I0051564

3D Printed Conducting Polymers

Conducting polymers are smart materials that possess unique and tuneable electrical, optical, and electrochemical properties. 3D printing technology is rapidly advancing, and using conducting polymers for this process can lead to many emerging applications as it can print complex structures cost effectively, though many challenges need to be overcome before this technology can be used on a large scale. *3D Printed Conducting Polymers* highlights the state of the art of these materials, the basics of additive printing, and the role of conducting polymers in additive manufacturing. It also discusses applications in energy, sensors, and biomedical areas.

- Covers fundamentals, synthesis, and various applications of conducting polymers.
- Discusses basics of energy devices, sensors, and materials technology for emerging applications.
- Explores new approaches for the synthesis of conducting polymers and composites for 3D print technology.
- Details future applications and challenges.

Offering direction to researchers and advanced students to better understand the chemistry and electrochemical properties of conducting polymers and technologies for 3D printing, this book advances the science and technology of this emerging field for readers in materials and chemical engineering, biotechnology, energy, and related disciplines.

Ram K. Gupta is an associate vice president for research and support and a professor of chemistry at Pittsburg State University. Gupta has been recently named by Stanford University as being among the top 2% of research scientists worldwide. Before joining Pittsburg State University, he worked as an assistant research professor at Missouri State University, Springfield, Missouri, then as a senior research scientist at North Carolina A&T State University, Greensboro, North Carolina. Dr. Gupta's research spans a range of subjects critical to current and future societal needs, including semiconducting materials and devices, biopolymers, flame-retardant polymers, green energy production and storage using nanostructured materials and conducting polymers, electrocatalysts, optoelectronic and photovoltaic devices, organic-inorganic heterojunctions for sensors, nanomagnetism, biocompatible nanofibers for tissue regeneration, scaffold and antibacterial applications, and biodegradable metallic implants. Dr. Gupta has mentored 10 PhD/post-doc scholars, 76 MS students, and 58 undergraduate/high school students. Dr. Gupta has published over 290 peer-reviewed journal articles (9600+ citations, 57 h-index, 207 i10-index); made over 420 national/international/regional presentations; chaired/organized many sessions at national/international meetings; written several book chapters (100+); worked as an editor for many books (50+) for the American Chemical Society, CRC, Springer, Elsevier, and so on; and received several million dollars for research and educational activities from external agencies. He also serves as Editor, Associate Editor, Guest Editor, and Editorial Board Member for various journals.

Smart 3D/4D Printing

Series Editor
Ram K. Gupta

Smart 3D Nanoprinting Fundamentals, Materials, and Applications
Edited By Ajit Behera, Tuan Anh Nguyen, Ram K. Gupta

3D Printed Conducting Polymers: Fundamentals, Advances, and Challenges
Edited By Ram K. Gupta

For more information about this series, please visit: https://www.routledge.com/Smart-3D4D-Printing/book-series/S3D4DP

3D Printed Conducting Polymers
Fundamentals, Advances, and Challenges

Edited by
Ram K. Gupta

CRC Press
Taylor & Francis Group
Boca Raton London New York

CRC Press is an imprint of the
Taylor & Francis Group, an **informa** business

Designed cover image: Shutterstock

First edition published 2025
by CRC Press
2385 NW Executive Center Drive, Suite 320, Boca Raton FL 33431

and by CRC Press
4 Park Square, Milton Park, Abingdon, Oxon, OX14 4RN

CRC Press is an imprint of Taylor & Francis Group, LLC

Library of Congress Cataloging-in-Publication Data
Names: Gupta, Ram K., editor.
Title: 3D printed conducting polymers : fundamentals, advances, and challenges / [edited by] Ram K. Gupta.
Description: First edition. | Boca Raton, FL : CRC Press, 2025. | Series: Smart 3D/4D printing | Includes bibliographical references and index. | Summary: "3D Printed Conducting Polymers highlights the state-of-the-art of these materials, the basics of additive printing, and the role of conducting polymers in additive manufacturing. It also discusses applications in energy, sensors, and biomedical areas. Offering direction to researchers and advanced students to better understand the chemistry and electrochemical properties of conducting polymers and technologies for 3D printing, this book advances the science and technology of this emerging field for readers in materials and chemical engineering, biotechnology, energy, and related disciplines"—Provided by publisher.
Identifiers: LCCN 2023054719 | ISBN 9781032541969 (hbk) | ISBN 9781032542621 (pbk) | ISBN 9781003415985 (ebk)
Subjects: LCSH: Electric batteries—Materials. | Conducting polymers—Synthesis. | Electric power supplies to apparatus—Materials. | Three-dimensional printing.
Classification: LCC TK2945.P65 A23 2025 | DDC 620.1/9204297028—dc23/eng/20240315
LC record available at https://lccn.loc.gov/2023054719

ISBN: 978-1-032-54196-9 (hbk)
ISBN: 978-1-032-54262-1 (pbk)
ISBN: 978-1-003-41598-5 (ebk)
ISBN: [Enter 13 digit eBook+ ISBN here if applicable] (eBook+)

DOI: 10.1201/9781003415985

Typeset in Times LT Std
by Apex CoVantage, LLC

Dedicated to Dr. Shawn Naccarato for his inspiration, innovation, and motivation. Thank you for championing research and educational advances.

Contents

Preface

Conducting polymers are some of the smart materials used in many applications. They possess many unique and tuneable electrical, optical, and electrochemical properties. The variation in synthesis provides many unique architectural characteristics, making them very suitable for energy and sensor applications. Their properties can be further improved by doping and making composites. A fundamental understanding of the charge-transport mechanism in conducting polymers and their composites is very important to use them for emerging applications. On the other hand, 3D printing technology is rapidly advancing, and using conducting polymers for this process can lead to many emerging applications, as they can print complex structures cost-effectively. However, there are still many challenges that need to be overcome before using this technology on a large scale.

This book provides state-of-the-art knowledge in conducting polymers, the basics of additive printing, and the role of conducting polymers in additive manufacturing. The applications of conducting polymers in the energy, sensor, and biomedical areas are attractive. This book covers the basics and workings of many energy generation and production devices. This book also explores new approaches for the synthesis of conducting polymers and composites for 3D print technology. The future applications and challenges in 3D print technology are also explored. All the chapters are by experts in these areas, making this a suitable textbook for students and researchers.

Ram K. Gupta, Professor
Department of Chemistry
National Institute for Materials Advancement
Pittsburg State University
Pittsburg, Kansas 66762, USA

Contributor List

Jazib Ali
University of Rome
Tor Vergata, Rome, Italy

Elyor Berdimurodov
National University of Uzbekistan
Tashkent, Uzbekistan
Akfa University
Tashkent, Uzbekistan

Nora Chelfouh
Université de Montréal
Succursale Centre-Ville, Montréal, Canada

George Z. Chen
University of Nottingham
Nottingham, UK

Yuanyuan Chen
Technological University of the Shannon
Athlone, Co. Westmeath, Ireland

Xiang Chu
Southwest Jiaotong University
Chengdu, P. R. China

Allen Davis
Pittsburg State University
Pittsburg, Kansas, United States

Mickaël Dollé
Université de Montréal
Succursale Centre-Ville, Montréal, Canada

Manon Faral
Université de Montréal
Succursale Centre-Ville, Montréal, Canada

Begum Nadira Ferdousi
Ahsanullah University of Science and Technology
Tejgaon, Dhaka, Bangladesh

Anjali Gupta
Pittsburg State University
Pittsburg, Kansas, United States

Ram K. Gupta
Pittsburg State University
Pittsburg, Kansas, United States

Ahmad Hussain
The University of Lahore
Sargodha, Pakistan

Mohy Menul Islam
University of Dhaka
Dhaka, Bangladesh

Md. Mominul Islam
University of Dhaka
Dhaka, Bangladesh

Nawishta Jabeen
Fatima Jinnah Women University
Rawalpindi, Pakistan

Tingting Jiang
Wuhan University of Science and Technology
Wuhan, P. R. China

Xinglin Jiang
Southwest Jiaotong University
Chengdu, P. R. China

S. Kalaiarasi
A.P.C. Mahalaxmi College for Women
Thoothukudi, Tamilnadu, India

Selcan Karakuş
Istanbul University-Cerrahpaşa
Avcilar, Istanbul, Turkiye

P. Karpagavinayagam
V.O. Chidambaram College
Thoothukudi, Tamilnadu, India

Ismail Koyuncu
Istanbul Technical University
Maslak, Istanbul, Turkey

Alok Kumar
Nalanda College of Engineering
Bihar Engineering University
Bihar, India

Ashish Kumar
Nalanda College of Engineering
Bihar Engineering University
Bihar, India

Audrey Laventure
Université de Montréal
Succursale Centre-Ville, Montréal, Canada

Sunil Kumar Baburao Mane
Khaja Bandanawaz University
Kalaburagi, Karnataka, India

G. Manjunatha
Shri Siddhartha Institute of technology
Tumkur, Karnataka, India

Mohammad Mazloum-Ardakani
Yazd University
Yazd, Iran

Wei Ni
ANSTEEL Research Institute of Vanadium & Titanium (Iron & Steel)
Chengdu, China

Nidhi
Lovely Professional University
Phagwara, Punjab, India

Mehmet E. Pasaoglu
Istanbul Technical University
Maslak, Istanbul, Turkey

Mayakkumar Patel
Pittsburg State University
Pittsburg, Kansas, United States

Pratik Patel
Pittsburg State University
Pittsburg, Kansas, United States

Tristan Perodeau
Université de Montréal
Succursale Centre-Ville, Montréal, Canada

Alexandra Robinson
Pittsburg State University
Pittsburg, Kansas, United States

Hamideh Mohammadian Sarcheshmeh
Yazd University
Yazd, Iran

Norazuwana Shaari
Universiti Kebangsaan Malaysia
Selangor, Malaysia

Naghma Shaishta
Khaja Bandanawaz University
Kalaburagi, Karnataka, India

Praveen K. Sharma
Lovely Professional University
Phagwara, Punjab, India

Esmaeil Sheibani
University of Isfahan
Isfahan province, Isfahan, Iran

Ling-Ying Shi
Sichuan University
Chengdu, China

Felipe M. de Souza
Pittsburg State University
Pittsburg, Kansas, United States

V. Sreeja
Vellalar College for Women (Autonomous)
Tamilnadu, India

Aasma Tabassum
The University of Lahore
Sargodha, Pakistan
Nanjing University of Science and Technology
Nanjing, China

Ramesh C. Thakur
Himachal Pradesh University
Summer Hill Shimla, Himachal Pradesh, India

Vahid Vatanpour
Istanbul Technical University
Maslak, Istanbul, Turkey
Kharazmi University
Tehran, Iran

C. Vedhi
V.O. Chidambaram College
Thoothukudi, Tamilnadu, India

Ajaz A. Wani
Universiti Kebangsaan Malaysia
Selangor, Malaysia

Bo Xu
Nanjing University of Science and Technology
Nanjing, China

Emre Yılmazoğlu
Istanbul University-Cerrahpaşa
Avcilar, Istanbul, Turkiye

Haitao Zhang
Southwest Jiaotong University
Chengdu, P. R. China

1 3D-Printed Conducting Polymers
An Introduction

S. Kalaiarasi, V. Sreeja, P. Karpagavinayagam, and C. Vedhi

1.1 INTRODUCTION

Conducting polymers are attractive substances applied for a wide range of uses, including energy storage, flexible electronics, and bioelectronics. However, traditional methods of conducting polymer fabrication, such as ink-jet printing, screen printing, and electron-beam lithography, have hampered rapid innovation and broad applications of conducting polymers. For conducting polymer three-dimensional (3D) printing, a high-performance 3D-printable conducting polymer ink based on poly (3,4-ethylenedioxythiophene): polystyrene sulfonate (PEDOT:PSS) has been presented. As an outcome of the improved printability, conducting polymers can be easily formed into high-resolution and high–aspect ratio microstructures, which can then be integrated with other materials such as insulating elastomers using multi-material 3D printing. The conducting polymers created by 3D printing can also be converted into highly conductive and soft hydrogel microstructures.

1.1.1 Properties and Applications

Conductive polymers are guaranteed as antistatic components and have been used in commercial displays and batteries. According to the literature, they are also promising in organic solar cells, printed electronic circuits, organic light-emitting diodes, actuators, electrochromism, supercapacitors, chemical sensors, chemical sensor arrays, and biosensors [1], as well as flexible transparent displays, electromagnetic shielding, and possibly as a replacement for the popular transparent conductor indium tin oxide. Microwave-absorbent coatings, especially radar-absorptive coatings on stealth aircraft, are another application. Conducting polymers are rapidly gaining popularity in new applications due to their raised processability, improved electrical and physical properties, and lower costs. With their higher surface area and better dispersibility, the new nano-structured forms of conducting polymers contribute to this field. According to studies, nanostructured materials conducting polymers in the form of nanofibers and even nanosponges have considerably greater capacitance [2, 3].

PEDOT and polyaniline have achieved a few large-scale uses due to their abundance of solid and consistent dispersions. While PEDOT is primarily used in

DOI: 10.1201/9781003415985-1

antistatic applications and as a transparent conductive layer in the form of PEDOT: PSS dispersions (PSS = polystyrene sulfonate), polyaniline is frequently employed in electronic board manufacturing facilities in the final finish to protect the copper from corrosion and to prevent solderability [4]. Polyindole is also gaining popularity for a variety of applications due to its higher redox activity, thermal stability, and slower degradation than competitors polyaniline and polypyrrole.

Conducting polymers (CPs) are organic materials that are currently used in a variety of applications such as energy storage, flexible electronics, and bioelectronics. They have the distinct advantage of combining conducting or semiconducting properties in addition to unique optoelectronic properties, with some organic polymer properties. CPs are a type of macromolecule with unique physical and chemical characteristics; their mechanical properties differ significantly from those of thermoplastic polymers. The majority of CPs are insoluble, infusible powdery materials that are difficult to process. Only in a few cases does the inclusion of dissolving chemical features to the primary link backbone allow the use of traditional thin film deposition methods, such as solvent casting or spin-coating, to process CPs into functional devices. As a result, applying additive manufacturing (AM) techniques to conducting polymers has proved more challenging than applying AM methods to other polymer families [3].

Various AM technologies have become known as exciting manufacturing techniques over the past years; their main benefit is the ability to fabricate multi-material objects. Metals, polymers, ceramics, and other materials can be 3D printed with various sections in a single printing process to meet specific requirements such as chemical, mechanical, thermal, and electrical properties. Several papers were published over the past decade concerning metals or ceramic materials 3D printed alongside organic polymers. In fact, a significant migration of equipment from traditional two-dimensional (2D) thin films to shape variable three-dimensional structures has started to occur in the design of electronic devices.

1.1.2 3D Printing

Due to its design versatility and capacity to effectively construct advanced structures, 3D printing has attracted investigators from a wide range of subjects in the past few years [5, 6]. Design ideas can be turned into products using direct digital manufacturing without the use of dies or molds. Subtractive lithography-based processing and patterning processes are currently employed for the fabrication of most devices. They are capable of fabricating intricate structures at high rates yet frequently generate significant amounts of waste [7, 8]. Aside from these benefits, additive manufacturing techniques enable the integration of modules into a single element with less assembly required [9]. These advantages are appealing to a wide range of industries, such as biomedical devices, the aerospace sector, power, and vehicles [10, 11]. Figure 1.1 depicts a 3D direct ink writing printer.

1.1.3 Properties of 3D-Printed Conducting Polymers

3D-printed conducting polymers have electrical conductivities that exceed 155 S/cm in the dry state and 28 S/cm in the hydrogel state, which is comparable to previously

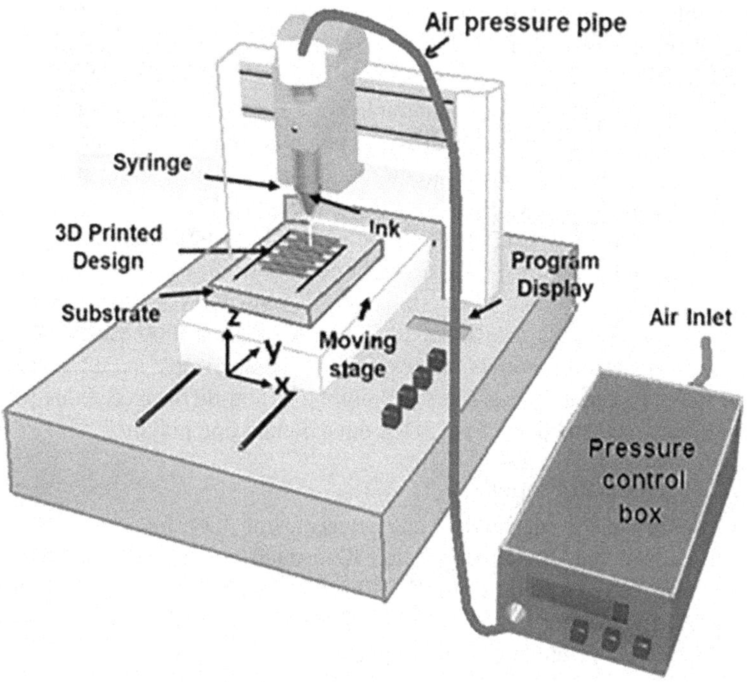

FIGURE 1.1 Direct ink writing printer in 3D representation.

established powerful conducting polymers [12–14]. Notably, a smaller nozzle diameter results in improved electrical conductivity for the created conducting polymers, possibly due to shear-induced improvements in the PEDOT: PSS nanofibril alignment [15]. Mechanical bending with a maximum strain of 13% in the dry state (65 m radius of curvature with 17 m thickness) and 20% in the hydrogel state (200 m radius of curvature with 78 m thickness) is possible with no failure due to the flexibility of 3D-printed conducting polymers. To examine the effect of physical bending on electrical wiring performance, Gladman et al. characterized the electrical conductivity of 3D-printed conducting polymers (100-m nozzle, 1 layer) on versatile polyimide substrates as a function of bending radius and bending cycle. In both dry and hydrogel, the electrical conductivity of 3D-printed conducting polymers changes only slightly (less than 5%) across a wide range of tensile as well as compressible bending circumstances (radius of curvature, 1–20 mm). Furthermore, after 10,000 cycles of repeated bending, 3D-printed conducting polymers can maintain high electrical conductivity (over 100 S/cm in the dry state and over 15 S/cm in the hydrogel state).

1.1.3.1 Electrical Properties

The electrical properties of 3D-printed conducting polymers (100-m nozzle, 1 layer on Pt) are investigated using the technique of electrochemical impedance spectroscopy (EIS). The EIS data are fitted to the equivalent circuit model, where R*e* denotes the electrical resistance, Ri denotes the ionic resistance, Rc denotes the total ohmic resistance of the electrochemical cell assembly, and CPEdl and CPEg denote the constant phase elements (CPEs) that correspond to the double-layer ionic capacitance and geometric capacitance, respectively [16, 17]. The semicircular Nyquist plot shape indicates the presence of comparable ionic and electronic conductivity in the 3D-printed conducting polymer hydrogels, which is supported by the extracted fitting parameters of the equivalent circuit model, which demonstrate similar magnitudes of ionic and electronic resistances (Ri = 105.5 and Re = 107.1).

The cyclic voltammetry (CV) results show that 3D-printed conducting polymers (100-m nozzle, 1 layer on Pt) have a height control storage capability (CSC) when compared to typical metallic electrode materials such as Pt, with remarkable electrochemical stability (less than 2% reduction in CSC after 1000 cycles). The CV of 3D-printed conducting polymers also exhibits broad and stable anodic and cathodic peaks at varying potential scan rates [18], implying non-diffusional redox processes and electrochemical stability of the 3D-printed conducting polymer.

1.1.3.2 Mechanical Properties

Nanoindentation of the mechanical characteristics of 3D-printed conducting polymers was carried out. In the dry state, 3D-printed conducting polymers have a relatively high Young's modulus of "1.5–0.31 GPa", which is comparable to previously reported values for dry PEDOT: PSS [19]. In contrast, Young's modulus, or the modulus of 3D-printed performing plastics in the hydrogel state, is reduced by three orders of magnitude to "1.1–0.36 MPa", which is comparable to soft elastomers such as polydimethylsiloxane (PDMS) (Young's modulus, 1–10 MPa). The softness of 3D-printed conducting polymer hydrogels can provide beneficial long-term biomechanical interactions with biological tissues, which may be especially useful in bioelectronic devices and implants [20–22].

1.1.4 3D Printing of Conducting Polymer Devices

Enabled by superior 3D printability and properties, 3D printing of conducting polymer ink can offer a promising route for easy and streamlined fabrication of high-resolution and multi-material conducting polymer structures and devices. Highly reproducible 3D printing of conducting polymers in high resolution allows the rapid fabrication of over 100 circuit patterns with less than 100 μm feature size on a flexible polyethylene terephthalate (PETE) substrate by a single continuous printing process with a total printing time of less than 30 min. The resultant 3D-printed conducting polymer electronic circuits exhibit high electrical conductivity to operate electrical components such as a light-emitting diodes (LEDs). This programmable, high-resolution, and high-throughput fabrication of conducting polymer patterns by 3D printing can potentially serve as an alternative to ink-jet and screen printing with a higher degree of flexibility in the choice of designs based on application demands [23, 24].

1.1.5 3D-Printable Conducting Polymer Ink

Conducting polymers are typically used in the form of a liquid monomer or polymer solution whose fluidity prevents their direct use in 3D printing [12, 20, 25].

In order to endow the rheological properties required for 3D printing to conducting polymers, Y. Wang et al. developed a simple process to convert a commercially available PEDOT:PSS aqueous solution to a high-performance 3D printable ink. The pristine PEDOT:PSS solution exhibits a dilute dispersion of PEDOT: PSS nanofibrils with low viscosity (below 30 Pa s). Inspired by the 3D printability of concentrated cellulose nanofiber suspensions [14, 26], B. Lu et al. hypothesized that a highly concentrated solution of PEDOT:PSS nanofibrils can provide a 3D printable conducting polymer ink due to the formation of entanglements among PEDOT:PSS nanofibrils (Figure 1.2). First isolated PEDOT:PSS nanofibrils by lyophilizing the pristine PEDOT:PSS solution. In order to avoid excessive formation of PEDOT-rich crystalline domains among PEDOT:PSS nanofibrils due to slow ice crystal formation during lyophilization at high temperature [27], we perform lyophilization in cryogenic conditions (i.e., frozen in liquid nitrogen). The isolated PEDOT:PSS nanofibrils are then re-dispersed with a binary solvent mixture (water:DMSO = 85:15 v/v) to prepare concentrated suspensions.

With increasing concentration of the PEDOT: PSS nanofibrils, the suspensions gradually transit from liquids to thixotropic 3D-printable inks due to the formation of reversible physical networks of the PEDOT: PSS nanofibrils via entanglements within the solvent. The small-angle X-ray scattering (SAXS) and rheological characterizations quantify microscopic and macroscopic evolutions of the conducting polymer ink with varying concentrations of the PEDOT: PSS nanofibrils, respectively. The SAXS characterizations show that the average distance between PEDOT-rich crystalline domains L (d-spacing calculated by the Bragg expression $L = 2\pi/q_{max}$) decreases with an increase in the concentration of the PEDOT: PSS nanofibrils (16.1 nm for 1 wt% and 7.0 nm for 10 wt%), indicating closer packing and a higher degree of interactions between the adjacent PEDOT: PSS nanofibrils in more concentrated inks.

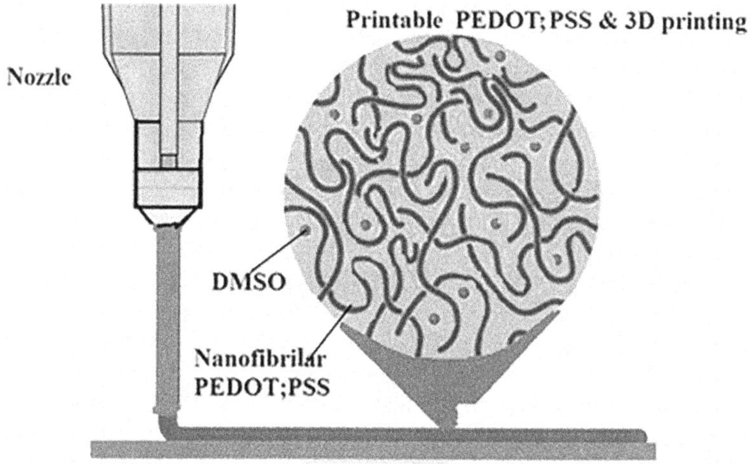

FIGURE 1.2 Printable conducting polymer ink (PEDOT: PSS).

The 3D printing process requires a computer-aided design (CAD) model sliced into a series of 2D digital models by specialized software. Computers manage the 3D printer, which realizes objects by adding each 2D layer onto the prior layers (i.e., layer-by-layer). In this process, a series of layers of materials is laid down in complex shapes or structures. This 'bottom up' manufacturing, conceptually different from traditional manufacture through 'cutting and drilling', can efficiently reduce the cost of production and prevent the waste of expensive advanced materials, meanwhile simplifying the design of microstructures and reducing post-processing (i.e., shaping, heating processing, and assembly). This technology has been extensively applied to jewelry, footwear, industrial design, pre-product models, engineering, automotive, aerospace, biomedical applications, dental, and medical industries. The two main technologies used for 3D printing are: (1) extrusion, the most common and probably the simplest 3D printing technique, where the main printing material is a plastic filament, which is heated, melting in the printing head of the 3D printer, and (2) laser stereolithography (SLA) or direct laser printing (DLP), which use photochemical processes by which light causes chemical monomers and oligomers to cross-link together to form polymers. Among the different 3D printing technologies, the photochemical approach is attractive since the choice of materials can be more ample, and the photopolymerization reactions of monomers is a tool with intrinsic environmental, economic, and production benefits. In SLA or DLP, printable materials are mainly composed of mono/multifunctional monomers, photoinitiating systems, and various additives. When the resin is exposed to UV light, the photoinitiators decompose into free radicals that react with the monomers, causing them to form cross-linked polymer chains. Absorbing additives are commonly used to adjust the horizontal and vertical resolutions.

1.1.6 3D Printing of Conducting Polymers

The superior printability of conducting polymer ink allows various advanced 3D printing capabilities, including printing of high-resolution, high–aspect ratio, and overhanging structures. The favorable rheological properties of the conducting polymer ink further enable the fabrication of multi-layered high–aspect ratio microstructures (100-μm nozzle, 20 layers, as well as overhanging features. 3D-printed conducting polymer structures can readily be converted into dry and hydrogel forms without loss of the original microscale structures, owing to the constrained drying (while attached to the substrate) and swelling property of the pure PEDOT:PSS hydrogels [15]. Furthermore, 3D-printed conducting polymer hydrogels exhibit long-term stability in physiological wet environments without observable degradation of microscale features (e.g., high aspect ratio and overhanging structures) after storage in PBS for 6 months.

1.1.7 Additive Manufacturing Techniques for 3D Printing of Electrically Conductive Polymer Composites

Despite recent developments in electrical CPCs, traditional manufacturing procedures are still mostly used in the fabrication of structures and devices based on conductive polymer composites [28, 29]. These manufacturing and fabrication methods are often limited to low-resolution (>100 mm), in-plane patterning (low aspect ratio), and intricate and expensive procedures [30]. Compared to traditional methods, 3D printing gives the potential to create microscale or even nanoscale structures in a way that is programmable, simple, and flexible while allowing for creative freedom in three dimensions. The technologies for the additive manufacture of polymers can be categorized as follows: vat photopolymerization (VP), material jetting (MJ), binder jetting (BJ), material extrusion (ME), and powder bed fusion (PBF). The basic and material requirements of each technique are different, and the choice of process would rely on the kind of conductive polymeric material employed and the required electrical qualities of the printed component.

3D printable conducting polymer ink can be readily incorporated into the multi-material 3D printing processes together with other 3D printable materials. For example, we fabricate a structure that mimics a high-density multi-electrode array (MEA) based on multi-material 3D printing of the conducting polymer ink and an insulating polydimethylsiloxane ink with a total printing time of less than 30 min. The 3D-printed MEA-like structure shows a complex microscale electrode pattern and a PDMS well that are comparable to a commercially available MEA fabricated by multi-step lithographic processes and post-assembly [31].

1.1.8 Main Additive Manufacturing and 3D Printing Technologies Used for Conducting Polymers

3D printing is the manufacturing of a structure with a specific design using computer-aided design and computer-aided manufacturing (CAM) software. Depending on the source, 3D printing methods can be classified as follows:

1. inkjet printing that uses controlled pulses for material deposition,
2. extrusion-based printing, where the source can be considered a mechanical movement,
3. electrohydrodynamic printing, which employs a controlled electric field for the deposition process, and
4. light-based printing, where lasers or light-emitting diodes are used for the curing/printing process [32, 33].

This section collects the 3D printing processing of the current most relevant conducting polymers.

1.1.8.1 Inkjet Printing

Inkjet printing operates through the same mechanism as inkjet office printers, which means that material droplets are ejected from a cartridge due to the pressure generated from the formation and collapse of microbubbles inside the nozzle. The bubbles can be generated from a thermal, piezoelectric, or electromagnetic stimulus, and the material in the form of droplets is deposited on a surface [34]. In contrast to extrusion-based printing, inkjet inks should be not viscous to ensure proper deposition and minimize the shear forces experienced by the material when it is ejected from

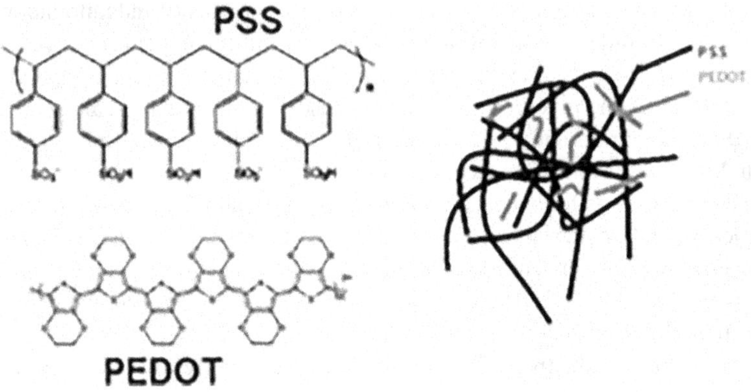

FIGURE 1.3 Inkjet printing using PEDOT: PSS.

the nozzle. Therefore, a viscosity lower than 100 mPa·s is recommended for inkjet inks [32]. This opens the possibility for CPs to be formulated into solvent- or water-based inks for inkjet printing. However, the low solubility of CPs brings difficulties in ink formulations. Nowadays, PEDOT is the most successful commercial CP in the (bio)electronics field due to its inherent properties, such as high conductivity, optical transparency in the form of thin films, and thermal and electrochemical stability [33, 35]. Moreover, PEDOT properties can be tuned using counterions and secondary dopants as well as by polymer blending, processing, and posttreatment methods [36] The most successful commercial PEDOT material is an aqueous dispersion of poly(3,4-ethylenedioxythiophene) and poly(styrenesulfonate) [34, 37]. The different processing methods of this PEDOT:PSS dispersion (Figure 1.3) and the combination with other polymers and conducting fillers allow for tuning the electrical, conducting, mechanical, and biological properties of the resulting materials [38, 39].

1.1.8.2 Extrusion-Based Printing

Extrusion-based printing consists of the layer-by-layer deposition of a material through a movable nozzle, which follows a specific shape previously programmed using software. There are two main extrusion printing methods that differ in the way to drive down the polymer. On the one hand, fused deposition modeling (FDM) uses a polymer in the form of a filament moved throughout a gear mechanism straight to a hot end, where the polymer is melted. On the other hand, direct ink writing (DIW) uses polymers that are semi-melted, in solutions or pastes, which are driven down by the action of current, air, pistons, or screws. Both FDM and DIW methods require the employment of polymers with a specific rheological behavior and viscosity values lower than 104 Pa·s for low shear rates ($10^{-1}s^{-1}$) and 101 Pa·s for high shear rates (102 s^{-1}) to be printable as well as to retain the desired shape after printing [40–45].

1.1.8.3 Electrohydrodynamic Printing

Electrohydrodynamic printing (EHD) is based on the deposition of a material dissolved in a polarizable liquid, which is subsequently evaluated for accessibility using an electric field that is typically placed between the nozzle and the grounded substrate. The EHD printing method possesses high resolution and overcomes the limitation related to the nozzle in the inkjet printing methodology. It can be used in pulsating or jet mode, creating dots or continuous fibers, so the deposition modulates the resolution at the micro- or nanoscale domain.

The printing quality is affected by the ink properties, such as viscosity, surface tension, electrical conductivity, or dipole moment, and the process-related factors, including the applied voltage, pressure, and flow rate. Overall, at low applied voltages and low ink viscosity, a dripping mode is observed at the apex of the Taylor cone. By increasing the voltage and keeping a low ink viscosity at low flow rates, the droplet size is much smaller than the nozzle size, giving rise to microdripping mode. An increase of the flow rate under these latter conditions makes the ink eject like a column, generating spindle mode. The employment of high-viscosity inks and high voltages generates a thin liquid jet at the apex of the cone known as cone-jet mode. In this way, by increasing the voltage and flow rate up to very high values, the unstable and uncontrollable multijet mode will be achieved. Among these jetting modes, microdripping and cone-jet provide the required printing process controllability for precision manufacturing. Moreover, electrodes can be located around the nozzle, which controls the deposition trajectory, achieving a sub-micrometer printing resolution.

1.1.8.4 Light-Based 3D Printing

Light-based 3D printing is based on the photopolymerization of a prepolymer or monomer in a liquid state placed inside a vat through partially controlled solidification in a specific shape, forming the 3D structure [46]. Two main methods are employed, stereolithography and digital light processing. SLA photocures resin by a laser beam controlled under a deflection mirror, and the liquid is solidified on the surface where the light spot was scanned. Regarding DLP, a digital micromirror device (DMD) formed by millions of mirrors is used to directly project a 2D image onto the photosensitive material. Moreover, SLA occurs in the top part of the vat, photocuring the resin point-by-point through the laser beam, whereas in the case of DLP, the light source projects the entire slice in the bottom part of the vat, where the photopolymerization takes place [47]. Besides these two methods, selective laser sintering (SLS) can be considered a light-based printing technique, where a photo-cross-linkable prepolymer in a powder is mixed with other polymers, metals, or ceramics to form composites that can be sintered by the action of a high-power laser [48]. However, CPs are rarely obtained by photopolymerization, which limits the applicability of light-based printing to inks where the CP is dispersed with a light-sensitive curable material.

1.1.9 Challenges

An important challenge in the field of additive manufacturing or 3D printing (Figure 1.4) is the lack of advanced polymer materials and available nanocomposites to match the performance and fabrication requirements. Thus, research efforts at academic, government, and industrial institutions have been focused on developing

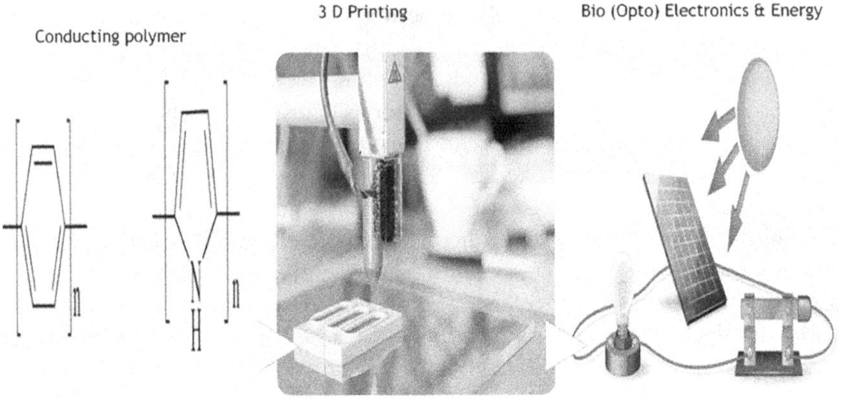

FIGURE 1.4 Challenges of 3D-printed conducting polymers.

new high-performance materials, improving the efficiency and speed of the process, and widening the range of its properties and applications [49]. The introduction and development of novel polymers, additives, and other thermoplastic composites for AM aims to extend the impact of AM on the fabrication of end-product components [50] and real part replacement.

A common issue in using commodity thermoplastic products in AM is their lack of strength and functionality as high load-bearing parts. The use of engineering plastics and high-performance polymers (HPPs) is desirable, but the cost and requirements for more demanding processing conditions (such as higher temperatures) present a challenge, especially if high build volumes are desired. Common matrix materials used in AM are thermoplastic material formulations. Traditional polymer blending has been used to combine the strength of two polymers or reinforce a weaker polymer with a high-performance polymer. It can also be done with the use of compatibilizers, telechelic copolymers, plasticizers, and other synergistic additives [51]. In a review by Dizon et al. (2017), it was emphasized that the mechanical properties of 3D-printed parts are indeed affected by the intrinsic material properties (before printing), the chosen manufacturing method, and build orientation, among many other factors. Thus, it is essential to constantly focus on improving the intrinsic properties of polymer materials from the molecular to the macroscopic level [52].

1.2 CONCLUSION

Three-dimensional printing methodologies have experienced explosive growth in the last decades; such growth has stimulated the interest of research groups. Among different CPs used for 3D printing, PEDOT is the most studied, especially the commercial dispersion PEDOT:PSS, which has resulted in the preferred material for different authors. The additive manufacturing of CPs presents great opportunities in the design of new devices and applications in the (bio)electronic field. However, CPs show important limitations in terms of processability. Therefore, their combination with other polymers and nano additives that improve their processability while keeping high conductivity is needed for their processing by additive manufacturing methods. In this regard, the 3D printing of CPs has started to be explored very recently and portrays a field with a huge number of possibilities and applications in the future.

REFERENCES

[1] M. Tebyetekerwa, S. Yang, S. Peng, Z. Xu, W. Shao, D. Pan, S. Ramakrishna, M. Zhu, Unveiling polyindole: Freestanding as-electro spun polyindole nanofibers and polyindole/carbon nanotubes composites as enhanced electrodes for flexible all-solid-state supercapacitors, Electrochim. Acta., 247 (2017) 400–409.

[2] M. Tebyetekerwa, Z. Xu, W. Li, X. Wang, I. Marriam, S. Peng, S. Ramakrishna, S. Yang, M. Zhu, Surface self-assembly of functional electro-active nanofibers on textile yarns as a facile approach towards super flexible energy storage, ACS Appl. Energy Mater., 1 (2017) 377–386.

[3] W. Zhou, J. Xu, Progress in conjugated polyindoles: Synthesis, polymerization mechanisms, properties, and applications, Polym. Rev., 57 (2016) 248–275.

[4] H.S. Nalwa ed, Handbook of Nanostructured Materials and Nanotechnology, 5 (2000) 501–575, New York, USA: Academic Press.

[5] S.C. Ligon, R. Liska, J. Stampfl, M. Gurr, R. Mülhaupt, Polymers for 3D printing and customized additive manufacturing, Chem. Rev., 117 (2017) 10212–10290.

[6] L.J.Y. Tan, W. Zhu, K. Zhou, Recent progress on polymer materials for additive manufacturing, Adv. Funct. Mater., 30 (2020) 54.

[7] J.A. Lewis, B.Y. Ahn, Three-dimensional printed electronics, Nature, 518 (2015) 42–43.

[8] J. Zhao, J. Gao, W. Li, Y. Qian, X. Shen, X. Wang, X. Shen, Z. Hu, C. Dong, Q. Huang, L. Cao, Z. Li, J. Zhang, C. Ren, L. Duan, Q. Liu, R. Yu, Y. Ren, S.C. Weng, H.J. Lin, C.T. Chen, L.H. Tjeng, Y. Long, Z. Deng, J. Zhu, X. Wang, H. Weng, R. Yu, M. Greenblatt, C. Jin, A combinatory ferroelectric compound bridging simple ABO_3 and A-site-ordered quadruple perovskite, Nat. Commun., 2 (2021) 747.

[9] H. Yuk, B. Lu, S. Lin, K. Qu, J. Xu, J. Luo, X. Zhao, 3D printing of conducting polymers, Nat. Commun., 11 (2020) 1604.

[10] S. Yuan, Y. Zheng, C.K, Chua, Yan, Q.K. Zhou, Electrical and thermal conductivities of MWCNT/polymer composites fabricated by selective laser sintering, Compos. Appl. Sci. Manuf., 105 (2017) 203e213.

[11] J.H.M. Wong, R.P.T. Tan, J.J. Chang, B.Q.Y. Chan, Zhao, J.J.W. Cheng, Y. Yu, Y.J. Boo, Q. Lin, V. Ow, X. Su, J.Y.C. Lim, X.J. Loh, K. Xue, Injectable hybrid cross linked hydrogels as fatigue-resistant and shape-stable skin depots, Biomacromolecules, 23 (2022) 3698e3712.

[12] Y. Wang et al., A highly stretchable, transparent, and conductive polymer, Sci. Adv., 3 (2017) e1602076.

[13] Y. Liu et al., Soft and elastic hydrogel-based microelectronics for localized low voltage neuro modulation, Nat. Biomed. Eng., 3 (2019) 58–68.

[14] B. Lu et al., Pure PEDOT:PSS hydrogels, Nat. Commun., 10 (2019) 1043.

[15] A.S. Gladman, E.A. Matsumoto, R.G. Nuzzo, L. Mahadevan, J.A. Lewis, Biomimetic 4D printing, Nat. Mater., 15 (2016) 413–418.

[16] C. Hsu, F. Mansfeld, Concerning the conversion of the constant phase element parameter Y0 into a capacitance, Corrosion, 57 (2001) 747–748.

[17] V.R. Feig, H. Tran, M. Lee, Z. Bao, Mechanically tunable conductive interpenetrating network hydrogels that mimic the elastic moduli of biological tissue, Nat. Commun., 9 (2018) 2740.

[18] J. Heinze, B.A. Frontana-Uribe, S. Ludwigs, Electrochemistry of conducting polymers—persistent models and new concepts', Chem. Rev., 119 (2010) 4724–4771.

[19] U. Lang, N. Naujoks, J. Dual, Mechanical characterization of PEDOT: PSS thin films, Synth, Met., 159 (2009) 473–479.
[20] H. Yuk, B. Lu, X. Zhao, Hydrogel bioelectronics, Chem. Soc. Rev., 48 (2019) 1642–1667.
[21] S.I. Park et al., Soft, stretchable, fully implantable miniaturized optoelectronic systems for wireless optogenetics, Nat. Biotechnol., 33 (2015)1280–1286.
[22] S.P. Lacour, G. Courtine, J. Guck, Materials and technologies for soft implantable neuro prostheses, Nat. Rev. Mater., 1 (2016): 16063.
[23] H. Sirringhaus et al., High-resolution inkjet printing of all-polymer transistor circuits, Science, 290 (2000) 2123–2126.
[24] C. Zhu et al., Stretchable temperature-sensing circuits with strain suppression based on carbon nanotube transistors, Nat. Electron. (2018) 183–190.
[25] E. Bihar et al., Inkjet printed PEDOT: PSS electrodes on paper for electro-cardiography, Adv. Healthc. Mater., 6 (2017) 1601167.
[26] K. Hong, S.H. Kim, A. Mahajan, C.D. Frisbie, Aerosol jet printed, Sub-2 V complementary circuits constructed from P—and N-type electrolyte gated transistors, Adv. Mater., 26 (2014) 7032–7037.
[27] B. Nazari, V. Kumar, D.W. Bousfield, M. Toivakka, Rheology of cellulose nanofibers suspensions: Boundary driven flow, J. Rheol., 60 (2016)1151–1159.
[28] L. Mendoza, W. Batchelor, R.F. Tabor, G. Garnier, Gelation mechanism of cellulose nanofibre gels: A colloids and interfacial perspective, J. Colloid Interf. Sci., 509 (2018) 39–46.
[29] A.G. Guex et al., Highly porous scaffolds of PEDOT: PSS for bone tissue engineering, Acta Biomater., 62 (2017) 91–101.
[30] Z. Huang, P. Luo, H. Zheng, Z. Lyu, Sulfur-doped graphene promoted $Li_4Ti_5O_{12}$@C nanocrystals for lithium-ion batteries, J. Alloys Compd., 908 (2022) 164599.
[31] P. Gong, D. Wang, C. Zhang, Y. Wang, Z. Jamili-Shirvan, K. Yao, X. Wang, Corrosion behavior of TiZrHfBeCu(Ni) high-entropy bulk metallic glasses in3.5 wt. % NaCl, npj Mater. Degrad., 6 (2022) 77.
[32] S.J. Lee, T. Esworthy, S. Stake, S. Miao, Y.Y. Zuo, B.T. Harris, L.G. Zhang, Advances in 3D bioprinting for neural tissue engineering, Adv. Biosyst., 2 (2018) 1700213.
[33] C. Mota, S. Camarero-Espinosa, M.B. Baker, P. Wieringa, L. Moroni, Bioprinting: From tissue and organ development to in vitro models, Chem. Rev., 120 (2020) 10547–10607.
[34] S.C. Ligon, R. Liska, J. Stampfl, M. Gurr, R. Mülhaupt, Polymers for 3D printing and customized additive manufacturing, Chem. Rev., 117 (2017) 10212–10290.
[35] M.J. Donahue, A. Sanchez-Sanchez, S. Inal, J. Qu, R.M. Owens, D. Mecerreyes, G.G. Malliaras, D.C. Martin, Tailoring PEDOT properties for applications in bioelectronics, Mater. Sci. Eng. R., 140 (2020) 100546.
[36] D. Mantione, I. Del Agua, A. Sanchez-Sanchez, D. Mecerreyes, Poly(3,4-ethylenedioxythiophene) (PEDOT) derivatives: Innovative conductive polymers for bioelectronics, Polymers, 9 (2017) 354.
[37] D. Mantione, I. Del Agua, W. Schaafsma, M. ElMahmoudy, I. Uguz, A. Sanchez-Sanchez, H. Sardon, B. Castro, G.G. Malliaras, D. Mecerreyes, Low-temperature cross-linking of PEDOT:PSS films using divinyl sulfone, ACS Appl. Mater. Interfaces, 9 (2017) 18254–18262.
[38] H.Yamato,M.Ohwa,W.Wernet,Stabilityofpolypyrroleandpoly(3,4-ethylenedioxythiophene) for biosensor application, J. Electroanal. Chem., 397 (1995) 163–170.
[39] Y.T. Tseng, Y.C. Lin, C.C. Shih, H.C. Hsieh, W.Y. Lee, Y.C. Chiu, W.C. Chen, Morphology and properties of PEDOT:PSS/soft polymer blends through hydrogen bonding interaction and their pressure sensor application, J. Mater. Chem. C., 8 (2020) 6013–6024.
[40] Y. Yang, H. Deng, Q. Fu, Recent progress on PEDOT:PSS based polymer blends and composites for flexible electronics and thermoelectric devices, Mater. Chem. Front., 4 (2020) 3130–3152.

[41] X. Yu, T. Zhang, Y. Li, 3D printing and bio-printing nerve conduits for neural tissue engineering, Polymers, 12 (2020) 1637.

[42] V.G. Rocha, E. Saiz, I.S. Tirichenko, E. García-Tuñón, Direct ink writing advances in multi-material structures for a sustainable future, J. Mater. Chem. A, 8(31) (2020) 15646–15657.

[43] J. Malda, J. Visser, F.P. Melchels, T. Jüngst, W.E. Hennink, W.J.A. Dhert, J. Groll, D.W. Hutmacher, 25th anniversary article: Engineering hydrogels for bio-fabrication, Adv. Mater., 25 (2013) 5011–5028.

[44] Y. Zhang, G. Shi, J. Qin, S.E. Lowe, S. Zhang, H. Zhao, Y.L. Zhong, Recent progress of direct ink writing of electronic components for advanced wearable devices, ACS Appl. Electron. Mater, 1 (2019) 1718–1734.

[45] J. Kim, R. Kumar, A.J. Bandodkar, J. Wang, Advanced materials for printed wearable electrochemical devices: A review, Adv. Electron. Mater, 3 (2017) 1600260.

[46] A. Ambrosi, M. Pumera, 3D-printing technologies for electrochemical applications, Chem. Soc. Rev., 45 (2016) 2740–2755.

[47] Y. Han, J. Dong, Electrohydrodynamic printing for advanced micro/nano manufacturing: Current progresses, opportunities, and challenges, J. Micro Nano-Manuf., 6 (2018) 040802–1.

[48] D. Ahn, L.M. Stevens, K. Zhou, Z. Page, Rapid high-resolution visible light 3D printing, ACS Cent. Sci., 6 (2020) 1555–1563.

[49] A.C. Leon De, Q. Chen, N.B. Palaganas, J.O. Palaganas, J. Manapat, R.C. Advincula, High performance polymer nanocomposites for additive manufacturing applications, React. Funct. Polym., 103 (2016) 141–55.

[50] J. Frketic, T. Dickens, S. Ramakrishnan, Automated manufacturing and processing of fiber-reinforced polymer (FRP) composites: An additive review of contemporary and modern techniques for advanced materials manufacturing, Addit. Manuf., 14 (2017) 69–86.

[51] Y.C. Ching, C.H. Chuah, K.Y. Ching, L.C. Abdullah, A. Rahman, 5-Applications of thermoplastic based blends, in: Visakh, P.M., Markovic, G., and Pasquini, D., eds. Recent Developments in Polymer Macro, Micro and Nano Blends (2017) 111–129, Cambridge: Woodhead Publishing.

[52] J.R.C. Dizon, A.H. Espera, Q. Chen, R.C. Advincula, Mechanical characterization of 3D-printed polymers, Addit. Manuf., 20 (2017) 44–67.

2 Route for the Synthesis of Conducting Polymers and Their Characteristics

Mayankkumar L. Chaudhary, Pratik Patel, and Ram K. Gupta

2.1 INTRODUCTION

In most cases, polymers are considered electrical insulators, as they have been mostly used to insulate wires and make switches. However, it is not true that all polymers serve as insulators. Like metals, some polymers may conduct electricity. In 1930, conductively packed polymers were developed specifically to shield against corona discharge. New materials with exceptional qualities often inspire cutting-edge innovations in manufacturing and other fields. The realization in 1960 that the inorganic polymer polysulfur nitride is a metal was a significant development in the expansion of conducting polymers (CPs) [1]. Polysulfur nitride has a conductivity of around 10^3 cm^{-1} at room temperature. Because they provided unequivocal evidence for the existence of strong CPs and prompted the enormous effort needed to create additional polymeric conductors, these developments were of the utmost significance. CPs may be tailored to meet the needs of a variety of industries thanks to their adaptability and the fact that they can be made both lightweight and corrosion resistant.

Numerous novel scientific ideas and potential technological advancements have arisen from CP research. MacDiarmid, Shirakawa, and Heeger received the Nobel Prize in chemistry in 2000 for their work on CPs, demonstrating the significance of their contributions to the field. Poly(acetylene) (PA), poly(aniline) (PANI), poly(p-phenylene) (PP), poly(p-phenylene)vinylene (PPV), poly(3,4-ethylene dioxythiophene) (PEDOT), poly(furan) (PF), and other polythiophene (PTh) derivatives have attracted attention due to their extraordinary properties such as electrical characteristics and reversible doping [2, 3]. Nowadays, CPs as polymers (both natural and synthetic) are used extensively in modern society because of their processability, thermal stability, diversity of optical and mechanical characteristics, and relatively low-cost fabrication [4]. When compared to materials like metals and ceramics, polymers have a greater degree of flexibility in terms of their shapeability. But until the discovery of CPs, these uses were mostly in non-electronic fields. Since then, researchers from many walks of science have poured into the area, greatly accelerating progress in the field of CPs. The wide range of electrical conductivity like metals that CPs exhibit, together with the preservation of their polymeric mechanical characteristics, is what initially drew scientists to this material. Using various

DOI: 10.1201/9781003415985-2

counter-ions (dopants), researchers are able to manipulate the conductivity of CPs [5]. The fabrication circumstances are just as important as the monomer and dopant types in determining the conductivity of CPs. The unique electrical and optical characteristics of conductive polymers are often compared to those of inorganic semiconductors and metals.

Methods that are straightforward, flexible, and economical can be used to synthesis CPs. Using straightforward electropolymerization procedures, they may be easily built into supramolecular structures with a wide range of useful properties. Many kinds of CPs exist, and we may categorize them according to the nature of their electric charge, which can take the form of delocalized electrons, conductive nanomaterials, or ions. Various conjugated CPs will be covered, each with its own set of characteristics and synthesis strategy. PA, PTh, polypyrrole (PPy), polyphenylene, and PANI will be discussed as the five primary forms of CPs. Conjugated CPs, conducting nanoparticles (NPs), carbon-based NPs, and alloys are all examples of conductive composite polymers [6]. The many routes to CP synthesis (Figure 2.1) and their resulting characteristics are discussed in this chapter.

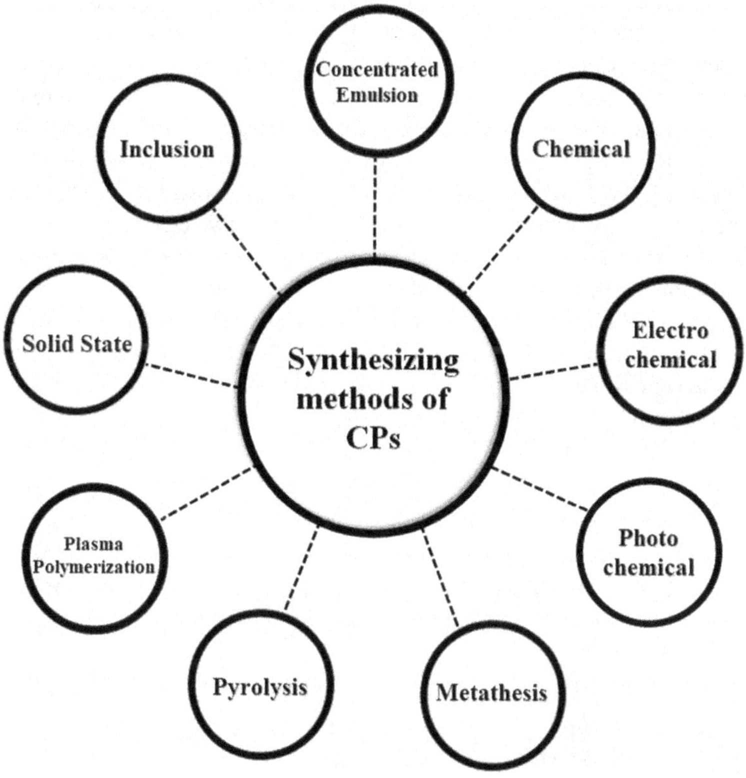

FIGURE 2.1 Different methods of synthesizing conducting polymers.

2.1.1 Conduction Mechanisms

Because the carbon atoms in the polymer backbone are sp^2 hybridized, the alternating single- and double-bond structure (conjugated backbone) can be electrically conductive. A continuous π-bond may be formed between carbon atoms in the z direction thanks to the p orbitals' parallelism, providing a channel for charge carriers to travel down the polymer chain. There must be charge carriers inside the polymer structure for it to be considered a conducting polymer. Most conjugated polymers made from organic compounds lack charge carriers on their own. To make the polymers conductive, it is necessary to add charge carriers from the outside. There are two types of electron acceptors and electron donors that can be used to tune the electrical properties of CPs. Doping refers to the procedure by which a conducting polymer is infused with charge carriers. To understand how the doping process modifies the electrical structure of CPs, band theory is used. Inorganic semiconductors often have their electronic structure described using band theory.

The band gap is the energy difference between the lowest unoccupied band (called the conduction band) and the highest occupied band (called the valence band). Band gap size in insulators is huge compared to semiconductors. Since there is no band gap between the conduction band and the valence band in metals, electrons can freely go from one band to the other. Doping CPs with a dopant either partially fills the conduction band or the valence band or creates polarons. One may use molecular orbital theory to back up band theory. Two new molecular orbitals can be formed when a p orbital from one carbon atom combines with another p orbital from a different carbon atom. Each of the two new orbitals has a different energy level relative to the original p orbital, with one having a lower energy level. Bonding molecular orbital refers to the orbital with the lowest energy level. Anti-bonding molecular orbitals are those with a greater energy than bonding molecular orbitals. Two electrons, one from each of the two carbon atoms, fill the bonding molecular orbital because electrons there are more stable. Because these two electrons are in the same orbital, they can form a π bond. (A π bond is a member of the carbon–carbon double bond.) The double bond consists of two bonds (one σ and one π). The HOMO of the bonding molecule orbital is the greatest possible, while the LUMO of the anti-bonding molecular orbital is the lowest possible. The valence band refers to the HOMO energy level, whereas the conduction band refers to the LUMO energy level. The band gap energy, or the energy at which electrons are activated, is the difference in kinetic energy between the HOMO and LUMO levels. Figure 2.2 represents the repeating units of a CP unit.

2.2 ROUTES FOR SYNTHESIZING CONDUCTING POLYMERS

2.2.1 Chemical Method

Chemical synthesis allows for the creation of CPs via the oxidation/reduction of monomers and subsequent polymerization of such monomers. One of its benefits is the potential for inexpensive mass manufacturing. Improved output and product quality from the oxidative polymerization process have been the focus of numerical research aimed at optimizing the production process. The application of electrochemical techniques is not mandated by the principles of the chemical route [7]. Poly (3-hexylthiophene), for instance, is a well-known and extensively researched CP

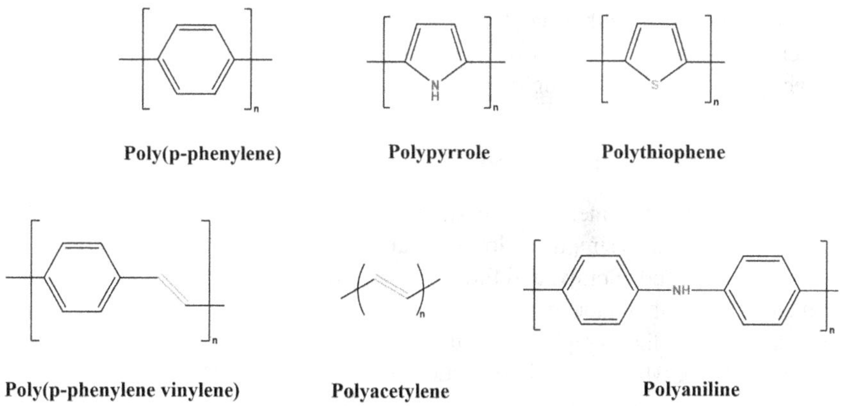

FIGURE 2.2 Chemical structures of repeating units of different types of conducting polymers.

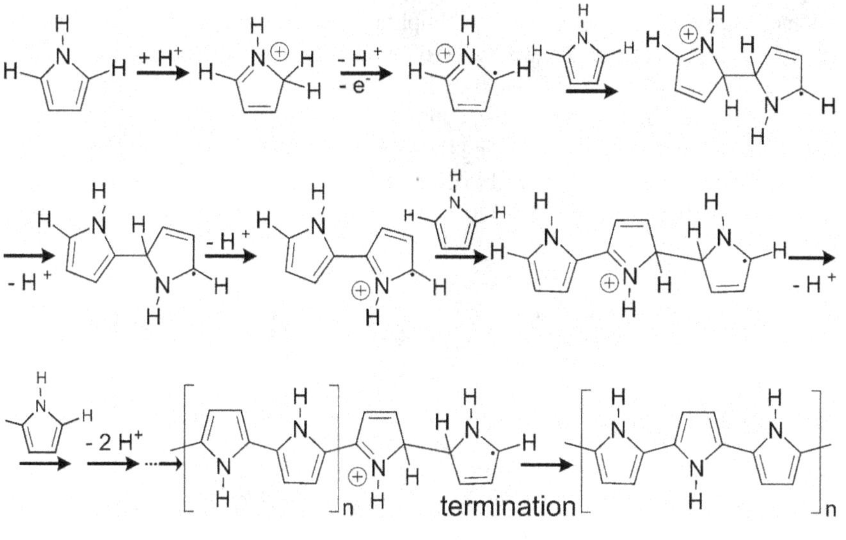

FIGURE 2.3 Chain polymerization mechanism of pyrrole. Adapted with permission [8]. Copyright (2016), American Chemical Society.

that is virtually always synthesized synthetically. While chemical methods exist for preparing PPy and PANI, the electrochemically generated variations often exhibit superior conductivity and mechanical qualities. Figure 2.3 shows the chain polymerization mechanism of pyrrole [8]. Stability is the primary need following conjugation for a chemical polymerization to occur. Oligomers and low-molecular-weight polymers must be reactive and soluble enough to polymerize to achieve high molecular weight polymerization. As the monomer and reactive polymer concentrations decrease, the possibility of heterogeneous polymerization proceeding as planned if an oligomer precipitates out of solution increases. If the molecular weight of the entanglement is not reached during the chemical polymerization, the process will end prematurely, and a mechanically unstable coating will be left on the walls of the reaction vessel. On the other hand, chemical polymerization allows for the selective production of cation radicals at the appropriate position on the monomer, provided that the system is soluble enough.

2.2.2 Electrochemical Method

Among the several documented techniques of synthesis, electrochemical synthesis of CPs is crucial due to its simplicity, low cost, ability to be accomplished in a single section glass cell, reproducibility, and thickness and homogeneity of the generated films. Anodic oxidation of suitable electroactive functional monomers is the most common electrochemical method employed in the manufacture of electrochemical conjugated polymers (ECPs), while cathodic reduction is utilized far less frequently. Doping of counterions because of oxidation occurs concurrently with polymer film production in the former. The polymerization potential of monomers oxidized to polymers is often greater than the polymerization potential of charged oligomeric intermediates. By alternating chemical and electrode reaction stages, electropolymerization of an electro-active monomer like pyrrole or thiophene was simplified. For instance, during the initial electrode reaction stage of thiophene electrooxidation, a radical cation is formed, which is then cleared by an anodic peak of high positive potential [9]. In the subsequent chemical reaction stage, the radical cation reacts with the monomer, resulting in the protonated dimer of a radical cation. Then, during the electrode reaction step, the radical cation's protonated dimer is electro-oxidized to di-cation. Figure 2.4 shows the electrochemical polymerization mechanism of PANI and associated morphology [10].

FIGURE 2.4 Electrochemical formation of PANI and its morphology at various magnifications. Adapted with permission [10]. Copyright (2010), American Chemical Society.

2.2.3 Photochemical Method

In both commercial and academic settings, chemical and ECP approaches have been the primary means of identifying polymers [11]. Though photochemical preparation has been extensively examined over the past two decapods, reports have indicated that the approach offers minimal benefits, even though it is a fast, cheap, and environmentally friendly option. Some CPs can benefit from using this technique during construction. Polymerization from pyrrole to PPy, for instance, has been achieved through visible light irradiation with either as the photosensitizer or a suitable electron acceptor. At present, horseradish peroxidase is used to kick off oxidative free radical coupling processes that polymerize aniline in the presence of hydrogen peroxide. When compared to chemical and electrochemical methods, polymerizing aniline at mild conditions is more plausible.

2.2.4 Metathesis Method

In chemistry, metathesis refers to a process in which two compounds undergo a reaction in which their constituent parts are switched. Ring-opening metathesis of cyclo-olefins is one type of metathesis polymerization, acyclic and cyclic alkynes are others, and metathesis of diolefins is still another. Metathesis between 1,2-dihydroquinoline and aniline was investigated by Evans *et al.* [12]. Masuda *et al.* [13] investigated metathesis polymerization synthesis and characteristics of polymers based on acetylene that are commonly conjugated.

2.2.5 Concentrated Emulsion Method

Water, latex particles, and monomer droplet phases may be distinguished in the emulsion polymerization approach, making it a heterophase polymerization technique. Radical polymerization is the primary process at play. Bulk and solution polymerization are examples of one-segment methods, where the monomer as solvent and the initiator are both present in the same part of the arrangement [14]. There is no incompatibility between the produced polymer and the monomer or the solvent until the polymer undergoes substantial change. A micelle-forming surfactant, an initiator that can be dissolved in water, and a water-insoluble monomer are the main components of this technique. The monomer-swollen micelles and latex particles are the primary sites of polymerization, as opposed to the monomer itself in suspension polymerization. Because the process begins with an emulsion of monomer droplets in water, but ends with a spread of latex particles, the term "emulsion polymerization" is misleading. Small monomer droplets, which are the actual sites of polymerization in microemulsions, are the norm in this process. When the monomer is water soluble, it is possible to do inverse emulsion polymerizations, in which the continuous segment is organic. Only modacrylic works of art may benefit from its application in the acrylic fiber business. Figure 2.5 shows the formation of PPy using the emulsion polymerization method.

2.2.6 Inclusion Method

Composite materials may be made at the atomic or molecular level by inclusion polymerization. This polymerization method, then, holds the key to developing very

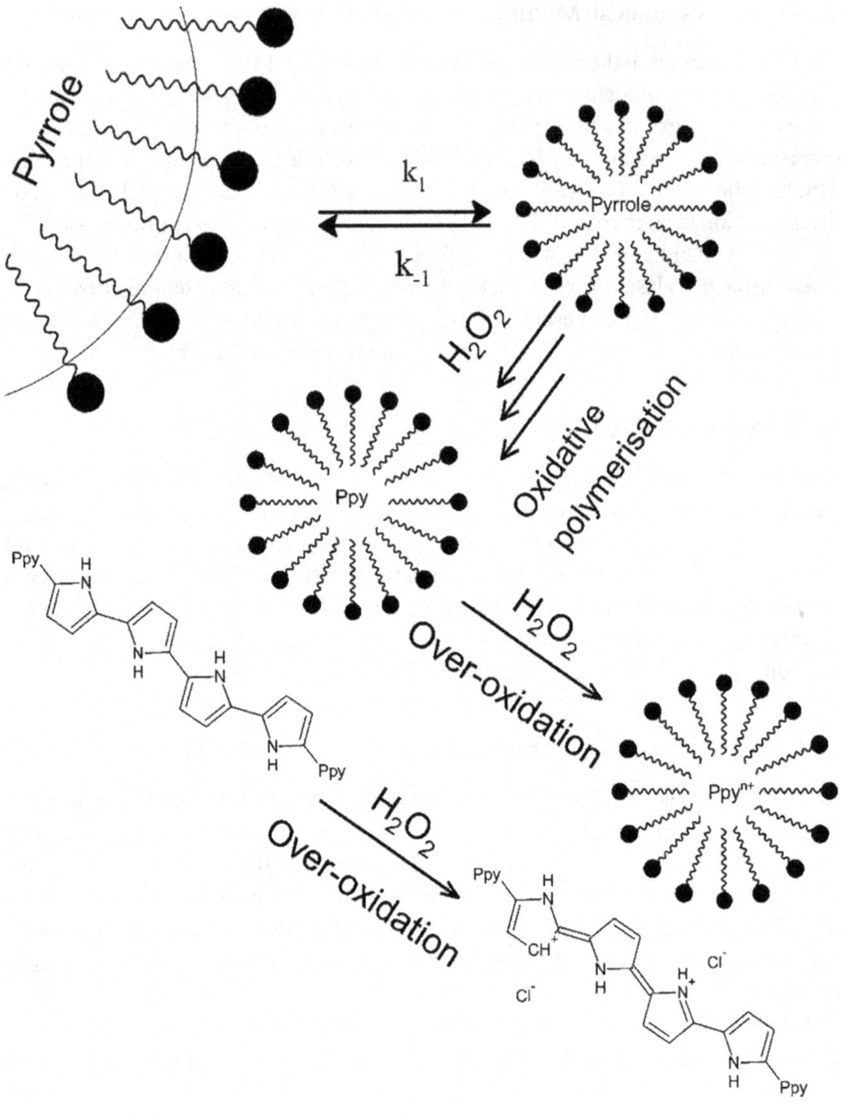

FIGURE 2.5 Formation of PPy nanoparticles using emulsion polymerization method. Adapted with permission [15]. Copyright (2011), American Chemical Society.

promising low-dimensional composite materials. For instance, a molecular wire might be created by including an electroconductive polymer. Composites of these polymers with organic hosts have been created based on inclusion. Not only should this polymerization not be seen from the standpoint of stereoregular polymerization, but Miyata *et al.* [16] also asserted that it may be recognized as a common space-dependent polymerization. Conventional solution and bulk polymerizations were ignored by the authors.

2.2.7 Solid-State Method

In solid-state polymerization, the lengths of polymer chains are increased by heat in an atmosphere devoid of oxygen and water, either by means of a vacuum or the elimination of the by-products of the processes using an inert gas. Temperature, pressure, and the migration of by-products from the pellet's inside to its outside regulate the reaction. It is a crucial step used to improve polymers' mechanical and rheological characteristics before injection blow molding, and it comes after melt-polymerization [17]. This technique has several applications in the manufacturing sector, particularly in the creation of bottle-grade polyethylene terephthalate (PET), films, and high-tech industrial fibers. Most of the issues with traditional polymerization methods may be sidestepped, and the process is much simpler and cheaper to implement in industry thanks to solid-state polymerization.

2.2.8 Plasma Polymerization

Thin films may be made from a variety of organic and organometallic precursors using a cutting-edge technology called plasma polymerization. The high degree of cross-linking in plasma-polymerized films makes them insoluble, thermally durable, chemically inert, and mechanically robust. Furthermore, such films are highly coherent and cling to a variety of substrates, including common polymers, glass, and metal [18]. In recent years, their widespread usage in a variety of applications from perm selective membranes and protective shells to medicinal materials and electrical, optical, and adhesion supporters has been largely attributable to their remarkable qualities.

2.2.9 Pyrolysis Method

Heating an organic substance to temperatures above its decomposition temperature results in a process known as pyrolysis. In industries as diverse as plastics and rubber manufacturing, dentistry, ecological shelter, and failure testing, it has proven itself a valuable method for studying and detecting organic polymeric compounds. Since no extensive sample preparation is required, this approach is ideal for studying extremely tiny sample sizes directly. However, pyrolytic degradation plays a crucial role in the ultimate assignment of the structure, and spectroscopic approaches can yield a detection of the monomeric species present. Extensive use of pyrolysis gas chromatography for the investigation of synthetic and natural polymers is now commonplace. Conjugated monomers undergo chemical polymerization when they react with an excess of an oxidant in a suitable solvent, such as acid. Since the polymerization occurs suddenly, constant stirring is required. The next method is electrochemical polymerization, which involves submerging both the counter and reference electrodes into a solution of diluted monomer and electrolyte. When the correct voltage is applied, the polymer thin film begins to deposit almost immediately on the working electrode.

2.3 PROPERTIES OF CONDUCTING POLYMERS

It is possible to employ CPs for a variety of purposes, since they possess a wide range of properties (Figure 2.6).

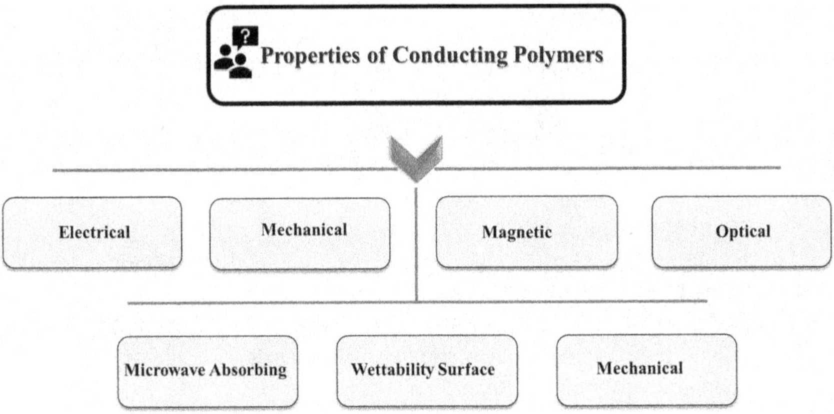

FIGURE 2.6 Properties of conducting polymers.

2.3.1 Electrical Properties

Doping percentage, polymer chain arrangement, conjugation length, and sample purity all play a role in determining a polymer's conductivity. Due to their molecular composition, ECPs may lack long-range organization. Due to their molecular structure, polymers generate electrical mobility around their macromolecules. Polymers and inorganic semiconductors require distinct methods for achieving high conductivity. The generation of self-localized excitons, such as solitons, polarons, and bipolarons, is linked to the enhanced conductivities that depend on doping in the polymers. These particles form because of a robust contact between charges throughout the chain, which is achieved via doping. While doping produces mostly polarons in CPs with non-degenerate ground states like cis-PA, PPy, PTh, or PPV, degenerate ground-state CPs like trans-PA have charged solitons as the charge carriers. Together, these polarons create spinless bipolarons that are used as the charge carriers. Polymers are cheap and versatile, making them an ideal material for electrical conductors, and their molecular structure lends itself to being engineered to have the appropriate qualities. The synthesis of highly conductive films of PA (CH) x derivatives was also investigated by the Chiang *et al.* [19], while Leung *et al.* [20] determined the electrical and optical characteristics of PA copolymers by calculation. Tanaka *et al.* [21] looked into the temperature dependency of the optical and electrical characteristics of doped PA. One author looked into what happens to PA's electrical characteristics due to impurities and heat isomerization [22]. Matsushita *et al.* [23] described how the pressure affects the electrical resistivity of iodine-doped PA. Carbon nanotube PANI composites with desirable electrical characteristics were produced by Long *et al.* [24]. Huang *et al.* [25] created self-assembled PANI carbon nanotubes and highly conductive PANI nanocables. Sarma *et al.* [26] investigated the oxidizing and reducing agent production of Au nanoparticle-conductive PANI composites. The poor energy conductivity of PF-doped PPy was shown by Chapman *et al.* [27]. The density of charge carriers, their mobility, the direction in which they move, the presence of doping materials, and temperature are all factors that influence electrical conductivity.

2.3.2 Magnetic Properties

There is a lot of curiosity about the magnetism of CPs because of their unique magnetic characteristics and potential technological uses. When looking for nanomaterials to include in a polymer matrix, the structural traits and magnetic properties of transition metal oxide nanoparticles are crucial. PANI/NiO conductors were studied in 2012 by Nandapure *et al.* [28] for their magnetic and transport characteristics. It was also shown through research that increasing the nickel oxide weight percentage in PANI lowered the conductivity and increased the magnetization of the PANI/nickel oxide nanocomposites. Yan *et al.* [29] developed a method for producing CP/ferromagnet films using an anodic-oxidation process. p-dodecyl benzenesulfonic acid sodium salt was used as a surfactant and dopant in a chemical process to create a PPy composite with ferromagnetic behavior. High saturation magnetization and other ferromagnetic behaviors were demonstrated by the final composites' magnetic characteristics. $NiFe_{1.95}Gd_{0.05}O_4$ composite particles including varying amounts of ferrite/PANI have been successfully generated using in situ polymerization, Aphesteguy *et al.* [30] used a range of ferrite to aniline ratios to demonstrate the change in magnetic characteristics with increasing PANI content.

After PANI coating, researchers looked at how the material's magnetic and conductive characteristics changed. Saturation magnetization drops dramatically, whereas coercivity remains almost the same. Using PANI and the acceptor molecule tetracyanoquinodimethane, Liu *et al.* [31] reported the synthesis and characterization of a new form of polymer. At a Curie temperature greater than 350 K and a maximum saturation magnetization of 0.1 J/Tkg, the polymer is predicted to be ferri- or ferro-magnetic, according to magnetic simulations. Zhang and colleagues [32] prepared and characterized ferromagnetic PANI that conducted electricity when a magnetic field was applied to the material.

2.3.3 Optical Properties

The optical properties of CPs are important to the development of an understanding of the basic electronic structure of the material. Due to their usefulness in nanophotonic devices, CPs' unique optical features have been the subject of much study [33]. Yoshino *et al.* [34] found that CPs doped with fullerenes and other molecular dopants exhibited novel optical properties. In research published in 2003 [35], novel di-substituted acetylene polymers [36] were shown to have enormous photoluminescence quantum efficiency in comparison to unsubstituted or monosubstituted PA [36]. Under a chiral nematic reaction environment, Akagi *et al.* [37] successfully synthesized helical PA $[(CH)_x]$ with left- and right-handed screw configurations. Electrical and optical characteristics of indium tin oxide/CP electrodes were studied by Lin *et al.* [38] to determine the impact of ammonium sulfide treatment, this treatment improves electrode performance by reducing trap-states and enhancing interfacial stability, offering potential efficiency gains in optoelectronic devices.

2.3.4 Mechanical Properties

Because of the wealth of data, they provide on charge-carrying groups and unpaired spins, CPs' magnetic characteristics have been the subject of much study. For a conducting polymer composite (CPC), Sulong *et al.* [39] looked at the mechanical characteristics of a carbon nanotube/graphite/polypropylene nanocomposite. The authors also analyzed the impact of filler content and chemical functionalization on the resultant CPCs' mechanical characteristics. Tensile and flexural testing reveal that functionalized CPC is stronger and more elastic than conventionally manufactured CPC. It has a flexural strength of 80 MPa and a tensile strength of 35 MPa at its maximum. Functionalized CPC is also more durable than unmodified CPC [40].

Machine and transverse mechanical characteristics of a uniaxially oriented PPV thin film have been calculated [41]. This film was manufactured by simultaneously thermally excluding and uniaxially extending the poly sulphonium salt precursor. These features vary in strength and other characteristics depending on the molecular orientation. Drawing in the machine direction causes a change in Young's modulus of 2.3 to 37 GPa, whereas drawing in the transverse direction causes a change of 2.3 to 0.5 GPa. Somanathan *et al.* [42] investigated the temperature dependence of the tensile characteristics of conducting poly (3-cyclohexyl thiophene) films. Hybrid proton CP combinations based on sulfonated polyetherketones were studied for their mechanical characteristics by Sgreccia *et al.* [43].

2.3.5 Microwave-Absorbing Properties

Since CPs are less dense, easily processible, and inexpensive, they have been explored as a novel microwave-interesting material. Ting determined the complex permittivity, complex permeability, and reflection loss in the microwave frequency range using the free space method [44]. In addition, this author demonstrated that PANI addition was helpful in achieving a high absorption across a wide frequency range. The performance of PANI nanocomposites including TiO_2 nanoparticles in microwave absorption were investigated by Phang *et al.* [45]. It is possible to quickly and easily create PANI nanocomposites reinforced with tungsten oxide nanoparticles and nanorods by use of the 6 surface-initiated polymerization process. Microwave absorbent shells on magnesium alloy were investigated by electrochemical impedance spectroscopy, as demonstrated by Guo *et al.* [46]. The microwave absorption capabilities of polystyrene polyvinyl pyrrolidone magnetic nanofiber were studied by Hosseini *et al.* [47] The coercive field was about 25 KV and saturation magnetization was 1.1 emu/g for nanofibers with a diameter of 30–40 nm measured at room temperature.

2.3.6 Wettability Surface

Material performance is significantly impacted by wettability. Surface wettability control is essential for many uses, such as in the production of non-stick coatings for cookware, the development of self-cleaning windows for the automotive and aerospace industries, the creation of watertight textiles, the development of anti-fingerprint or anti-reflective properties for optical instruments and mobile phones, the transportation of liquids, the separation of different substances, the adhesion of cells and bacteria, and so on. Preparation and surface wettability of triethoxyoctylsilane-modified TiO_2 nanorod films were studied by Xu *et al.* [48]. Darmanin *et al.* [49] investigated the wettability of nano porous, microporous, and micro/nanostructured arrangements of poly (3-alkyl-3,4-propylenedioxythiophene) fibrous structures. The wettability of poly(3-alkylthiophene) thin films was studied by Lin *et al.* [50]. The surface morphology and wettability of PPy were studied to see how redox-induced reformation affected them [51]. Researchers San *et al.* [52] looked at how changing the additive/binder and additive/filler ratios affected the wettability of polymer composite bipolar plates.

2.4 CONCLUSION

Processability, thermal stability, diversity of optical and mechanical qualities, and relatively low-cost production have led to widespread usage of CPs as polymers (both natural and synthetic) in modern life. In terms of their capacity to take on different forms, polymers excel above things like metals and ceramics. In terms of both scientific and technical interest, these materials have sparked a new path of inquiry in the field of materials science thanks to their adaptability in terms of both synthesis methods and characteristics, as well as their wide range of potential applications. This chapter offers a comprehensive and up-to-date overview of the methods used to synthesize and study these materials. The creation of CPs for use in diverse industries relies heavily on an in-depth comprehension of these features. Electrochemically created variants generally display greater conductivity and mechanical properties compared to those prepared using any of the chemical procedures for creating PPy and PANI. If the system is soluble enough, cation radicals can be produced selectively at the desired location on the monomer during chemical polymerization. Important features of electrochemical synthesis of CPs are their ease of use, cheap cost, repeatability, and the uniform thickness and composition of the produced films. Despite being a quick, low-cost, and eco-friendly choice, studies from photochemical approaches have shown that the strategy gives negligible advantages. To synthesize CPs for ring opening reactions, one can use a methodology called metathesis. The emulsion polymerization strategy used in the concentrated emulsion method is a heterophase polymerization method since it allows for the separation of water, latex particle, and monomer droplet phases. The process of inclusion polymerization allows for the creation of composite materials at the atomic or molecular level. Because of this polymerization technique, extremely promising low-dimensional composite materials may be created. Longer polymer chains can be produced in a solid state by applying heat in an environment free of oxygen and water, such as a vacuum or one created by purging the byproducts of operations with an inert gas. Injection blow molding happens after melt-polymerization and is essential for enhancing the mechanical and rheological properties of the polymer. Plasma polymerization thin films are a state-of-the-art method that may be used with a wide range of organic and organometallic precursors. Spectroscopic methods can identify the current monomeric species by pyrolytic destruction, which is a critical step in the final assignment of the structure. Nowadays, pyrolysis gas chromatography is widely used to study both synthetic and natural polymers. Batteries, solar cells, supercapacitors, medicinal applications, and electrothermic devices can all make extensive use of CPs due to their unique features.

REFERENCES

[1] J.L. Bredas, G.B. Street, polarons, bipolarons, and solitons in conducting polymers, Acc. Chem. Res. 18 (1985) 309–315.

[2] X. Lu, W. Zhang, C. Wang, T.-C. Wen, Y. Wei, One-dimensional conducting polymer nanocomposites: Synthesis, properties and applications, Prog. Polym. Sci. 36 (2011) 671–712.

[3] J. Stejskal, I. Sapurina, M. Trchová, Polyaniline nanostructures and the role of aniline oligomers in their formation, Prog. Polym. Sci. 35 (2010) 1420–1481.

[4] S. Kailasa, M.S.B. Reddy, M.R. Maurya, B.G. Rani, K.V. Rao, K.K. Sadasivuni, Electrospun nanofibers: Materials, synthesis parameters, and their role in sensing applications, Macromol. Mater. Eng. 306 (2021) 1–36.

[5] G. Tourillon, F. Garnier, Effect of dopant on the physicochemical and electrical properties of organic conducting polymers, J. Phys. Chem. 87 (1983) 2289–2292.

[6] T. Nezakati, A. Seifalian, A. Tan, A.M. Seifalian, Conductive polymers: Opportunities and challenges in biomedical applications, Chem. Rev. 118 (2018) 6766–6843.

[7] P.S. Sharma, A. Pietrzyk-Le, F. D'Souza, W. Kutner, Electrochemically synthesized polymers in molecular imprinting for chemical sensing, Anal. Bioanal. Chem. 402 (2012) 3177–3204.

[8] A.M. Bryan, L.M. Santino, Y. Lu, S. Acharya, J.M. D'Arcy, Conducting polymers for pseudocapacitive energy storage, Chem. Mater. 28 (2016) 5989–5998.

[9] A.L. Gomes, M.B. Pinto Zakia, J.G. Filho, E. Armelin, C. Alemán, J. Sinezio De Carvalho Campos, Preparation and characterization of semiconducting polymeric blends. Photochemical synthesis of poly(3-alkylthiophenes) using host microporous matrices of poly(vinylidene fluoride), Polym. Chem. 3 (2012) 1334–1343.

[10] G.-R. Li, Z.-P. Feng, J.-H. Zhong, Z.-L. Wang, Y.-X. Tong, Electrochemical synthesis of polyaniline nanobelts with predominant electrochemical performances, Macromolecules. 43 (2010) 2178–2183.

[11] A. Deronzier, J.C. Moutet, Polypyrrole films containing metal complexes: Syntheses and applications, Coord. Chem. Rev. 147 (1996) 339–371.

[12] P. Evans, R. Grigg, M. Monteith, Metathesis of aniline, Tetrahedron Lett. 40 (1999) 5247–5250.

[13] T. Masuda, S.M. Abdul Karim, R. Nomura, Synthesis of acetylene-based widely conjugated polymers by metathesis polymerization and polymer properties, J. Mol. Catal. A Chem. 160 (2000) 125–131.

[14] C.Y. Pan, Z.H. Chen, Y.L. Huang, K.L. Huang, Effects of water phase concentration on the emulsion polymerization of polyaniline, J. Cent. South Univ. Technol. (English Ed. 8 (2001) 140–142.

[15] K. Leonavicius, A. Ramanaviciene, A. Ramanavicius, Polymerization model for hydrogen peroxide initiated synthesis of polypyrrole nanoparticles, Langmuir. 27 (2011) 10970–10976.

[16] M. Miyata, F. Noma, K. Okanishi, H. Tsutsumi, K. Takemoto, Inclusion polymerization of diene and diacetylene monomers in deoxycholic acid and apocholic acid canals BT—Inclusion Phenomena in Inorganic, Organic, and Organometallic Hosts, in: J.L. Atwood, J.E.D. Davies (Eds.), Springer Netherlands, Dordrecht, 1987: pp. 249–252.

[17] P. Staiti, F. Lufrano, Design, fabrication, and evaluation of a 1.5 F and 5 V prototype of solid-state electrochemical supercapacitor, J. Electrochem. Soc. 152 (2005) A617.

[18] F. Arefi, V. Andre, P. Montazer-rahmati, J. Amouroux, Plasma polymerization and surface treatment of polymers, Pure Appl. Chem. 64 (1992) 715–723.

[19] C.K. Chiang, M.A. Druy, S.C. Gau, A.J. Heeger, E.J. Louis, A.G. MacDiarmid, Y.W. Park, H. Shirakawa, Synthesis of highly conducting films of derivatives of polyacetylene, (CH)x, J. Am. Chem. Soc. 100 (1978) 1013–1015.
[20] L.M. Leung, K.H. Tan, T.S. Lam, H. WeiDong, Electrical and optical properties of polyacetylene copolymers, React. Funct. Polym. 50 (2002) 173–179.
[21] J. Tanaka, K. Satake, T. Miyamae, Temperature dependence of optical and electrical properties of doped polyacetylene, Synth. Met. 101 (1999) 371–372.
[22] H. Tanaka, T. Danno, Effects of impurities and thermal isomerization on the electrical properties of undoped polyacetylene, Synth. Met. 17 (1987) 545–550.
[23] A. Matsushita, K. Akagi, T.S. Liang, H. Shirakawa, Effects of pressure on the electrical resistivity of iodine-doped polyacetylene, Synth. Met. 101 (1999) 447–448.
[24] Y. Long, Z. Chen, X. Zhang, J. Zhang, Z. Liu, Synthesis and electrical properties of carbon nanotube polyaniline composites, Appl. Phys. Lett. 85 (2004) 1796–1798.
[25] K. Huang, Y. Zhang, Y. Long, J. Yuan, D. Han, Z. Wang, L. Niu, Z. Chen, Preparation of highly conductive, self-assembled gold/polyaniline nanocables and polyaniline nanotubes, Chem.—A Eur. J. 12 (2006) 5314–5319.
[26] T.K. Sarma, D. Chowdhury, A. Paul, A. Chattopadhyay, Synthesis of Au nanoparticle-conductive polyaniline composite using H2O2 as oxidising as well as reducing agent, Chem. Commun. 2 (2002) 1048–1049.
[27] B. Chapman, R.G. Buckley, N.T. Kemp, A.B. Kaiser, D. Beaglehole, H.J. Trodahl, Low-energy conductivity of PF6-doped polypyrrole, Phys. Rev. B—Condens. Matter Mater. Phys. 60 (1999) 13479–13483.
[28] B.I. Nandapure, S.B. Kondawar, M.Y. Salunkhe, A.I. Nandapure, Magnetic and transport properties of conducting polyaniline/nickel oxide nanocomposites, Adv. Mater. Lett. 4 (2013) 134–140.
[29] F. Yan, G. Xue, J. Chen, Y. Lu, Preparation of a conducting polymer/ferromagnet composite ® lm by anodic-oxidation method, Synth. Met. 123 (2001) 17–20.
[30] J.C. Aphesteguy, P.G. Bercoff, S.E. Jacobo, Preparation of magnetic and conductive Ni-Gd ferrite-polyaniline composite, Phys. B Condens. Matter. 398 (2007) 200–203.
[31] R. Liu, H. Qiu, H. Zong, C. Fang, Fabrication and characterization of composite containing HCl-doped polyaniline and Fe nanoparticles, J. Nanomater. 2012 (2012).
[32] Z. Yang, H. Peng, W. Wang, T. Liu, Crystallization behavior of poly(ε-caprolactone)/layered double hydroxide nanocomposites, J. Appl. Polym. Sci. 116 (2010) 2658–2667.
[33] T.K. Das, S. Prusty, Review on conducting polymers and their applications, Polym.—Plast. Technol. Eng. 51 (2012) 1487–1500.
[34] K. Yoshino, K. Tada, K. Yoshimoto, M. Yoshida, T. Kawai, H. Araki, M. Hamaguchi, A. Zakhidov, Electrical and optical properties of molecularly doped conducting polymers, Synth. Met. 78 (1996) 301–312.
[35] D.B. Tanner, G.L. Doll, A.M. Rao, P.C. Eklund, G.A. Arbuckle, A.G. MacDiarmid, Optical properties of potassium-doped polyacetylene, Synth. Met. 141 (2004) 75–79.
[36] A. Fujii, R. Hidayat, T. Sonoda, T. Fujisawa, M. Ozaki, Z.V. Vardeny, M. Teraguchi, T. Masuda, K. Yoshino, Optical properties of disubstituted polyacetylene thin films, Synth. Met. 116 (2001) 95–99.
[37] K. Akagi, G. Piao, S. Kaneko, I. Higuchi, H. Shirakawa, M. Kyotani, Helical polyacetylene synthesized under chiral nematic liquid crystals, Synth. Met. 102 (1999) 1406–1409.
[38] Y.J. Lin, B.Y. Liu, Y.M. Chin, Effects of (NH4)2Sx treatment on the electrical and optical properties of indium tin oxide/conducting polymer electrodes, Thin Solid Films. 517 (2009) 5508–5511.
[39] A.B. Sulong, M.I. Ramli, S.L. Hau, J. Sahari, N. Muhamad, H. Suherman, Rheological and mechanical properties of carbon nanotube/Graphite/SS316L/polypropylene nanocomposite for a conductive polymer composite, Compos. Part B Eng. 50 (2013) 54–61.
[40] L.C.P. Almeida, Conducting polymers: Synthesis, properties and applications, Conduct. Polym. Synth. Prop. Appl. (2013) 1–358.

[41] J.M. Machado, M.A. Masse, F.E. Karasz, Anisotropic mechanical properties of uniaxially oriented electrically conducting poly(p-phenylene vinylene), Polymer (Guildf). 30 (1989) 1992–1996.
[42] N. Somanathan, G. Wegner, Mechanical properties of conducting poly(3-cyclohexyl thiophene) films, Polymer (Guildf). 37 (1996) 1891–1895.
[43] E. Sgreccia, M. Khadhraoui, C. de Bonis, S. Licoccia, M.L. Di Vona, P. Knauth, Mechanical properties of hybrid proton conducting polymer blends based on sulfonated polyetheretherketones, J. Power Sources. 178 (2008) 667–670.
[44] T.H. Ting, Y.N. Jau, R.P. Yu, Microwave absorbing properties of polyaniline/multiwalled carbon nanotube composites with various polyaniline contents, Appl. Surf. Sci. 258 (2012) 3184–3190.
[45] S.W. Phang, M. Tadokoro, J. Watanabe, N. Kuramoto, Microwave absorption behaviors of polyaniline nanocomposites containing TiO2 nanoparticles, Curr. Appl. Phys. 8 (2008) 391–394.
[46] Q. Guo, K. Du, X. Guo, F. Wang, Electrochemical impedance spectroscope analysis of microwave absorbing coatings on magnesium alloy in 3.5 wt.% NaCl solution, Electrochim. Acta. 98 (2013) 190–198.
[47] S.H. Hosseini, M. Sadeghi, Investigation of microwave absorbing properties for magnetic nanofiber of polystyrene-polyvinylpyrrolidone, Curr. Appl. Phys. 14 (2014) 928–931.
[48] C. Xu, L. Fang, Q. Huang, B. Yin, H. Ruan, D. Li, Preparation and surface wettability of TiO2 nanorod films modified with triethoxyoctylsilane, Thin Solid Films. 531 (2013) 255–260.
[49] T. Darmanin, F. Guittard, Wettability of poly(3-alkyl-3,4-propylenedioxythiophene) fibrous structures forming nanoporous, microporous or micro/nanostructured networks, Mater. Chem. Phys. 146 (2014) 6–11.
[50] P. Lin, F. Yan, H.L.W. Chan, Improvement of the tunable wettability property of poly(3-alkylthiophene) films, Langmuir. 25 (2009) 7465–7470.
[51] K.S. Teh, Y. Takahashi, Z. Yao, Y.W. Lu, Influence of redox-induced restructuring of polypyrrole on its surface morphology and wettability, Sens. Actuators, A Phys. 155 (2009) 113–119.
[52] F.G. Boyaci San, I. Isik-Gulsac, Effect of surface wettability of polymer composite bipolar plates on polymer electrolyte membrane fuel cell performances, Int. J. Hydrogen Energy. 38 (2013) 4089–4098.

3 Methods of Characterization of 3D-Printed Objects

Yuanyuan Chen

3.1 INTRODUCTION

The popularity of 3D printers has increased in the past decade, partially attributed to the expiration of significant patents and the availability of cost-effective equipment. The utilisation of 3D printing technology offers numerous benefits, including cost efficiency, expedited prototyping, minimised waste production, the ability to create lightweight structures, and decreased energy consumption, among others. According to a recent study, the global market for 3D printing products and services reached an estimated value of approximately 12.6 billion USD in the year 2020. Projections indicate that this market is poised to have a compound annual growth rate of 17% throughout the period spanning 2020 to 2023 [1]. The utilisation of 3D printing technology has been observed in various domains, including but not limited to medical devices and the automobile industry, electronics sector, construction field, and aerospace sector. This chapter will elucidate the primary techniques employed for characterising three-dimensional printed objects.

3.2 DIMENSIONAL ACCURACY TESTING

The assessment of dimensional accuracy holds significant importance in the realm of 3D-printed products. The primary approach for evaluating the precision of 3D-printed things involves comparing them with their corresponding digital models.

Callipers are widely utilised instruments for assessing dimensional correctness. The primary dimensions include height, width, length, diameter, and other relevant measurements. The assessment of linear dimensional correctness in 3D-printed objects involves the measurement of their dimensions along the X-direction, Y-direction, and Z-direction. Dimensional errors can be quantified by determining the disparity between the physical dimension and the corresponding dimension in the digital model. In their study, Carew et al. employed three different 3D printing technologies, fused deposition modelling (FDM), selective laser sintering (SLS), and stereolithography (SLA), to fabricate 3D-printed replicas of skeletal elements obtained from computed tomography (CT) scans. These replicas were utilised for the purpose of reconstructing forensic anthropology evidence. To assess the accuracy of the 3D-printed replicas, the dimensions of both the physical replicas and virtual 3D models were compared using callipers. The study revealed that every

DOI: 10.1201/9781003415985-3

printer generated replicas with mean differences falling within the range of ±1.2 mm. Furthermore, the printer utilising the SLS technique demonstrated the highest level of accuracy and produced prints that closely resembled the original CT scan in terms of aesthetics.

Nevertheless, a significant number of 3D-printed items possess intricate geometries, such as personalised medical apparatus, hence imposing limitations on the precision achievable through calliper measurements. The dimensional accuracy verification of complicated 3D-printed objects is frequently conducted using a coordinate measuring machine (CMM) (Figure 3.1a). The process involves the utilisation of a probe to detect discrete points on the surface of physical objects, hence enabling the measurement of their geometry.

In their study, Kacmarcik et al. conducted an analysis of the accuracy performance of two FDM 3D printers, the Ultimaker 2+ and ZEN3D. The samples were designed using SolidWorks, computer-aided design (CAD) software, incorporating various geometric elements such as cylinders, circles, cuboids, and squares. Subsequently, two FDM printers were employed to fabricate the designed samples. To assess the accuracy of the 3D-printed items, their dimensions and tolerances were measured using CMM scans. Finally, the variations in the dimensional parameters of the printed samples were determined. According to the findings, the Ultimaker 2+ printer exhibited superior dimensional accuracy in comparison to the ZEN3D printer [2].

Sandhu et al. conducted a study in which they utilised an FDM 3D printer to fabricate a drilling bit using acrylonitrile-butadiene-styrene (ABS) material. The researchers employed a CMM to assess any variations in the width and depth of the cut or groove generated by the ABS drilling bit. The findings of their investigation indicated that the 3D-printed drilling bit exhibited the ability to effectively machine soft materials while maintaining precise dimensional and topographic characteristics [3].

In order to enhance the precision of CMM, the integration of a micro tactile probe might be considered. This integration is particularly useful for accurately measuring intricate geometric features within the millimetre to micrometre scale, as depicted in Figure 3.1b [4]. The application of laser beams for ultra-high-precision measurement on the CMM was described, resulting in a measurement accuracy of ±6 μm [5].

Image analysis is another frequently employed method for evaluating the dimensional precision of 3D-printed items. This approach entails capturing visual representations of the 3D-printed artefact and juxtaposing them with the design model, subsequently computing the geometric disparities between the two.

Lin conducted research on the development of several resins specifically designed for UV digital light 3D printing. Additionally, Lin assessed the precision and fidelity of the printing materials by conducting a comparative analysis between the physical 3D-printed models and their corresponding digital models. The master digital model of a typodont molar was obtained by scanning it using an intraoral scanner. Subsequently, the digital molar model was replicated by printing it using formulated resins. The printed molars were then rescanned and compared to the master digital model in order to assess any deviations that occurred during the printing process. According to the paper, a 3D printing resin was developed using a composition consisting of 80% ethoxylated bisphenol A-dimethacrylate, 10% urethane

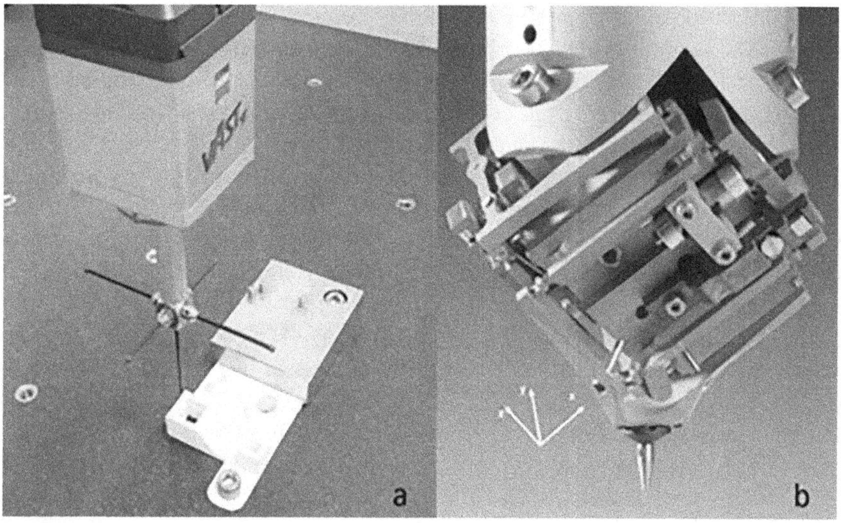

FIGURE 3.1 a) CMM for measuring 3D-printed objects. Adapted from [2]. Copyright the authors, some rights reserved; exclusive licensee IOP Publishing. Distributed under a Creative Commons Attribution License 4.0 (CC BY). b) Tactile microprobe for CMM. Adapted from [4]. Copyright the authors, some rights reserved; exclusive licensee MDPI. Distributed under a Creative Commons Attribution License 4.0 (CC BY).

dimethacrylate, and 10% triethylene glycol dimethacrylate. The resin was found to exhibit great precision and is deemed suitable for potential clinical applications in the future [6].

In their study, Fang et al. established an online methodology for image processing by incorporating geometric disparities into a closed-loop control system. The feedback mechanisms pertaining to the geometric disparities serve to rectify any flaws in the 3D printing procedure, hence facilitating the real-time detection and monitoring of errors [7]. In order to enhance the validity of the data collected, Straub devised multi-camera system and image processing software for the evaluation of access printing progress [8].

3.3 MORPHOLOGY AND SURFACE FINISH EVALUATION

The initial stage in examining the morphology and surface texture of 3D-printed objects involves doing an optical research. Certain faults can be readily detected, including over- and under-fill, loss of continuity, uneven layer marks, and irregular layers, among others.

An optical microscope, commonly referred to as a light microscope, employs optical principles to generate magnified images of minute sections of 3D-printed objects in order to detect any faults. In their study, Gkartzou et al. employed an optical microscope to examine the morphology of individual fibres that were extruded under various printing speeds, printing temperatures, and fibre widths [9].

The utilisation of a confocal microscope is frequently employed in the examination of surface roughness in 3D-printed objects. This instrument enables the optical segmentation of the surface, which is subsequently transformed into digital images and a topographic map through computational processing. By employing an algorithm, the computer is able to analyse the topographic map and derive a roughness parameter that effectively characterises the texture of the surface [10]. In their study, Polzin et al. employed a confocal microscope to examine 3D-printed items made from poly(methyl methacrylate) (PMMA). Their findings indicated that the introduction of wax infiltration led to a reduction in surface roughness of these things by as much as 50% [11].

The utilisation of an atomic force microscope (AFM) presents itself as a valuable instrument for the examination of surface roughness in the context of 3D-printed artefacts. The system functions based on the concept of surface sensing, employing a highly precise tip located on a micromachined silicon probe. The tip is affixed to a diminutive cantilever structure, so that when the tip comes into contact with the surface of the test material, the cantilever undergoes deflection, thereby providing an indication of the surface condition of the test material. In their study, Gorji et al. employed AFM as a means to investigate the surface roughness characteristics of stainless steel 316L powders, specifically in the context of powder bed fusion 3D printing [12].

The presence of air voids and pores within 3D-printed items is a widely seen phenomenon. The application of X-ray computed tomography (XCT) has been employed for the assessment of pore distribution and pore size within 3D-printed plastic and concrete structures. XCT employs a chromatic X-ray cone beam as a non-contact radiation medium in order to reveal the 3-dimensional geometric characteristics using a non-destructive methodology. In their study, Wang et al. employed XCT as a means to quantify the presence of air voids within FDM 3D-printed items. The dispersion and the size of the voids can be measured directly on the XCT images [13]. The void content or porosity of a sample can be measured by comparing the weight of wet samples and dried samples with Eq. (1):

$$Porosity = \frac{W_{wet} - W_{dry}}{W_{wet}} \times 100\% \qquad \text{Eq. (1)}$$

Mercury intrusion porosimetry (MIP) is a commonly employed technique for the characterisation of 3D-printed concrete materials. This method entails the application of pressure to introduce mercury into the pores of the sample, followed by the measurement of the volume of mercury utilised. The selection of mercury for the porosity study was based on its non-wetting characteristic, ensuring that it does not get absorbed by the test samples. Additionally, the need for external pressure to fill the spaces contributes to the high accuracy of porosity measurements [**14**]. The quantification of air voids present in 3D-printed concrete, mortar, and grout objects can be accomplished by adhering to the DIN EN 480–4 standard. This involves assessing the quantity of mixing water within the concrete, which can be segregated by analysing samples of fresh concrete both with and without admixtures.

Electron microscopy includes various techniques, including scanning electron microscopy (SEM), transmission electron microscopy (TEM), and field emission scanning electron microscopy (FESEM), which employ an accelerated electron beam rather than relying on light waves as in optical microscopy. The flow of electrons has both particle-like and wave-like properties, resulting in significant amplification and enhanced clarity. The utilisation of electron microscopy is frequently employed for the examination of the morphological characteristics of objects produced using three-dimensional printing techniques. Liu et al. utilised SEM to observe the voids between each layer of the items 3D printed via fused granule fabrication [15]. Nguyen et al. conducted a study wherein they generated a composite material for FDM 3D printing. They employed SEM to examine the flow direction or orientation of cellulose fibres within the polymer matrix during the 3D printing procedure [16].

3.4 MECHANICAL TESTING

Currently, there is a lack of universally accepted test protocols for the purpose of mechanically characterising things produced by 3D printing technology. Consequently, the mechanical properties of 3D-printed things have been evaluated by typical testing for objects produced using conventional methods. These tests include tension, compression, bending, fatigue loading, impact loading, and creep [17]. Table 3.1 presents the mechanical test standards utilised by the American Society for Testing Materials (ASTM) and the International Organisation for Standardisation (ISO) in the evaluation of 3D-printed products.

The mechanical parameters that are commonly evaluated for 3D-printed objects include Young's modulus, yield strength, ultimate strength, shear modulus, storage and loss modulus, elasticity, and elongation at break. Various mechanical properties can be ascertained through the utilisation of diverse testing methodologies. For example, the stress–strain relationship and Young's modulus of a material can be assessed through a tensile test. The shear modulus of a material can be determined by conducting either a static torsion test or a dynamic oscillatory rheometer. The hardness of a material can be evaluated using various hardness tests, such as the Brinell, Rockwell, and Vickers hardness tests. The toughness of a material can be measured through the Charpy test. The deformation behaviour of a material can be analysed using creep and fatigue tests. Metallographic analysis is commonly employed for the examination of metal 3D printing processes.

Ideally, it is envisioned that the mechanical qualities of 3D-printed things will closely resemble those of objects produced using traditional manufacturing processes. Nevertheless, the utilisation of the layer-to-layer production technique in 3D printing gives rise to interfacial weakness, resulting in anisotropy and heterogeneous mechanical characteristics. In their study, Ahn et al. conducted a comparison of the mechanical qualities between items produced using FDM printing and injection moulding. Their findings indicated that the FDM-printed objects exhibited around 70% and 85% of the tensile and compressive strength, respectively, when compared to the injection moulded products [18].

TABLE 3.1
Commonly Used ASTM and ISO Standards for Mechanical Property Characterisation

Standards	Type of Mechanical Testing	Test Material
ASTM D638	Tensile test	Plastics
ASTM D412	Tensile test	Vulcanised rubber and thermoplastic elastomers
ASTM D882	Tensile test	Thin plastic sheeting
ASTM D3039	Tensile test	Polymer composites
ISO 527	Tensile test	Plastics
ISO 37	Tensile test	Thermoplastic and vulcanised rubbers
ASTM D790	Flexural test	Plastics and electrical insulating materials
ISO 178	Flexural test	Rigid and semi-rigid plastics
ASTM D1938	Tear propagation resistance	Plastic film or sheeting
ISO 34	Tear test	Rubber, vulcanised or thermoplastics
ASTM D695–02a	Compression test	Rigid plastics
ASTM C109	Compressive test	Concrete
ISO 604	Compressive test	Rigid plastics
ASTM C348	Flexural test	Concrete
ASTM D3518	In-plane shear test	Polymer composites reinforced by fibres
ASTM D256	Izod impact test	Notched plastic
ASTM D6110	Charpy impact test	Notched plastic
ISO 180:2000	Izod impact test	Notched plastic
ISO 179–1:2010	Charpy impact test	Notched plastic
ISO 179–2:1997	Charpy impact test	Notched plastic
ASTM E384	Microindentation hardness test	Wide range of materials
ASTM D2240	Shore durometer hardness test	Elastomers
ASTM D785	Rockwell hardness test	Plastics
ISO 2039–1	Hardness test—ball indentation method	Plastics
ISO 2039–2:1987	Rockwell hardness test	Plastics
ASTM D2990	Tensile, compressive, flexural creep, and creep rupture	Plastics
ISO 899–1:2003C	Creep behaviour	Plastics
ISO 899–2:2003	Flexural creep behaviour by three-point loading	Plastics
ASTM D7791	Fatigue	Plastics
ASTM D3479	Fatigue	Polymer composites
ASTM D7774	Fatigue test under bending	Plastics
ISO 13003	Fatigue test	Fibre-reinforced plastics

The mechanical characteristics of three-dimensional printed items can be influenced by the materials utilised in the process of three-dimensional printing. Various types of reinforced 3D printing materials have been developed to enhance the mechanical properties of 3D-printed objects. These include short fibre-reinforced 3D printing material [19], long fibre-reinforced 3D printing material [20], continuous fibre-reinforced 3D printing material [21], and polymer composites [22].

The mechanical qualities of 3D-printed items have been seen to be influenced by the employed printing processes. The study conducted by Khosravani et al. examined the impact of raster orientations, namely the angle of the printing direction in relation to the loading direction, on the mechanical properties of 3D-printed things. The findings of the study indicated that as the raster direction increased, there was a corresponding drop in the strength of the 3D-printed products. According to the referenced study [23], the samples oriented at 0° exhibited the maximum level of strength, whilst the samples oriented at 90° demonstrated the lowest strength values. To clarify, optimal tensile properties are achieved when filaments are aligned longitudinally and parallel to the direction of loading, while the least favourable tensile properties are observed when samples are loaded along the build direction. This can be attributed to the inadequate bonding between layers [17].

In their study, Rankouhi et al. investigated the influence of layer thickness on the mechanical properties of 3D-printed objects. Their findings indicated that the specimens printed with a layer thickness of 0.2 mm demonstrated superior elastic modulus and ultimate strength when compared to those printed with a layer thickness of 0.4 mm. This disparity in mechanical properties can be attributed to the significantly reduced presence of air gaps between each bead or strand in the samples printed with a layer thickness of 0.2 mm, as opposed to those printed with a layer thickness of 0.4 mm [24].

Various 3D printing procedures yield distinct mechanical characteristics in the produced 3D-printed artefacts. The mechanical anisotropy of FDM is estimated to be roughly 50%, making it the highest among all available methods. The mechanical anisotropy of SLA is seen to be rather low, measuring at roughly 1%. Various parameters have been identified as potential influencers on the mechanical properties of SLA-printed items. These factors include curing wavelengths, annealing temperatures, resolution, and layer thickness, among others. Form Labs, a prominent producer in the field of SLA, recently released a white paper that investigates the influence of curing parameters on the mechanical characteristics of SLA-printed items. According to their findings, specimens exposed to a wavelength of 405 nm and subjected to a curing temperature of 60°C exhibited the most notable tensile strength and modulus [25]. The mechanical anisotropy of SLS is relatively low, estimated to be under 10%. Various parameters, such as energy density, laser power, scan spacing, and laser beam speed, have the potential to influence the mechanical properties of items manufactured using SLS technology.

3.5 CHEMICAL PROPERTY TESTING

The assessment of the chemical composition of the 3D-printed object typically involves examining the impact of 3D printing techniques and the ageing process on the objects, as well as exploring advancements in the development of new materials for 3D printing.

The Fourier transform infrared spectroscopy (FTIR) technique is employed to get an infrared spectrum that includes the absorption, emission, and photoconductivity characteristics of the substance under investigation. The characterisation of the chemical properties of 3D printing material is a widely employed practice. The detection of heat breakdown in thermoplastics during FDM 3D printing has been reported. The rise in absorption peaks for alkenes and aromatic chemicals can be attributed to heat deterioration in the context of ABS. The vibrational frequencies associated with alkenes and aromatic compounds are observed at wavenumbers of 3074, 1630, 910, 3033, 1496, and 698 cm^{-1} [26]. Polylactic acid (PLA) exhibits distinctive peaks at 1207 cm^{-1}, which can be attributed to the vibration of the alkyl-ketone chain, and at 920 cm^{-1}, which corresponds to the flexural vibration of the C-H bond. These peaks serve as indicators of the crystalline structure of PLA. However, the presence or absence of these peaks is influenced by the processing conditions. For instance, when fast cooling is employed during injection moulding, these peaks may vanish. Conversely, when a prolonged crystallisation process is carried out during annealing, these peaks may exhibit higher absorbance [27].

Raman spectroscopy is commonly employed for the purpose of characterising the structural properties of the material under investigation. This technique is based on the phenomenon of Raman scattering, wherein photons interact with molecular vibrations, so yielding a distinctive structural fingerprint of the molecules being analysed. The authors Roman et al. formulated inks containing lignin and graphene oxide for the purpose of 3D printing. To verify the presence of pyrolyzed carbon materials, Raman spectroscopy was employed, which revealed the typical bands at around 1350, 1580, and 2700 cm^{-1} [28]. Raman spectroscopy is frequently employed for the purpose of ascertaining the amorphous and crystalline characteristics of the substance under investigation. Trenfield et al. employed SLS 3D printing to fabricate tables containing hydroxypropyl cellulose. The researchers then utilised Raman spectroscopy as a means to quantitatively assess the amorphous composition of the 3D-printed pharmaceutical product [29].

The technique of nuclear magnetic resonance (NMR) utilises a robust and unchanging magnetic field to investigate the chemical composition of the specimen under examination. Kim et al. conducted a study in which they developed a bioink for digital light 3D printing using glycidyl methacrylate modified silk fibroin. To analyse the chemical properties of the bioink, proton NMR (H-NMR) spectroscopy was employed. The researchers measured the characteristic resonance of the methacrylate vinyl group ($\delta = 6.2–6$ and 5.8–5.6 ppm) and the methyl group of glycidyl methacrylate ($\delta = 1.8$ ppm). Additionally, they observed a decrease in the signal intensity of the lysine methylene group at $\delta = 2.9$ ppm, indicating a modification of the lysine residues present in the silk fibre [30]. Desai and Jagtap conducted a study in which they formulated a fibre reinforced resorcinol epoxy acrylate for use in SLA 3D printing. To assess the purity of the composite and confirm the occurrence of the chemical reaction, they employed carbon 13 nuclear resonance spectroscopy (13C-NMR). The analysis revealed specific resonation peaks at δ 132 ppm (C1) and δ 128 ppm (C2), which are indicative of the presence of acryloyl double bonds. Additionally, an absorbance at δ 167.5 ppm (C3) was observed, corresponding to the carbonyl carbon of the ester component. The presence of absorbance peaks at δ 68.5 ppm (C6) and δ 160 ppm (C7) can be attributed to the oxygen connection with carbon, which signifies the production of the ether group. This observation provides evidence for the excellent purity of the composite, as it suggests the absence or minimum presence of any by-products [31].

X-ray diffraction (XRD) is a technique that utilises a beam of incident X-rays to interact with the crystalline structures of materials, resulting in the diffraction of X-rays in certain directions. By quantifying the intensity of the diffracted X-ray beams, it becomes possible to assess the crystallinity of the materials. XRD is commonly employed in the field of 3D printing material development to ascertain the level of crystallinity and the crystal structure of the materials. For example, researchers have successfully generated composite filaments for FDM 3D printing by combining PLA with lignin. These PLA/lignin composite filaments have demonstrated notable antibacterial and antioxidant capabilities, making them promising candidates for many healthcare applications [32]. PLA exhibits a wide peak within the range of 10° to 25° at 2θ degrees, which is attributed to the semicrystalline characteristics of PLA. The presence of peaks at 2θ = 32° and 34.5° in the PLA/lignin composites suggests that the PLA experienced further crystallisation, which can be attributed to the nucleating properties of lignin (Figure 3.2) [33].

X-ray photoelectron spectroscopy (XPS) is a quantitative spectroscopic technique that operates on the principles of the photoelectric effect, specifically for surface analysis. When subjected to x-ray irradiation, atoms on the near-surface of a material undergo the ejection of photoelectrons. The measurement of the kinetic energy of a

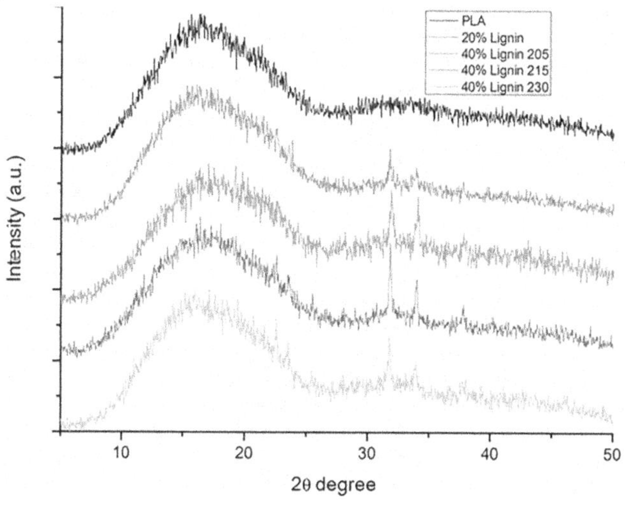

FIGURE 3.2 XRD patterns of PLA/lignin composites. Adapted from [33]. Copyright the authors, some rights reserved; exclusive licensee MDPI. Distributed under a Creative Commons Attribution License 4.0 (CC BY).

photoelectron that has been emitted can be conducted. XPS is frequently employed as a technique for quantifying the elemental composition in the context of material development for 3D printing. The researchers Hu et al. conducted a study in which they utilised a laser 3D printing technique to create objects by combining graphene and aluminium powder. To analyse the composition of the printed items, they employed XPS to quantify the presence of aluminium (Al), oxygen (O), and carbon (C) components on the surface [34].

Energy dispersive X-ray spectroscopy (EDX or EDS) is a widely employed analytical technique utilised for the purpose of elemental analysis or chemical characterisation of a given material. The X-ray beam engages in an interaction with the atoms present in the sample, resulting in the displacement of an electron from its shell, so creating a vacancy. Subsequently, an electron from a higher energy level fills this vacancy, leading to the release of energy. This emitted energy is then measured, and the energy levels of the resulting X-rays are used to establish a correlation with specific elements. In their study, Wang et al. successfully fabricated a magnetic covalent organic framework (COF) and bovine serum albumin (BSA) functionalised 3D-printed electrochemical biosensor. To validate the composition of the created material, the authors employed EDX analysis, which confirmed the presence of carbon (C), oxygen (O), nitrogen (N), sulphur (S), and iron (Fe) components [35].

Gel permeation chromatography (GPC) is a valuable analytical technique employed for the detection of chemical degradation through the quantification of alterations in molar mass. One example of a technique that employs the extrusion process is FDM 3D printing. This method requires the application of heat and shear force. The application of heat and stress to 3D-printed materials can result in a degree of deterioration and a reduction in molar mass. GPC has the capability to determine the molecular weight of materials before and after the 3D printing process. This enables the acquisition of conclusive data on material degradation, which is crucial for ensuring the quality of both the product and the manufacturing process. Furthermore, gas chromatography can be utilised to quantify the decrease in molar mass that occurs during the ageing process. This reduction is attributed to the degradation of chemical bonds resulting from factors such as fatigue, hydrolysis, oxidation, and exposure to UV radiation. Hence, GPC proves a valuable instrument for ascertaining the longevity of 3D-printed items and establishing appropriate storage parameters.

The assessment of particle emission in the atmosphere during 3D printing procedures has been conducted using a particle counter, scanning mobility particle sizer, and aerosol mass spectrometer. Research has indicated that the process of 3D printing results in the release of ultrafine particles. Among the commonly used materials, ABS emits a higher quantity of particles compared to PLA, primarily due to the elevated temperature required for printing. Furthermore, investigations utilising in vitro cellular assays have demonstrated the toxic nature of the particles emitted during the 3D printing process. Consequently, it is advisable to carry out 3D printing activities in well-ventilated environments, as recommended by previous studies [36].

3.6 THERMAL PROPERTY TESTING

Thermal analysis plays a crucial role in the context of hot melt extrusion-based 3D printing technology, namely in the case of FDM. The determination of the printing temperature and bed temperature during the 3D printing process is contingent upon the thermal parameters of the 3D printing material, including viscosity, glass transition temperature, melting temperature, and degradation temperature. Thermal analysis serves as a valuable characterisation tool in the development of innovative 3D printing materials, such as composites.

The printability or melt flow characteristic plays a crucial role in the FDM 3D printing process. If the printing qualities are not optimal, it becomes impossible to fabricate the object according to its intended design. The printability of a 3D printing material can be evaluated using a rheometer, a device that measures the viscosity of a material when subjected to external forces or stresses at elevated temperatures. An optimal printable substance should possess favourable shear-thinning properties to facilitate the process of melt extrusion. It is also crucial for the material to exhibit a very high zero-shear viscosity in order to provide the desired dimensional stability of the extruded melt subsequent to its discharge from the nozzle onto the printing bed. Nguyen et al. conducted a study to examine the printability of acrylonitrile-butadiene-styrene and high-impact polystyrene (HIPS) materials. Their findings on the optimal printing region are shown in Figure 3.3 [16].

Differential scanning calorimetry (DSC) is a widely employed technique in thermal analysis that enables the observation of material behaviour in response to controlled temperature variations, specifically with respect to heat flow rates. The provided data include the glass transition temperature, melting temperature, and crystallinity characteristics of a certain material. In their study, Damadzadeh et al. investigated the

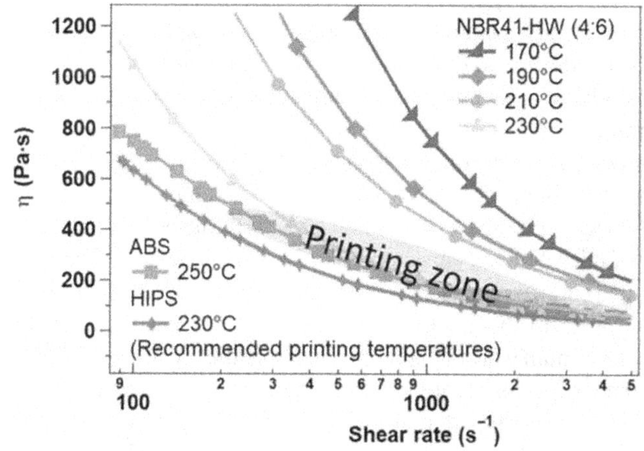

FIGURE 3.3 Viscosity of ABS, HIPS measured by rheology. Adapted from [**16**]. Copyright the authors, some rights reserved; exclusive licensee AAAS. Distributed under a Creative Commons Attribution License 4.0 (CC BY).

development of ceramic filler–reinforced composites using poly(lactic-co-glycolic acid (PLGA) and poly-L-lactic acid (PLLA) polymers for potential medical applications. The researchers observed that the incorporation of hydroxyapatite led to a reduction in the melting temperature of PLGA, as evidenced by the DSC curves. However, no significant impact on the melting temperature of PLLA was seen [37]. Composite filaments consisting of PLA and lignin were formulated specifically for FDM 3D printing. The results of the DSC analysis indicated that the incorporation of lignin led to a decrease in the glass transition temperature of PLA, from 71°C to 59°C. This reduction can be attributed to the enhanced intermolecular spacing and the introduction of stiff phenyl groups within the composite structure [33].

Thermogravimetric analysis (TGA) is a technique utilised to quantify the mass of a given sample in relation to both time and temperature. The primary purpose of its utilisation is to evaluate the thermal stability of a certain material. In their study, Wang et al. successfully fabricated nanocomposites for 3D printing by incorporating L-arg, graphite nanoplatelets, and polylactic acid. The inclusion of these materials resulted in a notable enhancement in thermal stability, as evidenced by a substantial increase of 60°C in the degradation temperature. This improvement in thermal stability not only expanded the processing window but also minimised the risk of thermal degradation occurring during the 3D printing procedure [38]. TGA is employed for the estimation of mineral content or deposition in the context of innovative material creation for 3D printing. In their study, Tanase-Opedal et al. conducted TGA on composites of PLA and lignin. They specifically focused on determining the residual mass of lignin when subjected to a temperature of 800°C. This temperature was chosen because lignin exhibited the formation of char, characterised by the presence of highly condensed aromatic structures, which enabled its retention at elevated temperatures [33].

3.7 OTHER APPLICATION-SPECIFIC TESTS

The utilisation of 3D printing technology has found widespread applications. Consequently, in addition to the conventional characterisation methods outlined earlier, it is imperative to take into account other characterisation methods that are relevant to the unique applications.

The utilisation of 3D printing technology has been implemented in the domain of microfluidics, including applications such as lab-on-a-chip systems and micro-thermal management, among others. The utilisation of 3D printing technology has facilitated the development of microfluidic connectors for microfluidic devices. The primary attribute of the 3D-printed connector is its dimensional correctness, as it has a direct impact on the fluid flow pathway. A further noteworthy characteristic of the 3D-printed fluid connector pertains to its maximum working pressure, denoting the highest level of pressure that the 3D-printed component can endure without experiencing any leakage. Pressure tests can be conducted using either liquid, typically water, or gas, typically compressed air or dry nitrogen. The pressure decreases inside the system or the development of bubbles upon submerging the objects in water serve as indicators of potential leaks (Figure 3.4) [39].

The utilisation of 3D printing technology for fabricating shape memory hydrogels has found significant applications within the domains of soft robotics, biomedicine, and sensing industries. The researchers Shiblee et al. created innovative hydrogels

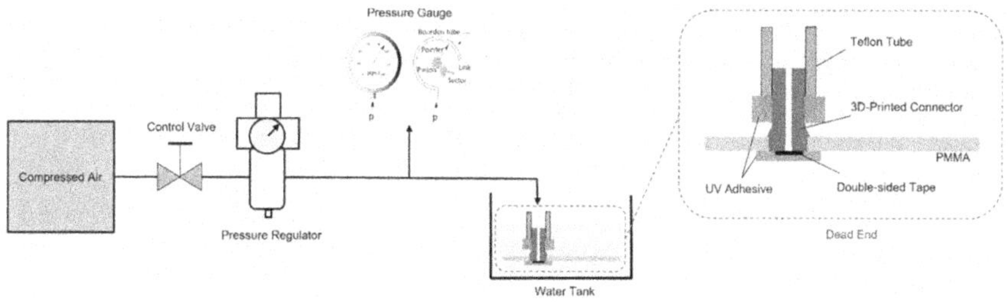

FIGURE 3.4 Pressure test of a 3D-printed connector for a microfluid device. Adapted from [39]. Copyright the authors, some rights reserved; exclusive licensee MDPI. Distributed under a Creative Commons Attribution License 4.0 (CC BY).

with shape memory properties specifically designed for use in SLA 3D printing. In addition to the conventional assessment of physiochemical, thermal, and mechanical properties, the shape memory property was rigorously assessed through the utilisation of a dynamic mechanical analyser employing a programmed thermomechanical technique. The samples were subjected to thermal equilibration at a temperature of 70°C for a duration of 10 minutes. Following this, a tensile load of 120 mN was applied to the samples. Subsequently, the samples were cooled to a temperature of –20°C, and the load was removed. The samples were then warmed to 70°C, and the recovery of the samples was observed [40].

The utilisation of 3D printing technology has been employed in the advancement of battery development. The researchers McOwen et al. successfully formulated 3D printing inks and used them to fabricate solid electrolytes for solid-state batteries. These solid electrolytes exhibited significantly reduced resistance in whole cell configurations, along with enhanced energy and power density. The process of material characterisation involved the use of various analytical techniques. XRD was employed to investigate the structure of the material, while a laser diffraction particle size distribution analyser was utilised to measure the particle size. Additionally, a rheometer was employed to determine the viscosity of the material, and SEM was utilised to examine the morphology of the material. Battery-related characterisation techniques were employed in this study. These techniques included the use of a potentiostat for conducting cell electroanalytical experiments, electrochemical impedance spectroscopy (EIS) for evaluating the ionic conductivity and interfacial charge transport of the developed material, and contact angle measurement for assessing the structural properties of the electrolyte–electrode interface [41, 42].

The utilisation of 3D printing technology presents numerous potential advantages to the field of biomedical science and industry. For example, the utilisation of 3D printing technology enables the production of porous scaffolds that possess meticulously regulated pore diameters, hence facilitating their application in the field of tissue engineering. Porous structures facilitate efficient mass transfer, hence influencing several biological processes such as cell migration, nutrition transportation, blood vessel infiltration, and tissue growth routes within the scaffold. Typically, conventional techniques are employed to characterise 3D-printed materials, including physiochemical analysis, porosity assessment, morphological examination, surface roughness evaluation, mechanical testing, and other related methodologies. Various specific studies are conducted to evaluate the suitability of 3D-printed scaffolds for tissue engineering purposes. These tests include assessments of hydrophilicity as well as in vivo and in vitro biological testing [43].

The measurement of hydrophilicity can be quantified by the use of a goniometer, an instrument specifically designed for determining water contact angles. The experimental procedure involves the application of a small quantity of de-ionised pure water onto the test sample's surface, resulting in the formation of a liquid droplet. Subsequently, the angle formed between the fluid and the surface is determined and referred to as the contact angle. When the angle measures less than 90°, the test material is classified as hydrophilic. Conversely, when the angle measures larger than 90°, the test material is classified as hydrophobic. In their study, Vaidya et al. conducted research on the development of polyhydroxybutyrate (PHB) composites with lignin for the purpose of FDM 3D printing. The authors observed that as the lignin content in the composites grew, the contact angle also increased. This finding suggests that the inclusion of lignin in the composites leads to enhanced hydrophobic properties [44].

The in vitro characterisation of 3D-printed objects for medical applications typically involves the utilisation of cells derived from several sources, including people, rabbits, rats, pigs, goats, and bovines. The objective of the in vitro experiments is to evaluate the biocompatibility and osteogenic capacity of the test specimen. The evaluation of in vitro cell biocompatibility include the examination of cell attachment, cell viability, and cell proliferation. Cell attachment refers to the phenomenon of adhesion between cells and a surface. This biological process includes various molecular interactions, such as ligand binding and intracellular signalling. Cell adhesion plays a pivotal role in facilitating cellular cohesion both among cells themselves and with their surrounding environment. This fundamental process is of utmost importance in preserving the integrity and functionality of tissues. The incorporation of minerals into the polymer matrix has been shown to enhance cell adhesion [45]. Cell viability refers to the quantification of living cells within a given population, while cell proliferation pertains to the assessment of cell division. The assessment of cell viability and proliferation serves as a reliable approach to determine whether the test substance exhibits the capacity to stimulate cellular growth. In their study, Chen et al. fabricated composites of PLA and halloysite nanotubes for the purpose of coronary stent application. The investigation involved assessing the vitality and proliferation of human umbilical vein endothelial cells, which demonstrated that the composites exhibited biocompatibility and were deemed suitable for utilisation in this specific application [46].

The authors Kim et al. formulated a bioink by modifying silk fibroin with glycidyl methacrylate for the purpose of digital light 3D printing. To evaluate the compatibility of the bioink with living cells, specifically NIH/3T3 fibroblasts, a live/dead experiment was employed. The cells were suspended within the bioink that was formulated and subsequently printed using a digital light 3D printer. Following the printing process, the cells were then cultured for a period of 14 days. The cells that were enclosed within the bioink exhibited a predominantly viable state. The long-term biocompatibility of the bioink that was created was evaluated by the utilisation of a digital light 3D printing process to produce a ring-shaped trachea composed of cartilaginous tissue. A 4-week in vitro investigation of the printed trachea revealed a remarkable proliferation of chondrocytes and the development of cartilage tissue [30].

The process of in vivo characterisation involves conducting tests and analyses inside the physiological context of a living organism. In their study, Moncal et al. utilised FDM 3D printing to fabricate bone scaffolds comprising composites of polycaprolactone, poly (D, L-lactide-co-glycolide), and hydroxyapatite. These scaffolds were subsequently implanted into rat calvarial defects to assess their in vivo performance. The findings of this investigation revealed that the composite scaffolds exhibited a significant level of newly mineralised bone tissue formation and degradation after 8 weeks of implantation, as reported in their publication [47].

3.8 CONCLUSION

The utilisation of 3D printing, a technology that is undergoing rapid development, has become prevalent across various industries. The methods employed for characterising 3D-printed things bear resemblance to those employed for conventionally created objects. In addition to the conventional testing and application-specific characterisation methods outlined in this chapter, it is noteworthy to see a growing inclination towards the integration of characterisation methods with the 3D printing process, enabling real-time monitoring of the 3D printing process. The application of a synchrotron imaging system to the laser 3D printing process using metal powder enables the real-time observation of the melting process and subsequent solidification of the powder [48]. ThermoFisher Scientific integrated rheology and Raman spectroscopy into the FDM 3D printing process in order to concurrently observe the impact of extrusion on the crystallisation kinetics and crystal structures of the 3D-printed items [49]. As the use of 3D printing technology continues to expand within the broader manufacturing environment, there will be ongoing advancements and refinement of the associated characterisation approaches.

REFERENCES

[1] Statista Research Department, Global 3D printing industry market size, Statista, 2021.
[2] J. Kacmarcik, D. Spahic, K. Varda, E. Porca, N. Zaimovic-Uzunovic, An investigation of geometrical accuracy of desktop 3D printers using CMM, IOP Conf. Ser. Mater. Sci. Eng. 393 (2018) 12085.
[3] K. Sandhu, G. Singh, S. Singh, R. Kumar, C. Prakash, S. Ramakrishna, G. Królczyk, C.I. Pruncu, Surface characteristics of machined polystyrene with 3D printed thermoplastic tool, Materials. 13 (2020) 2729.
[4] R. Thalmann, F. Meli, A. Küng, State of the art of tactile micro coordinate metrology, Appl. Sci. 6 (2016) 150.
[5] Y. Li, Y. Zhao, Z. Wang, C. Fang, W. Sha, Precision measurement method of laser beams based on coordinate measuring machine, IEEE Access. 7 (2019) 112736–112741.
[6] C.H. Lin, Y.M. Lin, Y.L. Lai, S.Y. Lee, Mechanical properties, accuracy, and cytotoxicity of UV-polymerized 3D printing resins composed of Bis-EMA, UDMA, and TEGDMA, J. Prosthet. Dent. 123 (2020) 349–354.
[7] T. Fang, M.A. Jafari, I. Bakhadyrov, A. Safari, S. Danforth, N. Langrana, On-line defect detection in layered manufacturing using process signature, Proc. IEEE Int. Conf. Syst. Man Cybern. 5 (1998) 4373–4378.

[8] J. Straub, Initial work on the characterization of additive manufacturing (3D printing) using software image analysis, Machines. 3 (2015) 55–71.
[9] E. Gkartzou, E.P. Koumoulos, C.A. Charitidis, Production and 3D printing processing of bio-based thermoplastic filament, Manuf. Rev. 4 (2017) 1.
[10] D.A. Lange, H.M. Jennings, S.P. Shah, Analysis of surface roughness using confocal microscopy, J. Mater. Sci. 28 (1993) 3879–3884.
[11] C. Polzin, S. Spath, H. Seitz, Characterization and evaluation of a PMMA-based 3D printing process, Rapid Prototyp. J. 19 (2013) 37–43.
[12] N.E. Gorji, R. O'Connor, D. Brabazon, X-ray tomography, AFM and Nanoindentation measurements for recyclability analysis of 316L powders in 3D printing process, Procedia Manuf. 47 (2020) 1113–1116.
[13] Z. Wang, L. Fuh, Effect of porosity on mechanical properties of 3D printed polymers: Experiments and micromechanical modeling based on X-ray computed tomography analysis, Polymers (Basel). 11 (2019) 1154.
[14] S. Yu, M. Xia, J. Sanjayan, L. Yang, J. Xiao, H. Du, Microstructural characterization of 3D printed concrete, J. Build. Eng. 44 (2021) 102948.
[15] H. Liu, K. Gong, A. Portela, Z. Cao, R. Dunbar, Y. Chen, Granule-based material extrusion is comparable to filament-base d material extrusion in terms of mechanical performances of printed PLA parts: A comprehensive investigation, Addit. Manuf. 75 (2023) 103744.
[16] N.A. Nguyen, S.H. Barnes, C.C. Bowland, K.M. Meek, K.C. Littrell, J.K. Keum, A.K. Naskar, A path for lignin valorization via additive manufacturing of high-performance sustainable composites with enhanced 3D printability, Sci. Adv. 4 (2018).
[17] J.R.C. Dizon, A.H. Espera, Q. Chen, R.C. Advincula, Mechanical characterization of 3D-printed polymers, Addit. Manuf. 20 (2018) 44–67.
[18] S.H. Ahn, M. Montero, D. Odell, S. Roundy, P.K. Wright, Anisotropic material properties of fused deposition modeling ABS, Rapid Prototyp. J. 8 (2002) 248–257.
[19] K. Korniejenko, M. Łach, S.Y. Chou, W.T. Lin, A. Cheng, M. Hebdowska-Krupa, S. Gadek, J. Mikuła, Mechanical properties of short fiber-reinforced geopolymers made by casted and 3D printing methods: A comparative study, Materials. 13 (2020) 579.
[20] J. Justo, L. Távara, L. García-Guzmán, F. París, Characterization of 3D printed long fibre reinforced composites, Compos. Struct. 185 (2018) 537–548.
[21] F. Mashayekhi, J. Bardon, V. Berthe, H. Perrin, S. Westermann, F. Addiego, Fused filament fabrication for polymers and continuous fiber-reinforced polymer composites: Advance in structure optimization and health monitoring, Polymers (Basel). 13 (2021).
[22] P. Parandoush, D. Lin, A review on additive manufacturing of polymer-fiber composites, Compos. Struct. 182 (2017) 36–53.
[23] M.R. Khosravani, F. Berto, M.R. Ayatollahi, T. Reinicke, Characterization of 3D-printed PLA parts with different raster orientations and printing speeds, Sci. Rep. 121. 12 (2022) 1–9.
[24] B. Rankouhi, S. Javadpour, F. Delfanian, T. Letcher, Failure analysis and mechanical characterization of 3D printed ABS with respect to layer thickness and orientation, J. Fail. Anal. Prev. 16 (2016) 467–481.
[25] Z. Zguris, How mechanical properties of stereolithography 3D prints are affected by UV curing, Formlabs White Pap. (2022). https://go.eacpds.com/acton/attachment/25728/f-069f/1/-/-/-/-/Formlabs%20How%20Mechanical%20Properties%20of%20SLA%20 3D%20Prints%20are%20affected%20by%20UV%20Curing.pdf
[26] S.U. Zhang, Degradation classification of 3D printing thermoplastics using fourier transform infrared spectroscopy and artificial neural networks, Appl. Sci. 8 (2018) 1224.
[27] F. Carrasco, P. Pagès, J. Gámez-Pérez, O.O. Santana, M.L. Maspoch, Processing of poly(lactic acid): Characterization of chemical structure, thermal stability and mechanical properties, Polym. Degrad. Stab. 95 (2010) 116–125.

[28] J. Roman, W. Neri, V. Fierro, A. Celzard, A. Bentaleb, I. Ly, J. Zhong, A. Derré, P. Poulin, Lignin-graphene oxide inks for 3D printing of graphitic materials with tunable density, Nano Today. 33 (2020) 100881.

[29] S.J. Trenfield, P. Januskaite, A. Goyanes, D. Wilsdon, M. Rowland, S. Gaisford, A.W. Basit, Prediction of solid-state form of SLS 3D printed medicines using NIR and Raman spectroscopy, Pharm. 14 (2022) 589.

[30] S.H. Kim, Y.K. Yeon, J.M. Lee, J.R. Chao, Y.J. Lee, Y.B. Seo, M.T. Sultan, O.J. Lee, J.S. Lee, S. Il Yoon, I.S. Hong, G. Khang, S.J. Lee, J.J. Yoo, C.H. Park, Precisely printable and biocompatible silk fibroin bioink for digital light processing 3D printing, Nat. Commun. 91. 9 (2018) 1–14.

[31] P.D. Desai, R.N. Jagtap, Synthesis and characterization of fiber-reinforced resorcinol epoxy acrylate applied to stereolithography 3D printing, ACS Omega. 6 (2021) 31122–31131.

[32] J. Domínguez-Robles, N.K. Martin, M.L. Fong, S.A. Stewart, N.J. Irwin, M.I. Rial-Hermida, R.F. Donnelly, E. Larrañeta, Antioxidant PLA composites containing lignin for 3D printing applications: A potential material for healthcare applications, Pharmaceutics. 11 (2019).

[33] M. Tanase-Opedal, E. Espinosa, A. Rodríguez, G. Chinga-Carrasco, Lignin: A biopolymer from forestry biomass for biocomposites and 3D printing, Materials. 12 (2019) 3006.

[34] Z. Hu, F. Chen, J. Xu, Q. Nian, D. Lin, C. Chen, X. Zhu, Y. Chen, M. Zhang, 3D printing graphene-aluminum nanocomposites, J. Alloys Compd. 746 (2018) 269–276.

[35] L. Wang, W. Gao, S. Ng, M. Pumera, Chiral protein-covalent organic framework 3D-printed structures as chiral biosensors, Anal. Chem. 93 (2021) 5277–5283.

[36] Q. Zhang, M. Pardo, Y. Rudich, I. Kaplan-Ashiri, J.P.S. Wong, A.Y. Davis, M.S. Black, R.J. Weber, Chemical composition and toxicity of particles emitted from a consumer-level 3D printer using various materials, Environ. Sci. Technol. 53 (2019) 12054–12061.

[37] B. Damadzadeh, H. Jabari, M. Skrifvars, K. Airola, N. Moritz, P.K. Vallittu, Effect of ceramic filler content on the mechanical and thermal behaviour of poly-l-lactic acid and poly-l-lactic-co-glycolic acid composites for medical applications, J. Mater. Sci. Mater. Med. 21 (2010) 2523–2531.

[38] Y. Wang, M. Lei, Q. Wei, Y. Wang, J. Zhang, Y. Guo, J. Saroia, 3D printing biocompatible l-Arg/GNPs/PLA nanocomposites with enhanced mechanical property and thermal stability, J. Mater. Sci. 2020 5512. 55 (2020) 5064–5078.

[39] Q. Xu, J.C.C. Lo, S.R. Lee, Characterization and evaluation of 3D-printed connectors for microfluidics characterization and evaluation of 3D-printed connectors for, Micromachines. 12 (2021) 874.

[40] M.D.N.I. Shiblee, K. Ahmed, A. Khosla, M. Kawakami, H. Furukawa, 3D printing of shape memory hydrogels with tunable mechanical properties, Soft Matter. 14 (2018) 7809–7817.

[41] D.W. McOwen, S. Xu, Y. Gong, Y. Wen, G.L. Godbey, J.E. Gritton, T.R. Hamann, J. Dai, G.T. Hitz, L. Hu, E.D. Wachsman, 3D-printing electrolytes for solid-state batteries, Adv. Mater. 30 (2018).

[42] C. Wang, K. Fu, S.P. Kammampata, D.W. McOwen, A.J. Samson, L. Zhang, G.T. Hitz, A.M. Nolan, E.D. Wachsman, Y. Mo, V. Thangadurai, L. Hu, Garnet-type solid-state electrolytes: Materials, interfaces, and batteries, Chem. Rev. 120 (2020) 4257–4300.

[43] C.N. Kelly, A.T. Miller, S.J. Hollister, R.E. Guldberg, K. Gall, Design and structure–function characterization of 3D printed synthetic porous biomaterials for tissue engineering, Adv. Healthc. Mater. 7 (2018) 1–16.

[44] A.A. Vaidya, C. Collet, M. Gaugler, G. Lloyd-Jones, Integrating softwood biorefinery lignin into polyhydroxybutyrate composites and application in 3D printing, Mater. Today Commun. 19 (2019) 286–296.

[45] J. Huang, J. Xiong, J. Liu, W. Zhu, J. Chen, L. Duan, J. Zhang, D. Wang, Evaluation of the novel three-dimensional porous poly (L-lactic acid)/nanohydroxyapatite composite scaffold, Biomed. Mater. Eng. 26 (2015) S197–S205.

[46] Y. Chen, A. Murphy, D. Scholz, L.M. Geever, J.G. Lyons, D.M. Devine, Surface-modified halloysite nanotubes reinforced poly(lactic acid) for use in biodegradable coronary stents, J. Appl. Polym. Sci. 135 (2018) 46521.
[47] K.K. Moncal, D.N. Heo, K.P. Godzik, D.M. Sosnoski, O.D. Mrowczynski, E. Rizk, V. Ozbolat, S.M. Tucker, E.M. Gerhard, M. Dey, G.S. Lewis, J. Yang, I.T. Ozbolat, 3D printing of poly(ε-caprolactone)/poly(D,L-lactide-co-glycolide)/hydroxyapatite composite constructs for bone tissue engineering, J. Mater. Res. 33 (2018) 1972–1986.
[48] C.L.A. Leung, S. Marussi, R.C. Atwood, M. Towrie, P.J. Withers, P.D. Lee, In situ X-ray imaging of defect and molten pool dynamics in laser additive manufacturing, Nat. Commun. 2018 91. 9 (2018) 1–9.
[49] C.D. Millholland, Combining rheology and Raman spectroscopy to analyze 3D printing polymers—advancing materials, ThermoFisher Sci. (2019). https://www.thermofisher.com/blog/materials/combining-rheology-and-raman-spectroscopy-to-analyze-3d-printing-polymers/

4 Electrochemical Properties of Conducting Polymers

Begum Nadira Ferdousi, Mohy Menul Islam, and Md. Mominul Islam

4.1 INTRODUCTION

Conducting polymers (CPs) are the subject of intensive research because of their exceptional qualities such as tunable electrical properties, high optical and mechanical capabilities, ease of synthesis and manufacturing, high environmental stability, and superior corrosion resistance to traditional inorganic materials [1–20]. In addition to these, CPs serve as key components of materials used in electrochemical charge storage [5, 15, 16], electrochromic devices [8, 12], transistors [11], light-emitting diodes [9, 10], actuators [13, 14], photovoltaic cells [17, 18], and sensors [19, 20]. However, the most crucial factor for progress in such emerging fields is achieving control of the electrical or electrochemical properties of CPs.

Different important CPs, as illustrated in Figure 4.1, have been studied continuously since the discovery of polyacetylene (**I**) in 1977 [1–5, 21]. Many methods have already been practiced to prepare CPs, including chemical oxidation, electrochemical polymerization, vapor phase synthesis, hydrothermal, solvothermal, template-assisted, electrospinning, self-assembly, and photochemical methods, the inclusion method, the solid-state method, and plasma polymerization [22–25]. The physicochemical properties of materials, including CPs, depend on the methods of their preparation and post-treatment regulating morphology and the chemical environment.

Recently, the 3D printing technique has received exceptional global attention, as it can create a myriad of high-resolution architectures from digital models [26–28]. The 3D printing technique has shown superiority [26–31] over conventional manufacturing techniques such as aerosol printing [29], ink-jet printing [27], screen printing [28], electrochemical patterning [30], and lithography [31]. Unlike conventional methods, the 3D printing process is effective for producing micro-scale structures in a programmable, simple, and flexible manner with design freedom in 3D space. This chapter focuses mainly on the electrochemical characteristics of CPs. The discussion is started with the development of CPs, which actually hints at how to retain the inherent characteristics of CPs, especially electrical conduction and redox behavior. The electrochemical properties, especially the electrical conduction, redox reactions, electrochromism, and electrical charge storage processes, of CPs are detailed. The electrochemistry of 3D-printed CPs is also discussed.

DOI: 10.1201/9781003415985-4

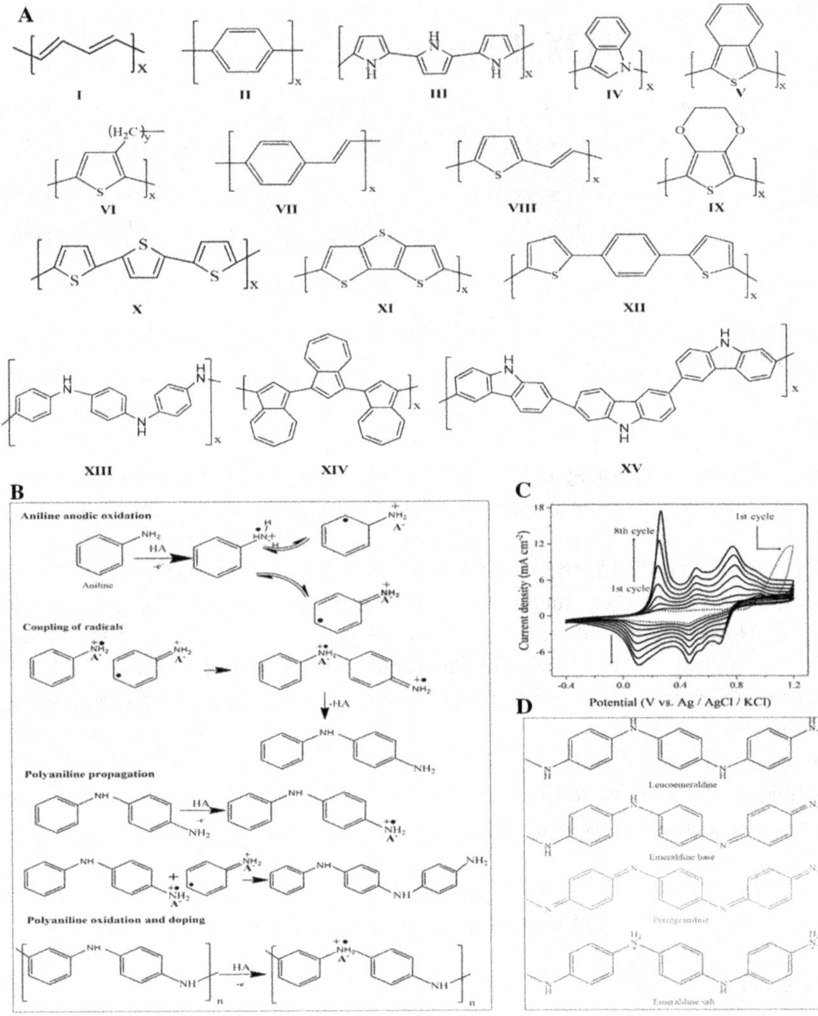

FIGURE 4.1 (A) Building units of different CPs: polyacetylene (I), poly-*p*-phenylen (II), polypyrrole (III), polyindole (IV), poly(isothionaphthalene) (V), poly(3-alkylthiophene) (VI), poly(phenylenevinylene) (VII), poly(thienylenevinylene) (VIII), poly(ethylenedioxythiophene) (PEDOT) (IX), polythiophene (X), poly(dithienothiophene) (XI), poly(dithienylbenzene) (XII), polyaniline (PAni) (XIII), polyazulene (XIV) and polycarbazole (XV). (B) Schematic presentation of mechanism of electrochemical polymerization of aniline. (C) Typical CVs recorded during the deposition of PAni on glassy carbon electrode from 0.05 M aniline in 1.0 M H_2SO_4 solution by continuous cycling of potential at the scan rate of 0.01 V s^{-1} [1–3]. (D) Different oxidation states of XIII.

4.2 ELECTROCHEMISTRY OF CPs

4.2.1 CPs AND ELECTROPOLYMERIZATION

The term 'polymer' was first introduced by Jacob Berzelius in 1832 [2]. Before the 1960s, polymers were generally considered plastic materials or macromolecular compounds due to their solid form. CPs have attracted a lot of interest because of their distinct optical and electrical features, which include the capacity to adjust the molecule structure [1–6]. CPs with interchangeable single (σ) and double (π) bonds in their carbon backbone (Figure 4.1) enable electron delocalization, which contributes to a range of electronic, electrical, electrochemical, and optical properties. In fact, the resistance of the charge/electron flow makes the conventional polymers inferior in technical uses, especially where conductivity is a prerequisite. CPs are a class of polymers with highly reversible redox activity and metallic characteristics that allow for connectivity. In a conductor, electrons can freely move and increase the possibility of charge transfer as the conduction band (CB) overlaps with the valance band (VB). It may be noted that the electrical conductivity of metals is 10^6 S cm^{-1}, while those for semiconductors in range from 10^3 to 10^{-9} Scm^{-1}, and an insulator is 10^{-22} S cm^{-1} [1–5, 32]. However, CPs are generally semiconducting in nature.

Electrochemical polymerization has been considered a simple, fast, and easy technique for the preparation of CPs. Chemical polymerization of aniline was studied in 1862, and later a detailed investigation was carried out by Mohilner *et al.* in 1962 [2, 33]. Dallollio *et al.* first reported the electropolymerization of pyrrole in 1968 [34], although chemical oxidative polymerizations of pyrrole have been reported since 1916 [3, 4]. Figure 4.1 illustrates the electropolymerization events of **XIII** (i.e., PAni) via the cyclic voltammetric method in which the first cycle shows a peak at 1.35 V corresponding to the oxidation of aniline [3, 8]. After the first cycle, two new anodic peaks at 0.5 and 0.9 V corresponding to the respective cathodic peaks appeared, and the anodic peak at 1.35 V disappeared. This indicates the initiation of the oxidative polymerization of aniline. After the second cycle, three redox pairs of cathodic and anodic peaks became visible, and a progressive increase in the peak currents took place due to the potential cycles being repeated. This is indicative of the build-up of surface-bound electro-active material as it occurs during electropolymerization of a monomer, that is, aniline on the surface of the electrode via the mechanism stated in Figure 4.1 [2–4]. However, all of these features are common for electropolymerization of CPs.

The most important characteristic of polymerization forming CPs is the selective position of monomers at which new bond formation takes place (see Figure 4.1 and Table 4.1). The initiation and coupling steps of polymerization in the case of PAni, for example, selectively occur so that polymer backbone that is finally formed can retain the mentioned π-electron delocalization. It has been proved by theoretical study that the initiation step of polymerization of common CPs creates a suitable chemical environment that rightly guides the propagation step in forming a polymer backbone with prerequisite π-electron delocalization [3–6]. The redox reaction of the backbone of CPs is characterized by the shifting of the oxidation potential of the monomer towards a negative potential (Table 4.1), as described later.

TABLE 4.1
Properties of Different Polymers and Chemical Environment of Their Monomers. (Adapted with permission [6]. Copyright (1986) Canadian Science Publishing))

Polymer			Monomer					Refs
Name	σ (S cm^{-1})	Oxidation potential (V)	Chemical structure	π-electron density on monomer	Primary electrophilic reactive sites(s)	Unpaired electron density of radical cation	Oxidation potential (V)	
Polypyrrole III	1–400	0.25–0.30	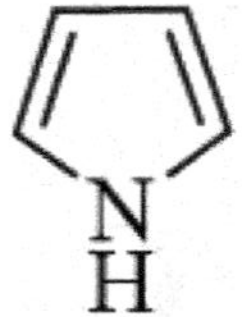	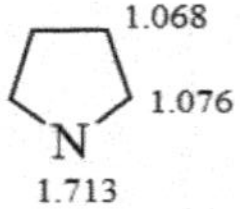	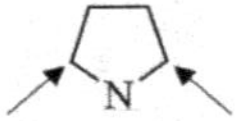	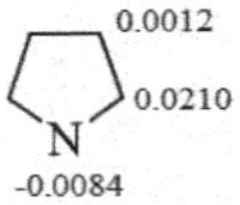	1.20	[4, 6, 35]
Polythiophene X	10–200	0.96		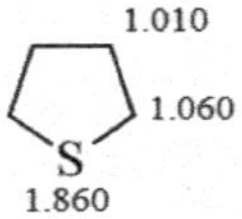	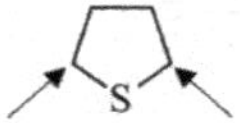		2.06	[6, 7, 36]
Polyindole IV		0.7–0.8	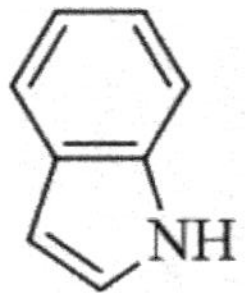	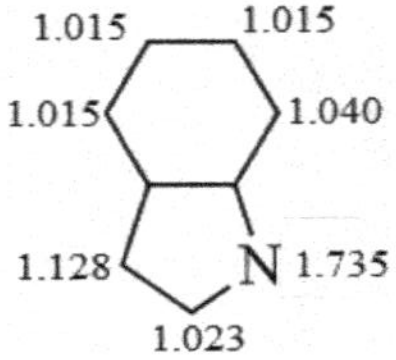	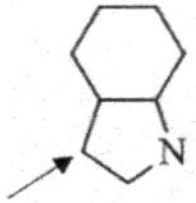	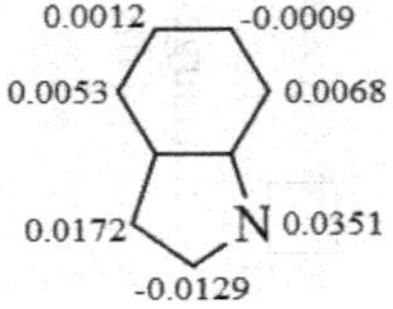	1.26	[6, 35]

Polyaniline XIII	1–400	0.2–0.3, 0.5–0.6, 0.8	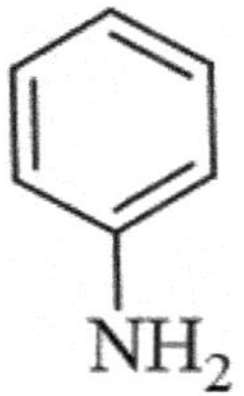	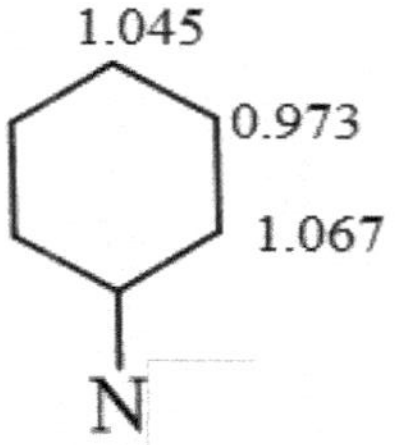	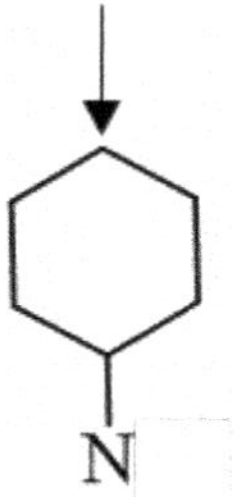	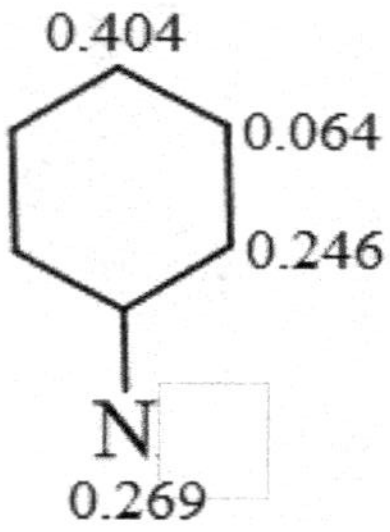	0.8–1.1	[6, 37]
Polyazulene XIV	10–1–10	0.15–0.43	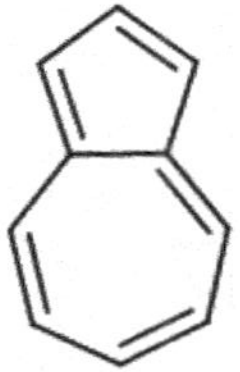	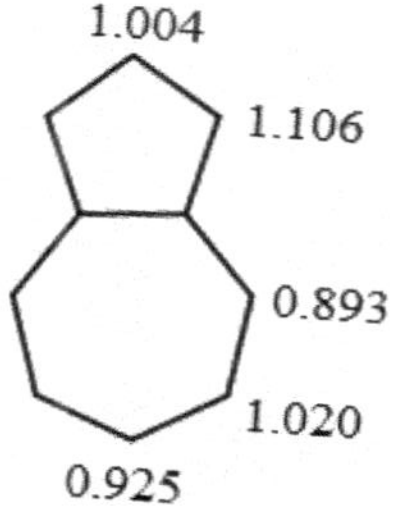	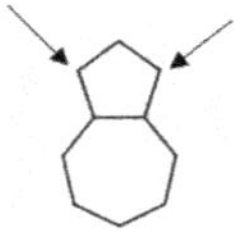	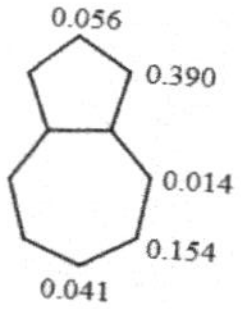	0.96	[6]
Polycarbazole XV		0.9	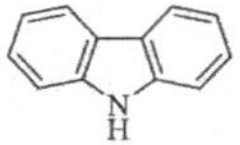	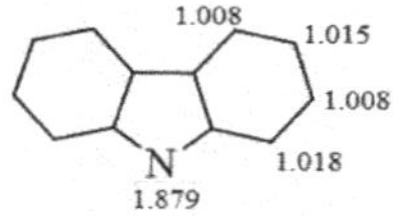	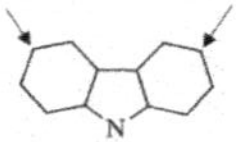	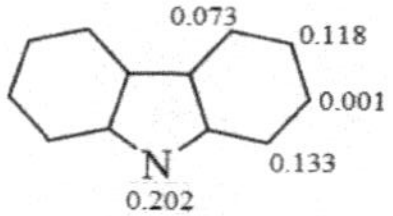	1.30	[6, 38]

4.2.2 Origin of Electrical Conductivity

The electrochemistry of polymers entirely depends on electrical conductivity, and, in fact, their application in the field of electrochemical systems is impossible without electrical conduction. The electrical conductivity of CPs remarkably varies depending on their chemical structures, as summarized in Table 4.1. It may be noted that energy band theory is adequate to clearly explain conduction by CPs, being organic materials. However, the common electronic features of pristine CPs arise from the existing conjugated single and double bonds along the polymer backbone (see Figure 4.1).

The principle of conduction in **I** with a simple chemical structure and high electrical conductivity is illustrated in Figure 4.2. The CPs like **I** conjugated single and double bonds contain a localized σ-bond that forms a strong chemical bond. On the other hand, each double bond possesses a localized π-bond [3–5, 21]. The π-bond between the first and second carbon atoms is transferred to the position between the second and third carbon atoms. In turn, the π-bond between the third and fourth carbon travels to the next carbon, and so on. As a result, the electrons in the double bonds migrate down the carbon chain, allowing electric current to pass via the conjugated double bonds. However, polymeric materials can not be made extremely conductive by conjugated linkages.

The extent of electrical conductivity of CPs belongs to the class of semiconductor of which the band gap energy, which is the difference between the lowest unoccupied

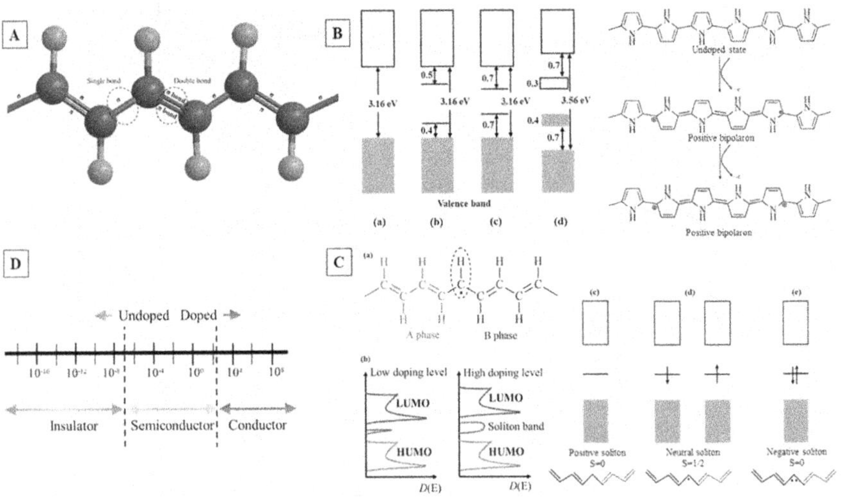

FIGURE 4.2 (A) The structure of I: The backbone contains conjugated double bonds. (B) Electronic bands and chemical structures illustrating (a) undoped, (b) polaron, (c) bipolaron, and (d) fully doped states of III. (C) (a) Schematic illustration of the geometric structure of a neutral soliton on a *trans*-form of I; (b) soliton band with light doping (left) and heavy doping (right) and the band structure of the *trans*-form of I containing (c) a positively charged soliton, (d) a neutral soliton, and (e) a negatively charged soliton. (D) General conductivity range of CPs [15, 21].

molecular orbital (LUMO) and highest occupied molecular orbital (HOMO), is the regulating factor of electrical conduction (Figure 4.2) [2, 3, 15, 21]. Two partially filled atomic orbitals of two identical atoms form two new MOs, of which one is known as VB and the other is an anti-bonding orbital with a higher energy called CB. When the energy gap between CB and VB is large, then electrons cannot participate in conduction or cross the gap. Consequently, the conduction of electricity does not take place, and the material is called an insulator. Several factors are identified to contribute to the band gap of CPs, including configuration of the conjugated bonds, π conjugated length, intermolecular interaction, alternation of bond length, energy-associated resonance, structures of acceptor and donor, effects of substituents, and so on [15]. In degenerate systems like **I**, solitons serve as the charge carriers. On the other hand, in both degenerate and non-degenerate systems like **I** and **III**, polarons and bipolarons act as the charge carriers [1, 2, 15, 21]. Through practicing redox reactions, as described later, one creates radical ions (polarons), dications, or dianions (bipolarons) in the polymer backbone as the charge carriers.

The use of dopants, which are tiny co-planner monomers, can increase electrical conductivity. Extensive research with the CPs was performed by Alan J. Heeger, who was awarded a Nobel Prize in 2000 [21]. Doping means a charge transfer through partial reduction (*n*-type) or partial oxidation (*p*-type) of CPs, as illustrated in Figure 4.2 [2, 15, 21]. These actually enable the movement of the positive or negative charge carriers or electrons along polymer chains, ensuring electrical conduction. Doping polymers with ions aids in the reduction of band gaps and hence enhances conductivity. This happens since the doping process extracts electrons from HOMO of the VB through oxidation or transfer of electrons to the LUMO by reduction.

p-type doping assists in moving the electron from the HOMO to the dopant species and consequently creates a hole in the polymer backbone. Conversely, *n*-type doping enables the electrons to move from the dopant species to the LUMO of the polymer. Hence, through doping, the density and mobility of charge carriers on the CPs backbone can be enhanced [2, 33, 34, 39, 40]. The electrical conduction in **III** is an outcome of *p*-type doping, as shown in Figure 4.2. The chain in **III** exhibits four distinct electronic band structures with different doping levels. The undoped state, **III**, behaves like an insulator with a large band gap of ca. 3.16 eV [15].

In creating a polaron structure, an electron is pulled from the neutral chain of **III**, called oxidation. This changes the benzenoid structure locally to form a quinoid structure of polymer and hence creates two localized electronic levels within the band gap wherein the unpaired electron takes room in the bonding state. In this way, further oxidation results in the removal of a second electron from the backbone of **III** that creates a doubly charged bipolaron. As a result, the energy gap decreases from 3.16 to 1.4 eV due to an overlap between bipolarons [2, 15]. It is worth mentioning that, besides chemical or any other process, doping of CPs can be achieved simply through electrochemical process, as described in the following.

4.2.3 Redox Properties of CPs

One of the most important properties of CPs is their redox reactions. Electrochemical techniques, particularly cyclic voltammetry, are efficient in exploring different features of redox properties of electroactive materials, including CPs. In a straightforward single-electron transfer process, reversible CVs are expected to exhibit fully symmetrical and mirror-image anodic and cathodic waves characterized by identical peak potentials and magnitudes of peak current [3]. CV responses of the redox reactions of different CPs are represented in Figures 4.1 and 4.3–5. As can be seen, the redox peaks of the CPs are generally observed at more negative potentials than those of the corresponding monomers (Table 4.1) [7]. Details about the electrochemical redox reactions of **XIII** (i.e., PAni) were discussed previously. This redox reaction is actually referred to as doping and dedoping processes regulating electrical conductivity. Among the different redox forms of **XIII** illustrated in Figures 4.1 and 4.4, only the emeraldine salt form is conductive [2, 8]. Therefore, through electrochemical techniques, one may switch a non-conducting state to a conductive state. Similar redox features can be noticed in the case of **III** (Figure 4.3). Even blending of two CPs, such as **XIII** with **III**, does not destroy these redox properties, that is, peaks. However, the different redox forms and the level of conductivity of **III** are illustrated in Figure 4.2, and the relevant discussion about the structural changes of its backbone was also given previously.

CPs with multiple redox active cites show a CV with multiple redox peaks; for example, the redox reactions of **VII** with six phenylenvinylene units exhibits at least seven redox states in the potential range between –2.0 and –3.0 V (Figure 4.3). In

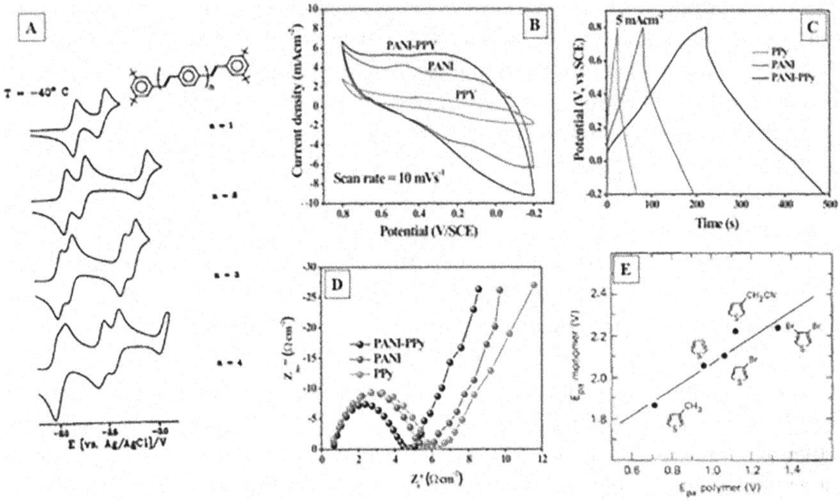

FIGURE 4.3 (A) CVs recorded for the redox reaction of VII (adapted with permission [3]. Copyright (2010) American Chemical Society). (B–D) Electrochemical responses of III, XIII, and their mixture deposited on stainless steel electrodes measured in 1.0 M H_2SO_4 solution. (B) CVs measured at the potential scan rate of 10 mV s^{-1}. (C) GCD curves recorded at 5 mA cm^{-2} and (D) Nyquist plots recorded from 10^3 to 10^{-3} Hz (adapted with permission [41]. Copyright (2012) Elsevier). (E) Oxidation potential of different thiophene derivatives as the monomers of XI vs. their respective polymers (adapted with permission [7]. Copyright (1983) American Chemical Society)

fact, there is a relationship between the length of the chain and the quantity of available redox states. Longer chain lengths lead to the overlapping of redox states across a wide range of potentials, as observed in the case of **VII** [3].

Further information about the redox behavior of CPs can be evaluated by diagnosing the CV with an established model. The peak current for a reversible system is expressed [3, 15] by

$$i = n^2 F^2 \Gamma \upsilon \left[\frac{\exp\theta}{RT(1+\exp\theta)} \right] \tag{1}$$

where $\theta = (nF/RT)(E - E^{\circ})$ and $\Gamma_T = (\Gamma_o + \Gamma_R)$ corresponds to the total surface covered by the reduced and oxidized states. n, F, R, and T are the number of electrons involved, Faraday constant, gas constant, and absolute temperature, respectively. υ is the potential scan rate employed for measuring the voltammogram. This equation is, in fact, applicable solely to monomolecular layers of which i varies linearly with υ, as expected for a surface-attached species [42]. On the other hand, a significant enhancement in the thickness of the film leads to the initiation of diffusion during the redox process. This results in a gradual alternation of peaks from mirror-symmetrical diagrams to the conventional asymmetrical shape, where i is proportional to $\upsilon^{1/2}$ [3, 4].

Substituents attached to the backbones of CPs affect the position of redox peaks. Electron-donating substituents favor oxidation, resulting in shifting the oxidation peak negatively, while electron-withdrawing substituents do the opposite. A linear correlation is observed in the case of **X** (Figure 4.3E), indicating that the substituted monomers and their corresponding polymers are composed of a closely interconnected system of π-electrons. Similar observations have been reported for **III** [7]. This linear correlation between the oxidation potentials of monomers and polymers suggests that the impact of substituents is predominantly of an electronic nature. Such a variation in oxidation potential (E) of CPs formed with substituted monomers may be associated with three factors: polar, steric, and mesomeric effects [6]. The Hammett-Taft equation can be used to evaluate their extent:

$$E = p_{\pi}\sigma^{+i} + S_h \tag{2}$$

where $p_{\pi}\ \sigma^{+i}$ refers to the polar mesomeric effects and S_h is the steric factor. This analysis would help to reveal the position of electrooxidation of the monomers, the elimination of a type of electron (σ or π) from the monomer, and the ease of oxidation of the respective monomers. The factor S_h signifies the level of steric influence on the electrooxidation reaction by the monomer units [6].

As described, the doping method undoubtedly influences the electrochemical characteristics of CPs. It is also crucial to remember that the shapes of CP matrices, particularly at the nanoscale level, can significantly affect their electrochemical properties. For example, different **XIII** nanostructures, namely nanorods, nanospheres, and nanofibers, have been investigated in the presence of SO_4^{2-} ions as a dopant [15]. Despite the similar shapes of CVs, the integrated area of each voltammogram increases in the order of nanospheres < nanorods < nanofibers. The variation of potential of anodic or cathodic peak as a function of the log of υ can be employed to determine kinetic parameters associated with the electrode reactions. Using so-called Laviron theory, the electron transfer coefficient (α) and electron transfer rate constant (k_s) can be calculated. The values of α have been found to be 3.6×10^{-1}–3.7×10^{-1}. The k_s value of 4.3×10^{-1} s^{-1} of nanofibers has been found to be higher than those of other forms: 3.1×10^{-1} s^{-1} for nanorods and 2.6×10^{-1} s^{-1} for nanospheres [15]. Introducing dopants generally induces the generation of charge carriers that is associated with alterations like the transition from benzenoid to quinoid in the geometric configuration of the conjugated polymer.

The formation of films of CPs has been reported to cause significant kinetic limitations through the combination of molecular motions of the backbone plus the penetration of ions and solvent into the film [18]. Diaz et al. studied in detail the effects of electrolyte on the redox reactions of **III** in acetonitrile solution using different inorganic and organic electrolytes possessing ions of various sizes [4, 34]. The variations of electrolyte ions have been found to generate multiple peaks and affect the kinetics of the reaction. The differences observed depend on the types of ions, and it has been suggested that the generation of multiple peaks resulted due to specific interactions between the polymer backbone containing the delocalized π-cation and the anion or the ion pair and/or triplet form of the electrolyte ions [4]. However, a linear relationship between i and υ with zero intercept has been obtained for all electrolytic solutions studied with only a difference in their slopes.

The mechanism of redox reactions of CPs can be explained by so-called bipolaron model, initially proposed by Brédas in the 1980s. In chemical terminology, bipolarons can be understood as di-ionic states ($S = 0$) that arise following oxidation or reduction from their neutral state, as described before [15, 21]. The transition from the neutral state to the bipolaron occurs through the intermediate polaron state (see Figure 4.2), exhibiting a spin of ½ identified with an electron spin resonance measurement. Consequently, this sequential transition aligns with the observed redox transitions in two-step redox systems of **III** (Figure 4.3B). It is worth mentioning that in contrast to conventional redox processes, the charging of the polymer induces an additional local distortion in the chain.

The initial stage of polaron formation requires a relaxation energy (E_{rel}), and the structural relaxation after ionization, on the other hand, involves the release of energy. This process requires a vertical Franck-Condon-like ionization energy [3]. The term E_{rel} means the bonding energy associated with the polaron. The occurrence of structural relaxation induces a localized distortion in the chain surrounding the charge. This distortion transforms the affected segments from a twisted benzoid-like structure to a quinoid-like one wherein the single bonds between the monomeric units become shorter and exhibit characteristics of double bonds.

When a second electron is extracted from the polymer segment, it does not form two polarons. Instead, it creates a bipolaron, which is expected to have a lower energy state than the polaron (Figure 4.2). The disparity in structural relaxation between the bipolaron and the polaron accounts for this phenomenon, with the former exhibiting significantly greater relaxation than the latter. The ionization energy necessary for removing a second electron shows a decrease, or the electron affinity for acquiring a second electron demonstrates an increase. Furthermore, it is postulated that the regionally distorted bipolaron state consists solely of a chain segment composed of four or five units. The energy increase observed in the bipolaron relative to two polarons is estimated to be approximately 0.4 eV [3]. In the context of redox energies, it is implied that the redox potentials associated with the formation of bipolarons should exhibit a notable decrease compared to the potential related to the construction of polarons. The model is predicated on the theoretical assumption that a linear and well-arranged polymer is infinitely long. This assumption leads to the emergence of a band structure in terms of electronic properties, as depicted in Figure 4.2.

4.2.4 Electrochemical Capacitors

An electrochemical capacitor, simply known as a supercapacitor, is an electrical energy storage system. Supercapacitors consist of two porous electrodes which are dipped in an electrolyte and separated by a separator. When potential is applied, the ions get separated and create an electrical double layer (EDL) structure at the interface. As a result, charges are stored at these interfaces [2, 5, 15, 42]. Alternatively, charge storage can be achieved by the combination of a redox reaction with EDL formation. EDL capacitors (EDLCs) and pseudocapacitors are the two main categories of supercapacitors. Compared to EDLCs, pseudocapacitors have a higher capacitance. However, CPs exhibiting pseudocapacitor behavior have attracted a lot of attention due to their high charge density, acceptable cycle stability, quicker charge-discharge ability with predicted low material cost, and so on [15, 41–43].

The charge storage capacity of supercapacitor materials can be assessed with cyclic voltammetry, chronopotentiometry (i.e., galvanostatic charge/discharge (GCD)), and electrochemical impedance spectroscopy (EIS). Relevant analyses of these primary responses provide the specific storage capacity per unit mass of material used [41–43]. Figure 4.3 shows examples of supercapacitor studies of CPs. The CVs of CPs represented in Figures 4.1, 4.3, 4.4, and 4.5 exhibit distinct characteristics, including initial sharp anodic wave(s) during the charging process. Reverse scan results in the emergence of cathodic wave(s) that undergo a possible shift, observed at the negative terminal of the plateau resembling a capacitance [3]. From the anodic or cathodic response of the CV, one may determine the specific capacitance that is pseudocapacitance and varies with the type of CP.

The typical GCD responses of **III**, **XIII**, and a blend of **III** and **XIII** are shown in Figure 4.3C. Generally, the time-dependent potential for EDLC exhibits a linear relationship. Conversely, a change in slope in the time-dependent potential signifies the characteristic behavior of pseudo-capacitance, which arises from electrochemical adsorption/desorption or redox reactions. However, in the present cases, the GCD responses exhibit deviations from the idealized triangular shape due to the redox reactions occurring in the backbones of **III**, **XIII**, and a blend of **III** and **XIII** modified stainless steel electrodes. Using the time required for discharging, specific capacitance can be determined [41].

Figure 4.3D displays the Nyquist plot for these electrodes. In this plot, Z' and Z" refer to the real and imaginary components, respectively. The semi-circle observed in the plot indicates the occurrence of faradaic reactions. The occurrence of a linear

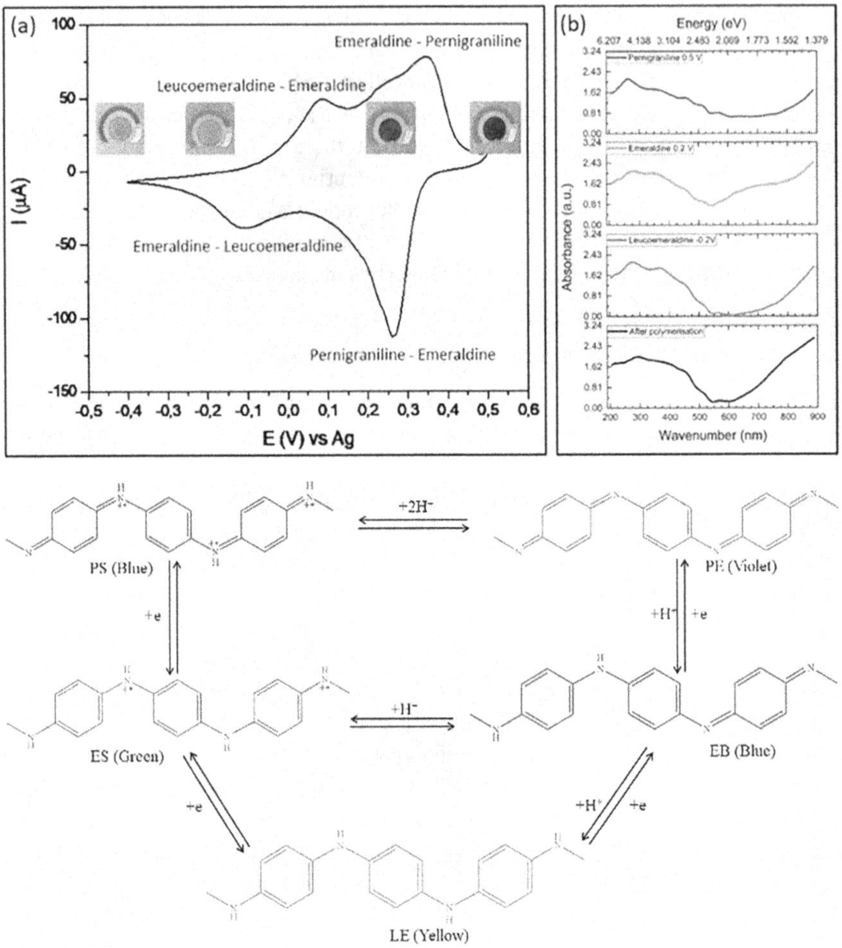

FIGURE 4.4 (A) (a) CV measured for XIII and (b) UV-visible spectra of the different oxidation forms of XIII generated by applying constant potentials of –0.2, 0.2, 0.5, and 0.3 V vs. Ag wire in air (adapted with permission [45]. Copyright (2020) MDPI). (B) The structures of different redox/protonation states and colors of XIII [2].

curve in the low-frequency region is ascribed to the diffusion-controlled process. The initial non-zero intercept at Z′ observed at the onset of the semi-circle exhibits high similarity across all curves. This intercept is attributed to the electrical resistance of the electrolyte. The resistance exhibited by the semi-circle is a result of the resistance of the active electrode material. By simulating the EIS response using a relevant circuit, as shown in the inset of Figure 4.5(d), one can determine the values of different components, including specific capacitance, solution resistance, and charge transfer resistance. It may be mentioned that the values of charge transfer resistance for **III**, **XIII**, and blend of **III** and **XIII** composite electrodes are 6, 5, and

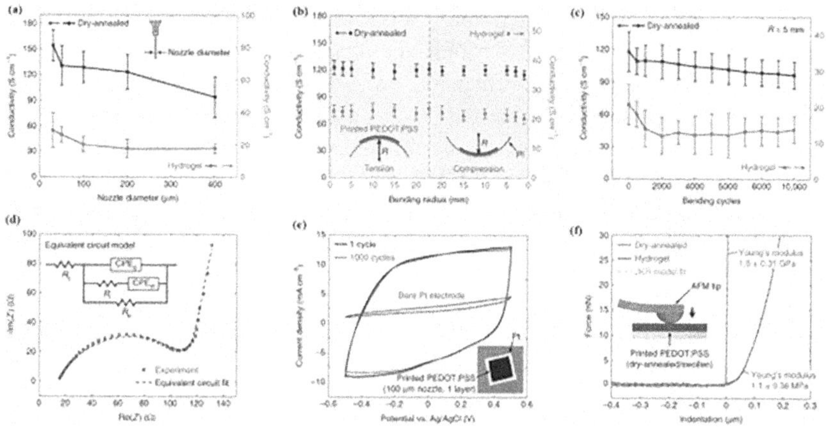

FIGURE 4.5 Properties of 3D-printed PEDOT:PSS CPs: (a, b) conductivity as a function of nozzle diameter for 3D-printed CPs in dry and hydrogel states where PI indicates polyimide substrate; (c), (d) Nyquist plot obtained from the EIS characterization for a 3D-printed CPs on Pt substrate simulated with corresponding equivalent circuit model, where symbols have their usual meaning; (e) CVs; and (f) nanoindentation fitted with JKR model (adapted with permission [26]. Copyright (2020) Springer Nature).

4 ohm cm^{-2}, respectively [26]. This observation suggests that making a composite with two CPs of **III** and **XIII** facilitates the faradaic process.

Further details about the supercapacitive behavior of CPs may be found elsewhere [5, 15, 16, 41, 43]. In addition, an idea of the charge storage capacity of some of the CPs listed in Figure 4.1 may be seen. The values of specific capacitance depend on various factors, including type of backbone of CP, method of preparation that may alter the morphology and microstate of CPs, and electrolytic media. Furthermore, the inherent capacitance of CPs can be enhanced by making composites with different materials, including other polymers, carbonaceous materials, and metal oxides [32]. Moreover, CPs have been advantageously used as a matrix phase of a composite to play a dual role such as to firmly hold the components of the disperse phase and ensure electrical conduction of the disperse phase of non-conducting/insulating materials that cannot singly be employed for electrochemical applications.

4.2.5 Electrochromism

Electrochromism of a material is a phenomenon involving changes in color or opacity in response to an electrical stimulus. Under applied potential or current, an electrochromic material may thus absorb electromagnetic radiation at specific wavelengths of ultraviolet, visible, or (near) infrared light. In the modern era, electrochromic materials have been used to control the amount of light and heat allowed to pass through a surface. In fact, the intriguing property of electrochromism renders CPs highly appealing for utilization in electrochromic applications, including electrochromic displays, smart windows, and rearview mirrors. The majority of CPs, including **X**, **XIII**, **XIV**, **XV**, polypyrene, polyanthracene, polyquinoline, and so on, exhibit electrochromic properties through electrochemical redox reaction. The chromophores of CPs exhibit variations in color across different redox states. The electrochromic characteristic of a CP, such as poly-*N*-methylpyrrole, a derivate of **III**, were first reported in 1981 [2, 4]. Later on, significant research efforts were dedicated to studying electropolymerized **X** and **XIII** [2, 41, 44].

In studying electrochromic properties, cyclic voltammetry is the primary experimental technique. The typical study of **XIII** is represented in Figure 4.4. As discussed, several forms of PAni are associated with pronation/deprotonation and insertion/de-insertion of anions. The presence of nitrogen atoms within the chain facilitates the injection of protons or anions to form radical cations. These redox/protonation states demonstrate intriguing variations in color, transitioning from a transparent yellow color to shades of green, blue, and violet, as can be monitored through *in situ* measurement of absorption spectra (see Figure 4.4).

The observed phenomenon of color change is thought to be contingent upon both the energy gap of the CPs and the presence of dopants. The phenomenon of electrochromism can be attributed to the insertion and de-insertion of dopant ions, which occurs through doping and de-doping mechanisms. Doping leads to a reconfiguration of the electronic structure of the polymers, causing a decrease in the energy gap associated with the π–π^* transition. The absorbance of CPs is influenced by the formation of sub-bands resulting from the presence of charge carriers, such as polarons and bipolarons. These sub-bands contribute to the modulation of the absorbance characteristics of CPs, resulting in perceptible color variations. Typically, the energy gap observed in undoped, pristine CP thin films exceeds 3.0 eV, resulting in a colorless and transparent appearance. In the doped state, thin films can exhibit the most pronounced absorption spectra within the visible region. When pristine CPs possess an energy gap of approximately 1.5 eV, they show a pronounced absorption of visible light and manifest vivid colors, with distinct contrasts in their undoped state. Nevertheless, the absorption wavelength shifts towards the near-infrared region following the doping process.

The electrochromic characteristics of CPs are contingent upon various factors, including their chemical composition, redox potential, temperature, and the pH level of the electrolyte solution. The switching time of color changes is significantly influenced by the migration speed of protons/dopant ions in and out of the polymer matrix, which can be controlled through tuning the morphology and microstructural characteristics of CPs, including their dimensions and porosity, to improve the rate of color transformation.

4.3 ELECTROCHEMISTRY OF 3D-PRINTED CPs

The electrochemical properties of CPs depend on their morphology and microstates that can be tuned by varying the method of preparation and post-treatment. Due to its greatly increased design freedom, structural complexity, and environmental sustainability, 3D printing offers promising future developments in the fields of CPs and organic electronics [26–31]. Because of his groundbreaking patent that he filed in 1984, Chuck Hull is largely considered as the "father" of 3D printing. However, the use of 3D printing to create CP-based materials is still in its infancy and is still fraught with difficulties.

CPs, mainly **III**, **XIII**, and the blend of **IX** and poly(styrene sulfonate) (PEDOT:PSS), have been studied extensively for 3D printing. Materials made of PEDOT:PSS are highly intriguing because they combine tunable electrical transport characteristics with outstanding mechanical qualities [26]. To achieve favorable rheological properties for 3D printing, a paste-like CP ink based on cryogenic freezing of aqueous PEDOT:PSS solution has been developed, followed by lyophilization and controlled re-dispersion in water and dimethyl sulfoxide mixture. The resulting CP ink has exceptional 3D printability, enabling the manufacture of CPs with high resolution (over 30 μm), high aspect ratio (over 20 layers), and great uniformity. The process dry-annealing of the 3D-printed CPs results in extremely conductive in nature with an electrical conductivity above 155 S cm^{-1} and flexible PEDOT: PSS 3D microstructures. In addition, dry-annealed 3D-printed CPs can be easily turned into a soft (Young's modulus <1.1 MPa), but highly conductive (up to 28 S cm^{-1}) PEDOT:PSS hydrogel by swelling in the wet condition [26].

The CV demonstrates a high charge storage capability of 3D-printed PEDOT: PSS (100-μm nozzle, 1 layer on platinum substrate) compared to metallic electrode materials such as platinum, with remarkable electrochemical stability (less than 2% reduction in CSC after 1000 cycles). The occurrence of broad and stable anodic and cathodic peaks measured at various υ suggests a non-diffusional redox process and electrochemical stability of the 3D-printed CPs. Bai et al. recently reported the printing of CP-based composite inks from a mixture of **XIII** and graphene oxide [28]. This 3D-printed composite has been studied for supercapacitor application.

Compared to high-performance CPs prepared in different ways, 3D-printed CPs can attain electrical conductivity as high as 155 S cm^{-1} in the dry state that decreases to 28 S cm^{-1} when it is transferred to a hydrogel state. Notably, a smaller nozzle diameter results in printed CPs with greater electrical conductivity, possibly as a result of shear-induced improvements in the alignment of the PEDOT: PSS nanofibrils. Because 3D-printed CPs are flexible, they can be bent mechanically without breaking, with a maximum strain of 13 and 20% in the dry and hydrogel states, respectively.

4.4 CONCLUDING REMARKS

Modern demands have been expanding the spectrum of research fields. CPs have been recognized as emerging materials for practical applications. Many modern applications of CPs are based on their electrochemical properties. The extent of these properties is regulated by the structure of the backbone that can be further tuned by doping with various techniques, including electrochemical approaches. Redox reactions enable us to tune the physicochemical properties of CPs, including chromism, as can be operated by electrochemical techniques like voltammetry. The excellence of CPs in storing electrical charge in a supercapacitor device supersedes other materials like carbon, metal oxides, and so on. Ensuring the electrical conductivity of insulating materials, CPs share their inherent conduction behaviors in composite with their non-conducting counterparts. Reforming the morphology and microstates of CP-based materials, electrochemical properties, including kinetics of redox reaction and charging/discharging at the interface, can be boosted. In addition, substituents of the monomer, chemical environment of the solvent, types of the electrolyte, etc., influence the redox properties, electrochromism, charging/discharging, and conductivity. Besides the conventional preparation techniques, 3D printing technology may be adopted in developing CP materials with tunable electrochemical properties.

REFERENCES

[1] N. Hall, Twenty-five years of conducting polymers, Chem. Commun. 7 (2003) 1–4.
[2] T. Ohsaka, A.N. Chowdhury, M.A. Rahman, M.M. Islam, Trends in polyaniline research, Nova Publishers, New York (2013).
[3] J. Heinze, B.A. Frontana-Uribe, S. Ludwigs, Electrochemistry of conducting polymers—persistent models and new concepts: Review, Chem. Rev. 110 (2010) 4724–4771.
[4] A.F. Diaz, J.I. Castillo, J.A. Logan, W.-Y. Lee, Electrochemistry of conducting polypyrrole films, J. Electroanal. Chem. 129 (1981) 115–132.
[5] A. Rudge, I. Raistrick, S. Gotesfeld, J.P. Ferrarist, A study of the electrochemical properties of conducting polymers for application in electrochemical capacitors, Electrochimica Acta. 39 (1994) 273–287.
[6] R.J. Waltman, J. Bargon, Electrically conducting polymers: A review of the electropolymerization reaction of the effects of chemical structure on polymer film properties, and of applications towards technology, Can. J. Chem. 64 (1986) 76–95.
[7] A.F. Diaz, Electrochemical studies of some conducting polythiophene film, J. Phys. Chem. 87 (1983) 1459–1463.
[8] R. Prakas, Electrochemistry of polyaniline: Study of the pH effect and electrochromism, J. Appl. Polym. Sci. 83 (2002) 378–385.
[9] O.L. Gribkova, O.D. Omelchenko, A.R. Tameev, D.A. Lypenko, A.A. Nekrasov, O.Y. Posudievskii, V.G. Koshechko, A.V. Vannikov, The specific effect of graphene additives in polyaniline-based nanocomposite layers on performance characteristics of electroluminescent and photovoltaic devices, High Energy Chem. 50 (2016) 134–138.
[10] P. Biju, X. Jining, K.A. Jose, K.V. Vijay, A new synthetic route to enhance polyaniline assembly on carbon nanotubes in tubular composites, Smart Mater. Struct. 13 (2004) N105.
[11] C.T. Kuo, W.H. Chiou, Field-effect transistor with polyaniline thin film as semiconductor, Synth. Met. 88 (1997) 23–30.
[12] K.Y. Shen, C.W. Hu, L.C. Chang, K.C. Ho, A complementary electrochromic device based on carbon nanotubes/conducting polymers, Sol. Energy Mater. Sol. Cells 98 (2012) 294–299.
[13] J.C. García-Gallegos, I. Martín-Gullón, J.A. Conesa, Y.I. Vega-Cantú, F.J. Rodríguez-Macías, The effect of carbon nanofillers on the performance of electromechanical polyaniline-based composite actuators, Nanotechnology 27 (2016) 015501.
[14] T. May, T. Van-Tan, M.S. Geoffrey, G.W. Gordon, Carbon nanotube and polyaniline composite actuators, Smart Mater. Struct. 12 (2003) 626–630.
[15] T.H. Le, Y. Kim, H. Yoon, Electrical and electrochemical properties of conducting polymers, Polymers 9 (2017) 150–182.
[16] X. Jiang, S. Setodoi, S. Fukumoto, I. Imae, K. Komaguchi, J. Yano, H. Mizota, Y. Harima, An easy one-step electrosynthesis of graphene/polyaniline composites and electrochemical capacitor, Carbon 67 (2014) 662–672.
[17] J. Lee, H. Kang, S. Kee, S.H. Lee, S.Y. Jeong, G. Kim, J. Kim, S. Hong, H. Back, K. Lee, Long-term stable recombination layer for tandem polymer solar cells using self-doped conducting polymers, ACS Appl. Mater. Interfaces 8 (2016) 6144–6151.

[18] E.M.J. Johansson, L. Yang, E. Gabrielsson, P.W. Lohse, G. Boschloo, L. Sun, A. Hagfeldt, Combining a small hole-conductor molecule for efficient dye regeneration and a hole-conducting polymer in a solid-state dye-sensitized solar cell, J. Phys. Chem. C 116 (2012) 18070–18078.
[19] J. Janata, M. Josowicz, Conducting polymers in electronic chemical sensors, Nat. Mater. 2 (2003) 19–24.
[20] H. Yoon, Current trends in sensors based on conducting polymer nanomaterials, Nanomaterials 3 (2013) 524–549.
[21] W.P. Su, J.R. Schrieffer, A.J. Heeger, Solitons in polyacetylene, Phys. Rev. Lett. 42 (1979) 1698–1701.
[22] M. Charles, Template synthesis of electronically conductive polymer nanostructures, Acc. Chem. Res. 28 (1995) 61–68.
[23] A.G. MacDiarmid, W.E. Jones, I.D. Norris, J. Gao, A.T. Johnson, N.J. Pinto, J. Hone, B. Han, F.K. Ko, H.E. Okuzaki, M. Llaguno, Electrostatically-generated nanofibers of electronic polymers, Synth. Met. 119 (2001) 27–30.
[24] Q. Tang, J. Wu, X. Sun, Q. Li, J. Lin, Shape and size control of orientated polyaniline microstructure by a self-assembly method, Langmuir 25 (2009) 5253–5257.
[25] K. Namsheer, C.S. Rout, Conducting polymers: A comprehensive review on recent advances in synthesis, properties and applications, RSC Adv. 11 (2021) 5659–5697.
[26] H. Yuk, B. Lu, S. Lin, K. Qu, X. Zhao, J. Xu, J. Luo, 3D printing of conducting polymers. Nat. Commun. 11 (2020) 1604–1612.
[27] H. Sirringhaus, High-resolution inkjet printing of all-polymer transistor circuits, Science 290 (2000) 2123–2126.
[28] R.S. Jordan, Y. Wang, 3D printing of conjugated polymers: Review, J. Polym. Sci. B Polym. Phys. 5 (2019) 1592–1605.
[29] K. Hong, S.H. Kim, A. Mahajan, C.D. Frisbie, Aerosol jet printed p-and n-type electrolyte-gated transistors with a variety of electrode materials: Exploring practical routes to printed electronics, ACS Appl. Mater. Interfaces 6 (2014) 18704–18711.
[30] S. Sekine, Y. Ido, T. Miyake, K. Nagamine, M. Nishizawa, Conducting polymer electrodes printed on hydrogel, J. Am. Chem. Soc. 132 (2010) 13174–13175.
[31] S. Wang, Skin electronics from scalable fabrication of an intrinsically stretchable transistor array, Nature 83(2018) 555.
[32] M.M. Islam, M.S. Islam, M.A.B.H. Susan, M.M. Islam, Conjugated polymers as the materials for supercapacitor electrode in book "Organic electrodes: Fundamental to advanced emerging applications", edited by R.K. Gupta, Springer, Cham, Switzerland (2023) 265–413.
[33] D.M. Mohilner, R.N. Adams, W.J. Argersinger, Investigation of the kinetics and mechanism of the anodic oxidation of aniline in aqueous sulfuric acid solution at a platinum electrode, J. Am. Chem. Soc. 84(19) (1962) 3618–3622.
[34] S. Sadki, P. Schottland, N. Brodlie, G. Sabouraud, The mechanism of pyrrole electropolymerization, Chem. Soc. Rev. 29 (2000) 283–293.
[35] R.L. Flurry, E.W. Stout, J.J. Bell, The inclusion of non-nearest neighbour β terms in Pariser-polypyrrole type S.C.M.O. calculations, Theorem. Chim. Acta 8 (1967) 203–211.
[36] A.J.H. Wachters, D.W. Davies, A theoretical study of the UV spectra of dithienyls and thiophene, Tetrahedron 20 (1964) 2841–2849.
[37] W.W. Focke, G.E. Wnek, Y. Wei, Influence of oxidation state, pH, and counterion on the conductivity of polyaniline, J. Phys. Chem. 91 (1987) 5813–5818.
[38] K. Hattory, Y. Wada, Molecular orbital calculation of the electronic structure of poly (N-vinylcarbazole), J. Polym. Sci. 13 (1975) 1863–1869.
[39] Q. Tang, J. Wu, X. Sun, Q. Li, J. Lin, Shape and size control of oriented polyaniline microstructure by a self–assembly method, Langmuir 25 (2009) 5253–5257.
[40] H.S. Nalwa, Handbook of organic conductive molecules and polymers, Wiley, New York, vol. 1–4 (1997).

[41] D.P. Dubal, S.V. Patil, G.S. Gund, C.D. Lokhande, Polyaniline-polypyrrole nanograined composite via electrostatic adsorption for high performance electrochemical supercapacitor, J. Alloys Comp. 552 (2013) 240–247.
[42] A.J. Bard, L.R. Faulkner, Electrochemical methods: Fundamentals and applications, 2nd Edition, Wiley, New York (2000).
[43] V.S. Jamadade, D.S. Dhawale, C.D. Lokhande, Studies on electrosynthesized leucoemeraldine, emeraldine and pernigraniline forms of polyaniline films and their supercapacitor behavior, Synth. Met. 160 (2010) 955–960.
[44] S.K. Dhawan, M.K. Ram, B.D. Malhotra, S. Chandra. Novel electrochromism phenomenon observed in polyaniline films, Synth. Met. 75(2) (1995) 119–122.
[45] A. Korent, K.Ž. Soderžnik, K.Ž. Rožman, In-situ spectroelectrochemical study of conductive polyaniline forms for sensor applications, Proceedings 56 (2020) 32.

5 Emerging Applications of 3D-Printed Conducting Polymers

Mayankkumar L. Chaudhary, Pratik Patel, and Ram K. Gupta

5.1 INTRODUCTION

3D printing offers capabilities to fabricate microscale structures in a programmable, easy, and flexible manner with freedom of design in 3D space. The development of 3D printing over time is shown in Figure 5.1 [1]. Oftentimes, 3D things are constructed via additive manufacturing (AM), also known as 3D printing, by adding material layers to achieve the required shape. Unlike traditional manufacturing methods like milling and moulding, AM is flexible and can immediately generate complicated structures from a digital model. Whenever a new component is made or the design of an existing part is altered, a new tool, mould, or jig is needed in conventional production. In contrast, 3D printing only requires digital product design and no actual hardware modifications. Manufacturing operations that formerly required a lot of manpower and time are being revolutionised by 3D printing. They help cut down on waste as compared to subtractive production techniques. In addition to its obvious benefits in manufacturing, AM also offers novel perspectives in other fields, including biomedicine, electrical energy storage, electronics, and robotics, thanks to its adaptability in the design of complicated geometries and its ability to accommodate individual preferences [2].

In fused filament manufacturing, a popular form of 3D printing, various fillers, such as metal and ceramic nanoparticles, are added to polymers and polymer-based composites to provide the desired material properties. Except for the directed-energy deposition technique, polymers are one of the most common feedstocks utilised in other 3D printing processes. Printing processes can be optimised for a wide variety of polymers, from thermoplastics to photopolymers, depending on the specific needs of the finished product [2]. Due to their poor mechanical qualities and generally insulating electrical and thermal behaviour, 3D-printed polymer goods are rarely employed for structural and functional parts. To address these shortcomings in mechanical performance, numerous studies have investigated the use of 3D printing for polymer composites with the incorporation of reinforcement components such nanoparticles and short or continuous fibres. Existing studies highlighted many difficulties with the printing and application of soft polymer materials, although this area of research is still in its infancy: (1) There are few forms of content that can be printed. Most

DOI: 10.1201/9781003415985-5

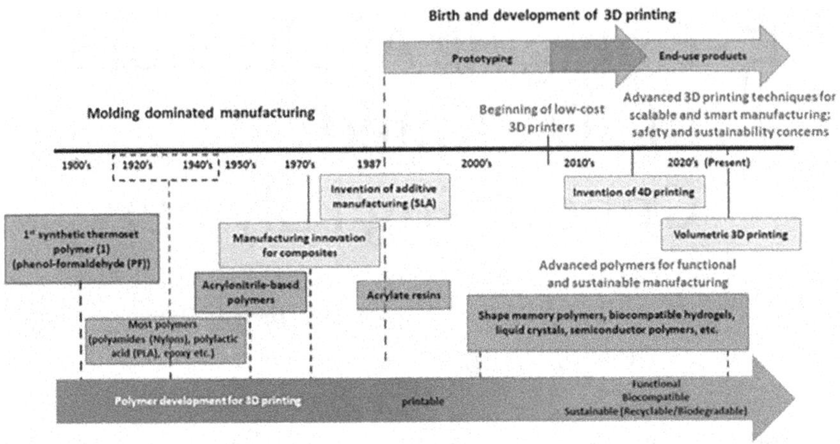

FIGURE 5.1 Timeline of 3D printing using polymers. Adapted with permission [2]. Copyright 2021, Elsevier.

commercial printing platforms still use soft polymer materials in one of three forms: polymer powders, thermoplastic monofilaments, or photocurable resins. This is even though some soft polymer materials with good qualities could not be printed. (2) Low print quality and speed, particularly with more common forms of 3D printing like direct ink writing. (3) Inadequate features. The complexity and variety of printed subjects are both low, limiting the usefulness of printed soft systems. Therefore, the material and techniques of 3D-printed soft materials must be studied in order to realise the enormous potential of 3D printing of soft polymer materials [3].

Given their unique polymeric nature, favourable electrical and mechanical properties, stability, and biocompatibility, conducting polymers have emerged as one of the most promising materials for a wide range of applications, including energy storage, flexible electronics, and bioelectronics [1, 4–6]. Traditional manufacturing techniques like ink-jet printing, screen printing, aerosol printing, electrochemical patterning, and lithography have been relied upon for the most part in the fabrication of conducting polymer structures and devices despite the limitations and challenges these methods present. Sensing performance in analytical and bioanalytical systems can be improved by using conducting polymers with desirable qualities such as high electrochemical activity, high electrical conductivity, high mechanical flexibility, high biocompatibility, and high environmental stability. Because of their extensive technological potential, conductive polymers (CPs) like polypyrrole (Ppy), polyaniline (PANI), polythiophene (PTH), and poly(3,4-ethylenedioxythiophene) (PEDOT) are widely used in a wide range of applications. These include not only sensors and biosensors but also rechargeable batteries, corrosion-preventing coatings, electromagnetic shielding, solar cells, and supercapacitors [7]. Compared to their existing inorganic equivalents, CPs have several desirable properties, including a wide range of chemical diversity, low density, flexibility, corrosion resistance, simple control over structure and morphology, and tuneable conductivity [8]. The properties of CPs have advanced, albeit not to the same extent as their metallic and inorganic semiconductor analogues. To improve upon their poor solubility, poor conductivity, and brittle long-term stability, materials with desirable qualities and novel application possibilities in industries as varied as electronics and energy devices can be created by judiciously combining CPs with other materials.

5.2 BASICS OF 3D PRINTING AND CONDUCTING POLYMERS

Stereolithography (SLA), selective laser sintering (SLS), fused deposition modelling (FDM), and polyjet modelling (PJM) are four common and effective methods of 3D printing. The evolution of each of these four phenomena is detailed in the following.

5.2.1 STEREOLITHOGRAPHY

One of the first 3D printing technologies was stereolithography, with a landmark discovery by Charles Hull that ultimately gave rise to 3D systems. As shown in Figure 5.2, this technique involves exposing a curable liquid to a pattern of light,

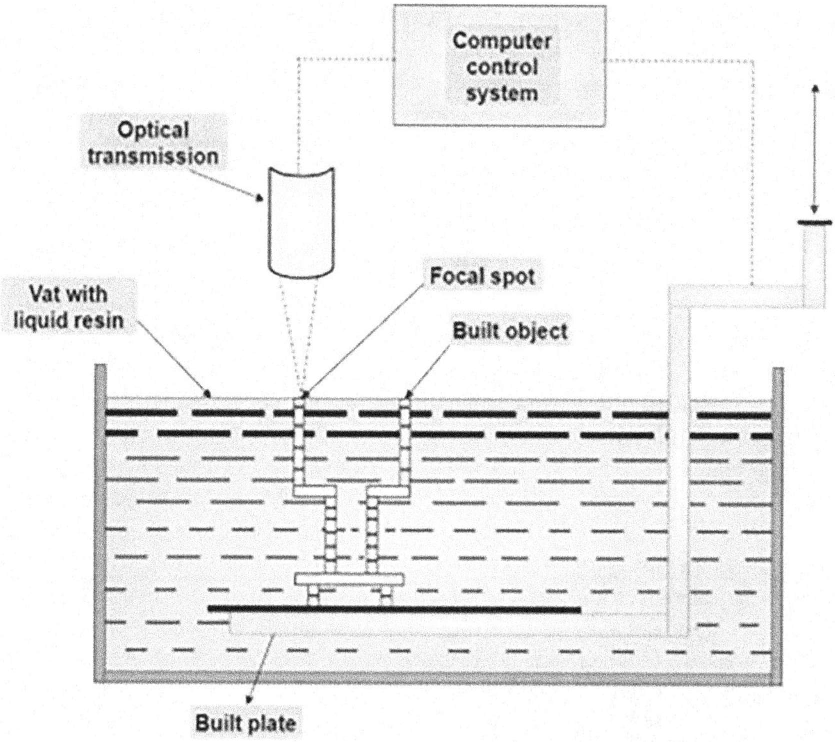

FIGURE 5.2 Schematic diagram of SLA. Adapted with permission [10]. Copyright 2017, Elsevier.

laser, or radiation to selectively draw and harden a cross-section of a planned object. When a cross-section has been cured, it is lowered below the liquid's surface so that it can backfill during the curing and bonding of future cross-sections. Prisms, mirrors, lenses, and movable gimbals are all viable options for the optical transmission path's design, as they allow for extremely fast and precise section tracing. The diameter of the concentrated beam of light can be made so narrow that the only thing limiting the tolerances is the amount of vibration in the chassis and the fluid. The layer thickness in the vertical direction and the resolution in the horizontal plane are both configurable in commercial systems (such those provided by 3D Systems, Form Labs, and others), with a resolution on the order of 20 m. [9] Because the fluid is being treated, the dimensions of the cured part are only constrained by the position and focus of the light source, making stereolithography the highest resolution 3D printing method. Despite stereolithography's reputation as a panacea for 3D printing, it has severe drawbacks. When using polymer resins, there are some post-process cleanings and curing of the parts that must be taken into consideration, as well as certain stability and ageing concerns for the resins themselves, which can be a pain for the operator.

5.2.2 Selective Laser Sintering

Selective laser sintering was developed by Carl Deckard of the University of Texas at Austin and later used to form DTM Corp. SLS involves layer-by-layer part fabrication using a scanning laser to fuse tiny particles (often on the order of 50 m in diameter). Every time a layer is sintered, the build volume is reduced by a few hundred micrometres and fresh powder is introduced into the annealing oven. To ensure that the fresh powder is evenly distributed around the build platform before the next layer is sintered, a roller moves back and forth across the build chamber. In most cases, the powder alone is sufficient to support the sintered cross-sections above without the need for any underlying solid support structures to be sintered because of how tightly the roller packs the powder. While Deckard was the first to use a laser with a wavelength of 10.6 m, SLS of ceramics and plastics had been done previously with the same laser. The insight that melting and residual stress govern SLS of plastics similarly to other polymer processing methods was a major discovery made by Deckard. As a result, Deckard prioritised keeping the build chamber close to the plastic's melting temperature to reduce residual stress and improve the SLS process's reliability. The build chamber's temperature regulation not only lessened the energy needed to sinter the particles but also minimised the thermal gradient and warpage that resulted from it. Wax (used in investment casting), nylon, polyvinyl chloride, polycarbonate, and ceramics are just some of the polymers that can be printed with SLS. To further enhance the tolerances and surface roughness of SLS-printed components, Deckard additionally designed the sintering laser to highlight the boundary of the printed cross-sections. To reduce residual stress and ensure that the part's characteristics are consistent throughout, he also scanned it during sintering from multiple angles.

5.2.3 Fused Deposition Modelling

AM techniques like FDM utilise molten fibres that are extruded and deposited to print stacks of 2D cross-sections to create complex 3D items (Figure 5.3). FDM

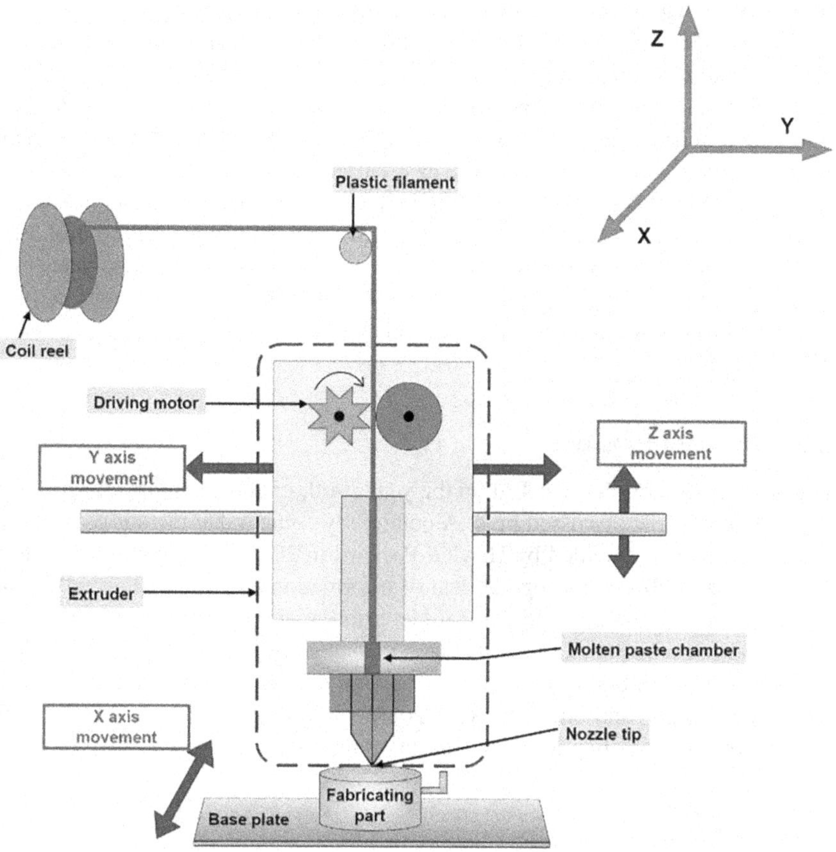

FIGURE 5.3 Schematic diagram of FDM. Adapted with permission [14]. Copyright 2017, Elsevier.

components have also been manufactured from metal and densely filled ceramics. The FDM system was created by Scott Crump, which eventually gave rise to Stratasys. Crump's first innovation required several driving wheels to deliver a continuous strand of feedstock into and through the nozzle. The solid, cooler part of the driven filament acts as a plunger to build up pressure in the liquefier's liquid material, making FDM possible. To printing larger components and structures, larger FDM machines have been designed that work by directly extruding thermoplastic pellets. Three areas of FDM technology have seen substantial development as of late. The first step was the realisation among researchers that control algorithms incorporating mechanistic models of the process dynamics can boost process and product performance with no extra expense. Using a "move-compiler" that considers the dynamics of the polymer melt flow, one example manages the filament drive and head traversal velocity simultaneously. The delay is reduced by overdriving the filament, and the flow rate is predicted using the flow's acceleration and compressibility. Next, the head velocity is set so that the depositing speed is proportional to the extrudate output. Expanding the range of materials that may be used with FDM has also been a major focus of recent progress, with the creation of both new materials and improved process conditions. In contrast to the closed build environments used by hobbyist systems for the deposition of polylactic acid plastic, the temperature of the build environment must be controlled to provide more robust procedures for use with engineering thermoplastics. This necessitates a closed build chamber, where the build plate, environment, and nozzle can all be independently controlled, a feature unique to commercial FDM machines. As a result of advancements in process control, FDM may now be used with a wide range of materials, from ABS and translucent ABS to nylon, PC, PPS, Ultem, and other compounds containing a wide variety of fillers [11–13]. FDM is more user friendly but produces lower-quality items than other 3D printing methods.

5.2.4 Polyjet Modelling

Developed by Object Methods Ltd. in the year 2000, PJM is the most modern of the 3D printing methods discussed here. A potentially comparable method, called multijet modelling, was revealed by Hewlett-Packard in 2015 [15]. However, neither its commercial availability nor a great deal of information about it are currently available. Both methods are extremely like inkjet printing, with the exception that instead of dropping ink onto paper, these 3D printers deposit tiny amounts of curable liquid photopolymer into a build tray. Figure 5.4 shows a print carriage with several print heads, curing units, and other subsystems used in polyjet modelling. Pieces of the object's cross-section are printed by depositing primary build material voxels one by

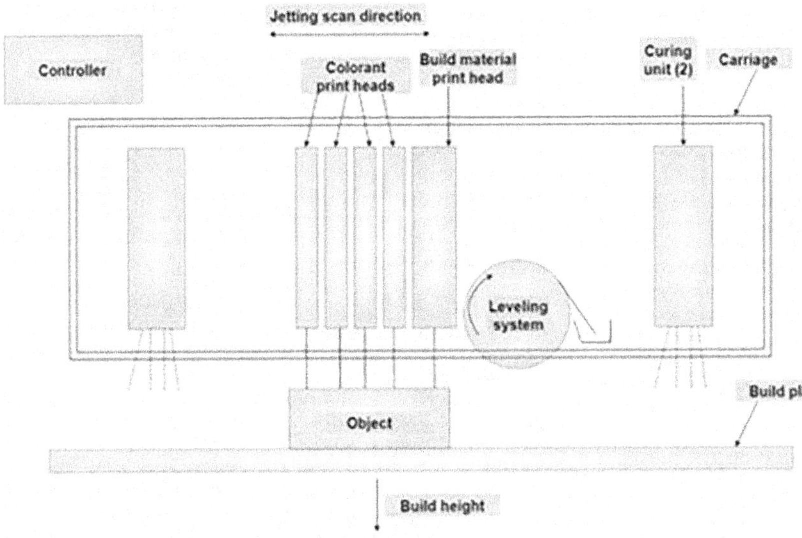

FIGURE 5.4 Schematic diagram of PJM. Adapted with permission [10]. Copyright 2017, Elsevier.

one through a print head. Multiple print heads can be used on the same voxel to distribute different colours or other additives. After the print material has been added, it may go through a levelling system, which consists of a roller and wiper, to remove any surplus. After printing, curing, and setting the build plate height for the subsequent print layer, the process is complete. Because of their short curing time following deposition, the photopolymers are quite like those used in stereolithography.

Common plastics like polythene may be the first things that come to mind when we hear the word "polymers." Imagine these plastics stuffed with conductors like metal or carbon particles, and you have the makings of a conducting polymer. In contrast, organic polymeric conductors (OPCs) are intrinsically conductive and do not contain any conductive fillers of any kind; they are also known as conducting polymers (CPs), conductive polymers, or conjugated conducting polymers (CCPs). CPs are distinct from common organic polymers like poly(ethylene), poly (vinylidene chloride) (Saran wrap), polyesters (used in textiles), and other plastics in that they are electrically conductive at room temperature and have many other properties due to their unique electronic properties, which distinguish them from the three broad material categories of insulators, semiconductors, and conductors. The use of conducting polymers in composite for various applications is shown in Figure 5.5.

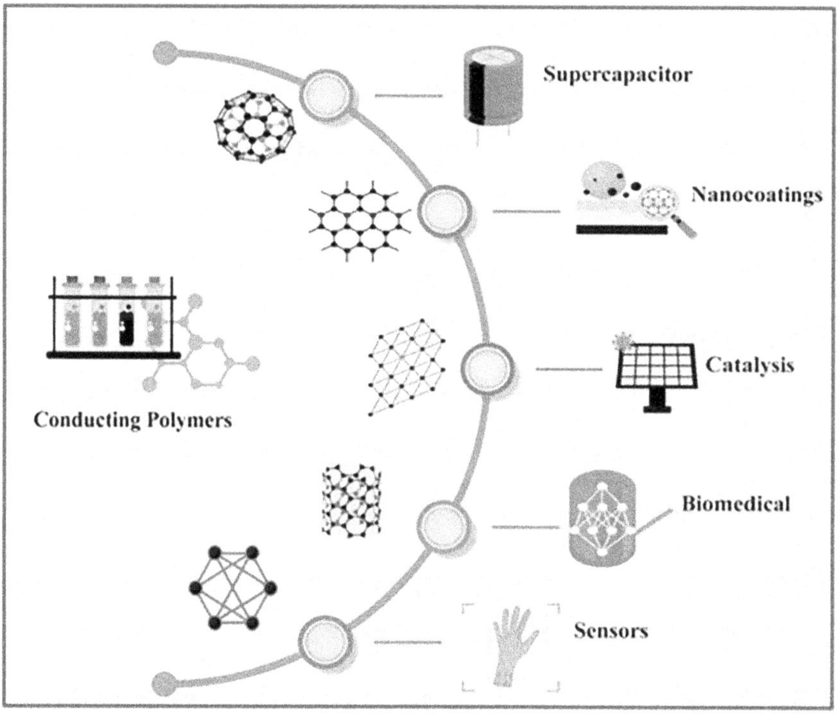

FIGURE 5.5 Illustration of potential applications for composites, including conducting polymers. Adapted with permission [16]. Copyright 2021, the authors. Published by the Royal Society of Chemistry. This article is licensed under a Creative Commons Attribution 3.0.

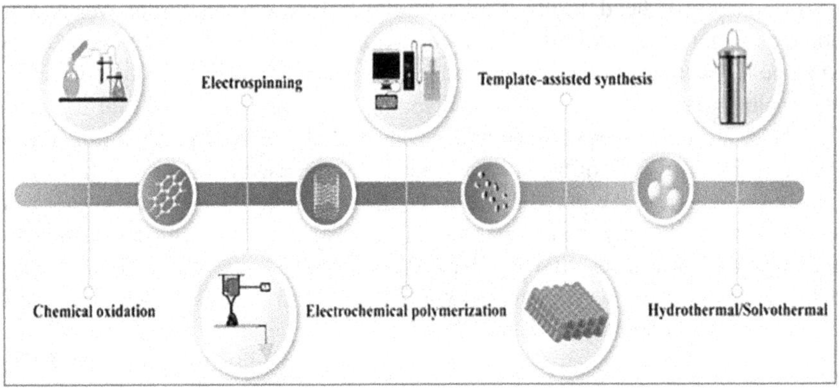

FIGURE 5.6 Schematic depiction of various approaches to the manufacture of conductive polymers. Adapted with permission [16]. Copyright 2021, the authors. Published by the Royal Society of Chemistry. This article is licensed under a Creative Commons Attribution 3.0.

Chemical oxidation, electrochemical polymerisation, vapour phase synthesis, hydrothermal, solvothermal, template-assisted, electrospinning, self-assembly, photochemical, inclusion, solid state, and plasma polymerisation were just some of the methods used to synthesise conducting polymers [16] (Figure 5.6). Catalytic polymerisation, non-catalytic polymerisation, catalytic polymerisation of other polymers, and precursor-assisted synthesis are all viable routes to synthesising polyacetylene. One of the simplest ways to synthesise polyaniline is via chemical oxidation, which involves reacting a monomer precursor of the corresponding polymer with an oxidising agent in the presence of a suitable acid at room temperature to produce the desired products. A similar procedure is used for composite preparation. Oxidising agents are commonly utilised, and some examples are ammonium persulfate, ammonium peroxy disulfate, ceric nitrate, ceric sulphate, and potassium bichromate. Physical properties are effectively modulated by conductivity, which varies depending on the pH of the acid dopant. When the pH is between 1 and 3, both the polymer and the composite have high conductivity [17].

5.3 EMERGING APPLICATIONS

5.3.1 Flexible Electronics and Wearable Devices

3D printing's benefits, such as rapid prototyping and design flexibility, can boost innovation across many fields. Polymer composites' potential as electronic components is greatly enhanced using feedstocks that contain conductive materials like carbon nanotubes and metal nanoparticles. Because 3D printing can manufacture custom-shaped multifunctional electronics, such as stretchable electronics with functional elements (e.g., sensors, circuits, and other embedded components [18], embedded electronics [18], and structural electronics [19]), it has been actively researched in recent years alongside the rise in popularity of wearable electronics, soft robots, and stretchable electronics. For instance, integrated 3D printing of carbon-based ink was used to create very stretchy sensors [19]. Mechanically resilient flexible sensors that can sense mechanical deformation via the change in electrical resistance can be made via embedded 3D printing inside the elastomeric matrix (Figure 5.7A). Soft strain sensors were produced utilising multicore-shell printing, marking yet another application of 3D printing to the manufacturing of electronic components. Coaxially aligned nozzles were used to extrude a mixture of a silicone elastomer and an ionically conductive fluid (made up of glycerol, sodium chloride, and polyethylene glycol) to create the sensors. The sensor's concept is shown in Figure 5.7B; the sensor can function as a capacitive sensor because of the dielectric elastomer and stretchable ionically conductive fluid. Wearable electronics, like garments and shoe insoles, can benefit from both examples of 3D-printed conductive elements implanted inside elastic polymers. More and more people are looking for miniaturised and individualised electronic components as they develop portable electronic devices. Customising the size and shape of structural electronics via 3D printing allows their manufacture for specific uses. For example, FFF is used to manufacture microelectronics such as components and circuits [20]. Microchannels are shown in Figure 5.7C, which depicts a 3D printed structure that was filled with liquid metal paste after the FFF printing process.

Printed devices can be shaped into a unified system that fits into any desired 3D architecture. Direct ink printing and 3D scanning of surface topologies have been demonstrated as a 3D printing method for the seamless integration of different materials on a 3D surface. Quantum dot light-emitting diodes (QD-LEDs) with full integration into active device components are shown printed in Figure 5.7D using

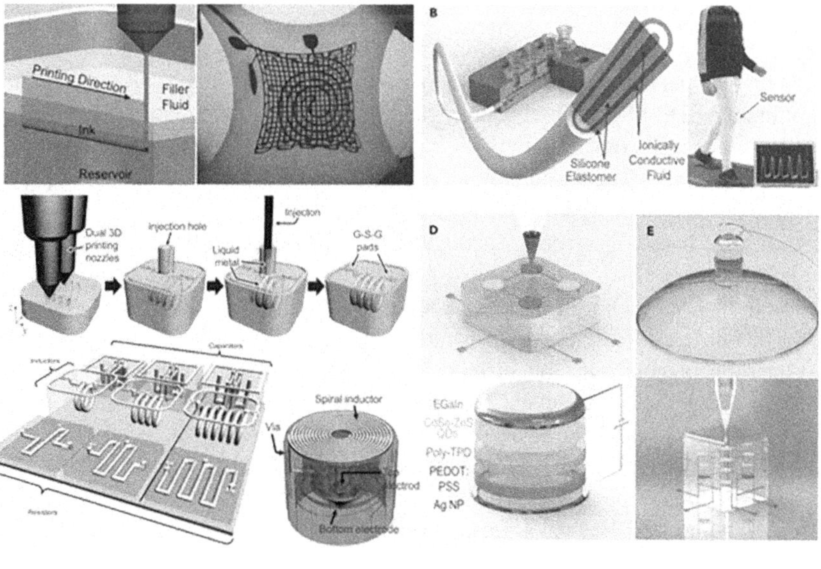

FIGURE 5.7 3D printing for electronics. Adapted with permission [2]. Copyright 2021, Elsevier.

direct ink printing. As shown in Figure 5.7E, when direct ink printing is combined with 3D scanning technology, materials can be printed on a variety of substrates, including curvilinear surfaces. This is made possible by the layer-by-layer process of 3D printing, which allows for the fabrication of encapsulated active electronics like multidimensional arrays of embedded QD-LEDs 3D. A variety of electromechanical connections and operations can be provided by assembling numerous units of advanced electronics into blocks, much like the printed structural electronics described previously. Bitbloxes are digital materials created on a personal 3D printer based on MJ.111 with some customisations. Each Bitblox is shorter than 1 cm and plugs together easily with standard thru-hole pins and sockets. Combining the Bitblox form factor with DC motors results in a cheap leadscrew actuator [2].

5.3.2 Supercapacitors

Many methods of efficiently storing electrical energy have been investigated since the invention of electricity. Electrochemical energy storage (EES) technology is a key component in the development of reliable energy infrastructure. Academic studies have traditionally centred on developing novel electrode, electrolyte, and separator materials for EES devices. Even more important than the bulk structure of EES devices, researchers have found, is the structure and multiple sizes of the components. In recent years, the architecture of EES devices has received a lot of research and development focus. Due to its ease of use, geometric versatility, and capacity to produce multi-scale structures, 3D printing has found widespread application in the field of energy storage. A micro-lattice cathode composed of sulphur copolymer-graphene (3DP-pSG) was created by employing DIW and a basic thermal treatment to create a copolymer between elemental sulphur and polymer [21]. Fast and precise manufacturing of a porous cathode structure that promotes ion movement is made possible by DIW technology, and the printing ink for DIW printing is made up of S particles, 1,3-diisopropenylbenzene, and condensed GO dispersion. In addition, the 3DP-pSG current density range of 50–800 mA/g demonstrates superior cycle performance compared to that of a 3D-printed sulphur-graphene cathode without polymer (3DP-SG), thanks to 3DP-pSG's strong covalent link between polymer and sulphur, which prevents the dissolution of polysulfides.

Graphene-based electrodes made using the FFF process are another option because they are efficient, cheap, and friendly to the environment. Printed 3D electrodes and a PVA-KOH gel electrolyte are used to fabricate a solid-state supercapacitor. The surface resistance of 3D-printed electrodes was decreased by sputtering a thin coating of gold onto the surface. Over a 100-cycle test, the printed electrode supercapacitor showed excellent capacitance performance over a potential range of +0.0–1.0 V at a scan rate of 50 mV/s. All-component 3D-printed lithium-ion batteries are another product that makes use of the extrusion method. LTO/GO ink was used to create the anode, whereas LFP/GO ink was used to create the cathode. After the electrodes had been annealed, a solid-state gel polymer ink containing a combination of poly(vinylidene fluoride)-co-hexafluoropropylene and alumina oxide (Al_2O_3) nanoparticles was injected in between them. The batteries had a capacity of 100 mAh/g after 10 cycles at a particular current of 50 mA/g and were printed using interdigitated architecture. At first, the battery could only charge and discharge at a rate of 117 and 91 mAh/g, but after 10 cycles, it had improved to 110 and 108 mAh/g, demonstrating approximately 100% columbic efficiency.

Utilisation of fossil fuel sources, such as natural gas and coal, is critical to the world economy. Many issues, both social and environmental, have arisen because of our dwindling supply of fossil fuels. Altering our energy infrastructure away from fossil fuels is essential in the face of climate change, global warming, and health concerns. For this reason, researchers have been hard at work on clean, high-power renewables like solar panels, wind turbines, and fuel cells. These days, supercapacitors are a hot commodity because of their potential applications in emerging industries like wearable tech and electric automobiles. Supercapacitor devices can store 1000 times more energy than a dielectric capacitor, making this the primary difference between the two. High energy and power density, as well as a long cycle life, are additional hallmarks of these devices. Based on how they store electricity, supercapacitors can be divided into three basic categories. Charge is accumulated at the interface of an electrode and electrolyte in an electrochemical double-layer capacitor (EDLC) by a non-faradic process. Capacitance results from a charge-separated double layer formed by electrochemistry. Electrodes for electrochemical double-layer capacitors often consist of carbon-based materials, including graphite, carbon nanotubes, and graphene, because of their high surface areas and the ease with which they can store energy [16]. Pseudocapacitors store their charges through redox reactions or intercalation, but their operation is based on a faradic mechanism. While EDLCs have a higher power density, pseudocapacitors have a higher energy density [22]. Pseudocapacitors typically use conducting polymers [23] (Figure 5.8) or transition

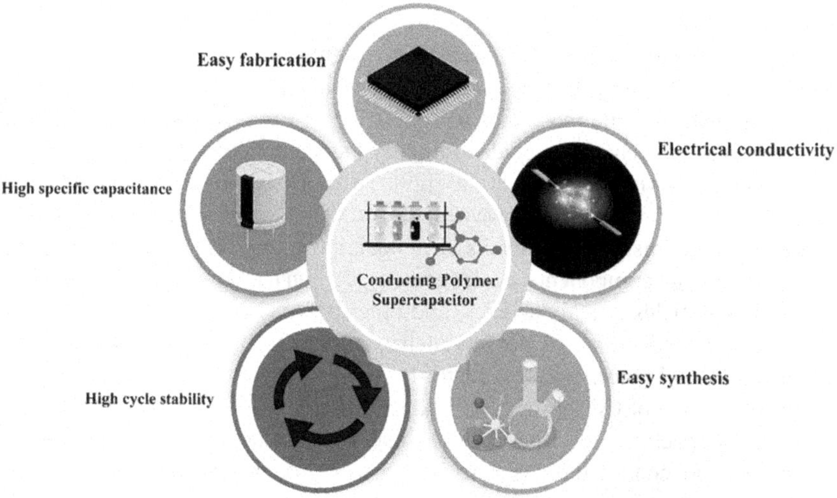

FIGURE 5.8 Schematic representation of basic properties of conducting polymer-based supercapacitors. Adapted with permission [16]. Copyright 2021, the authors. Published by the Royal Society of Chemistry. This article is licensed under a Creative Commons Attribution 3.0.

metal oxides as electrode materials due to their redox activity. Examples of such materials include Mn_3O_4 and Co_3O_4 The hybrid supercapacitor is an EDLC plus a pseudocapacitor. Pseudocapacitors have a high energy density and strong cycle stability, while EDLCs have a correct specific capacitance. Both the EDLC and pseudocapacitor benefits can be seen in hybrid capacitors.

Strong conductivity, structural flexibility, and chemical stability in aggressive electrolytes are hallmarks of pseudocapacitive CPs. Amazing new findings are discussed in this section, and their impact on the evolution of CP-based pseudocapacitive materials is discussed. CPs and metal oxides like MnO_2 [24] and NiO [25] are commonly used as electrode materials for pseudocapacitors, as are metal chalcogenides like TiS_2 [26], MoS_2 [27], and $Co_{0.85}Se$ [28] and metal hydroxides like $CoSn(OH)_6$ [28]. Because of their potential as pseudocapacitors, metal oxides have received a great deal of attention. However, their limited versatility, high price tag, and low conductivity limit their usefulness. While CPs like PANI and PPy exhibit great energy storage capacity, they also allow for the fabrication of flexible electrodes. Poor cycling stability, however, is a major limitation of CPs in actual use. Hybridisation and the production of well-designed nanostructures are two approaches that have been explored by several researchers as potential solutions to this problem. Power density, energy density, and cycling stability are all improved with these materials. An intriguing case in point is the study by Park et al. [28] of PANI nanostructures and the correlation between morphology and capacitance.

5.3.3 Sensors and Actuators

A sensor is a device that produces a measurable output signal from an input parameter (such as temperature, humidity, or chemical and biological species). Most sensors rely on a change in electrical or optical characteristics to generate an output signal. CPs are versatile sensing materials because of the reversibility of their conductivity, colour, and volume changes. It is necessary to enhance CP sensitivity and selectivity to a target analyte because they are occasionally unstable because of their interactions with organic molecules and moisture in the surrounding environment. Numerous attempts have been made to modify the features of CP sensors in order to increase their sensitivity and selectivity [29]. To alter their electrical, chemical, and microstructural properties, CPs have been doped with a wide range of functional heterogeneous materials, and the results have been intriguing [30]. In terms of heterogeneous functional components, chalcogenides, carbon nanotubes (CNTs), graphene, metal oxide particles, metal particles, and other polymers are of particular interest now. Utilising conductometry, potentiometry, amperometry, voltammetry, gravimetry, pH-based, and incorporated-receptor-based sensing modes have all contributed to the development of CP sensors [31, 32].

Gas/vapor species (such as NO_2, SO_2, and I_2) were detected by exposing first-generation, undoped CP film sensors in the target environment and measuring conductivity variations. This approach, however, was limited by its lack of selectivity due to the presence of numerous interferences in the surrounding environment. The chemical sensor created by Forzani et al. [33] uses amperometry and conductometry. To detect an analyte, the PANI-based dual-mode sensor altered the polymer's electrochemical current and conductance in two distinct but correlated ways. By adjusting the applied potential, this sensor was able to detect target analytes with greater selectivity in mixed samples and in the presence of relatively large amounts of unanticipated substances. To create CP chemical sensors that are both fast and selective, many scientists have constructed metal/CP nanocomposites. Selective methanol sensors based on a Pd/PANI nanocomposite were developed by Athawale et al. [34] After being exposed to methanol vapours in the air, the nanocomposite became extremely sensitive and selective. Analysis with Fourier transform infrared spectroscopy showed that the Pd nanoparticles catalysed the reduction of the imine nitrogen of PANI to an amine by methanol, hence contributing to the improved selectivity of the nanocomposite.

The size of PPy nanoparticles impacted chemical sensing behaviours in the detection of volatile organic chemicals and hazardous gases, with the goal of improving the efficacy of CPs in chemical sensors. Chemical fabrication of PPy nanoparticles in aqueous solution utilising a soft-template technique yielded particles with sizes of 20, 60, and 100 nm. Both conductivity and the surface-to-volume ratio were shown to improve as particle size was reduced to the nanometre range. The smallest PPy nanoparticle (20 nm)-based chemical sensor had the best sensing performance (high sensitivity, quick reaction time, reversibility, and reproducibility of responses). High sensing performance can be achieved through good contact with analytes, which is facilitated by a broad surface area. Also, highly sensitive and selective chemoreceptive sensors for monitoring volatile organic chemicals and hazardous gases in human breath have been developed using multidimensional PPy nanotubes with surface substructures. The highest sensitive recognition of ammonia by such sensors to date occurred at 10 ppm for gaseous ammonia. The main finding here is that the multidimensional nanotubes' unusual form benefited sensor application by increasing their effective surface area and enhancing their charge transport behaviour. In addition, a chemiresistive gas sensor was built using PPy nanotubes functionalised with carboxylic acid to detect the nerve agent dimethyl methyl phosphonate [35]. PPy nanotubes' sensitivity to dimethyl methyl phosphonate was improved by the presence of carboxylic groups on their surface via intermolecular hydrogen bonding. The outcome showed that chemiresistor sensitivity strongly depends on the level of carboxylic group functionalisation.

Researchers have been working to perfect tools for the quick and precise diagnosis of biological target species in response to the increasing human toll of disease. Due to their great compatibility with biological systems, CPs and CP composites have been used by a number of researchers for biosensor applications [36]. CP used a liquid-ion gated field-effect transistor (FET) setup to create a glucose biosensor based on enzyme-functionalised PPy nanotubes. By depleting or accumulating charge carriers in the nanomaterial bulk, 1D CP nanomaterials used as the conductive channel of FETs can deliver more sensitive responses than 2D materials. In order to use the CP nanoparticles for sensing in the liquid phase, they must be immobilised on electrode substrates. However, CPs are inappropriate for the traditional lithographic process because of the risks of chemical, thermal, and kinetic degradation. The surface-modified microelectrode substrate allowed for chemical immobilisation of the CP nanoparticles. The comprehensive surface modification procedures used to create the FET sensor substrate are shown in Figure 5.9.

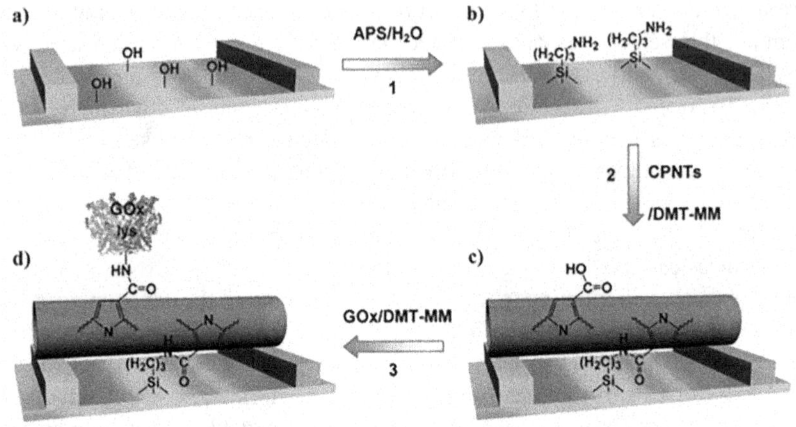

Reaction 1. Hydrolysis: $H_2N(CH_2)_3Si(OCH_3)_3 + 3H_2O \rightarrow H_2N(CH_2)_3Si(OH)_3 + 3CH_3OH$

Condensation: $H_2N(CH_2)_3Si(OH)_3$ + 3OH-substrate $\rightarrow$ $H_2N(CH_2)_3Si(O)_3$-substrate

Reaction 2. Condensation: pyrrole-COOH + $H_2N(CH_2)_3Si(O)_3$-substrate $\rightarrow$

pyrrole-$CONH(CH_2)_3Si(O)_3$-substrate

Reaction 3. Condensation : pyrrole-COOH + H_2N-lys-GOx $\rightarrow$ pyrrole-CONH-lys-GOx

FIGURE 5.9 Schematic illustration of the reaction steps for the fabrication of a field-effect transistor sensor platform based on carboxylated PPy nanotubes. Adapted with permission [36]. Copyright (2008), American Chemical Society.

As a first stage, two metal electrodes (the source and the drain) were contacted to carboxylate PPy nanotubes (CPNTs) that had been attached onto a microelectrode substrate via covalent bonds. The glucose oxidase (GOx) enzyme could then be covalently bound to the CPNT surface, rendering the CPNTs functional. The presence of glucose prompts the enzyme to alter the source–drain current. The glucose detection range of this FET sensor was from 0.5 to 20 mM, where it showed excellent sensitivity. Proteins, hormones, tastants, and odorants are just few of the biomolecules that have been detected using a similar principle. Since 1D CP nanomaterials can deplete or accumulate charge carriers in the nanomaterial bulk, they can provide more sensitive responses than 2D materials when used as the conductive channel of FETs. For sensing applications in the liquid phase, CP nanoparticles should be immobilised on electrode substrates. CPs are not suited for the typical lithography process since they are susceptible to damage from chemicals, heat, and motion. The surface of a microelectrode may be changed chemically, so CP nanoparticles could be immobilised there. Protein, hormone, and tastant detection are all examples of how this general idea has been used.

It is also usual practice to construct CP-based biosensors using a three-electrode electrochemical setup. The electrochemical detection of prostate-specific antigen (PSA) using a gold/PANI nanocomposite has been described.[37] Gold nanoparticles assembled themselves electrostatically on PANI nanowire electrodes, yielding the hybrid nanocomposite. The immobilisation of anti-PSA on the nanostructured gold/PANI composite surface improved charge transport and sensing characteristics. The PSA concentration was determined using differential pulse voltammetry by the immunochemical sensor. With a detection limit of 0.6 pg/mL and a wide linear range in the calibration curve, this sensor demonstrated excellent sensitivity, selectivity, and reproducibility in its measurements. In order to identify the breast cancer susceptibility gene (BRCA1), Hui et al. [38] recently created an electrochemical DNA sensor based on polyethylene glycol (PEG)ylated PANI nanofibers. Because of its low cost, lack of toxicity, and high hydrophilicity, PEG has found widespread employment in a variety of fields, from heavy industry to the medical field. PEGylation is the process of attaching a PEG structure to another molecule via covalent bonding. PEGylated PANI nanofibres retained their conductivity and enormous surface area, which is rather remarkable. Both in simple protein solutions and in complex human blood samples, they demonstrated remarkable antifouling abilities. This innovative biomaterial has great potential for use in biosensors, as the DNA sensor demonstrated extremely high sensitivity to BRCA1 over a linear range of 0.01 pM to 1 nM.

Oxygen monitoring was made easier with the use of the earliest electrochemical sensors in the 1950s. For the detection of flammable and dangerous gases in outer space, modern electrochemical sensors have proven highly effective. Toxic gases in the PEL range can now be detected by miniature kinds of very selective electrochemical sensors, which have recently come into the spotlight. Despite superficial similarities in appearance, electrochemical sensors for gas sensing perform very diverse tasks. Different sensors have different specificities, reaction times, sensitivities, and lifetimes, all of which affect how well they operate. Hydrophobic porous membranes, used in low-concentration gas sensors, allow more molecules of gas to flow through, improving the sensor's sensitivity through increased signal generation. However, water molecules from the electrolyte can escape through this design and into the surrounding environment. The loss of water molecules through the porous membrane also suggests that a highly sensitive electrochemical sensor will have a short lifetime. The selectivity of an electrochemical gas detector relies on the specific characteristics of the sensor itself, the kind of the gas being detected, and the concentration of the gas being detected. Just as the chemical reactivity of a certain target gas is affected by the electrolyte and electrode material used in the sensing process, so too is the sensitivity of the reading. Electrochemical sensors with high sensitivity use an electrical current to react to a target gas [39]. The interaction with the target gas is at the heart of how electrochemical sensors function, with the resulting electrical signal being directly proportional to the concentration of the gas being sensed.

An electrochemical sensor has two electrodes: one for sensing (the "working" electrode) and one for "countering," which are separated by an electrolyte. The gas first diffuses through a hydrophobic barrier before entering the electrode surface through a tiny capillary-like hole. This strategy is being investigated with the hope that it will allow more gas molecules to reach the sensor, resulting in a stronger electrical signal, while limiting electrolyte loss from the sensor electrode. A change in oxidation or reduction state occurs on the sensor electrode surface because of the gas diffusing through the barrier. Current between an electrochemical cell's anode and cathode is proportional to the concentration of gas molecules in the cell's environment. Because of the current that is generated by this process, this electrochemical sensor is also known as a miniature fuel cell or an amperometry gas sensor. Since electrochemical reactions are constantly taking place on the surface of the sensing electrode, the potential at that electrode is not stable. The electrode's effectiveness naturally declines over time as a result. Consequently, the functionality of the sensing electrode can be enhanced by employing a reference electrode. In a sensing-electrode–reference-electrode setup, the sensing electrode is subjected to a stable fixed potential, while the reference electrode is tasked with maintaining the sensing electrode's fixed potential in the absence of current flow. The sensing electrode's fixed potential determines the sensor's sensitivity to a given gas.

High costs, a lack of donors, and immunological rejection continue to be issues for tissue or organ transplantations in clinical medicine using current procedures (such as auto-transplantation, xenotransplantation, and implantation of artificial organs). By using biocompatible polymers, such as synthetic polymers (poly(ethylene glycol) diacrylate, PVA, poly(D,L-lactic-co-glycolic acid), PCL, and PLA) and natural polymers (gelatin methacrylate [GelMA], alginate, and collagen), 3D printing provides the opportunity to rapidly create personalised tissues and organs [40]. Polymers used in 3D bioprinting, in contrast to those used in traditional printing, need to be biocompatible and have strong mechanical and structural qualities. Engineering bone requires scaffolds with the right structure (shape, pore size, and porosity), material qualities (biocompatibility and degradability), mechanical properties, and desirable cellular interactions. Nano-hydroxyapatite/poly(ethylene glycol) (PCL) was used in SLS to create bone tissue. SLS permits quick prototyping of complicated geometrical scaffolds without the use of organic solvents, which can cause inflammation and harmful byproducts, making it distinct from traditional processes for scaffold production like solution casting and injection moulding.

Using biodegradable polymers and living cells in a bioprinting process has led to significant progress in biomedical applications. Bio-inks and an extrusion-based bioprinter have been used to manufacture vascularised bone tissue, for instance. Bone-like 3D structures were created, complete with osteogenic and vasculogenic niches. Human umbilical vein endothelial cells isolated from blood and human mesenchymal stem cells (hMSCs) isolated from bone marrow were combined with the GelMA bio-inks to generate a sturdy vascular system. In addition, a gradient of vasculogenesis was made by adding varying amounts of vascular endothelial growth factor (VEGF) to functionalised GelMA bio-ink. The central section is made up of a rapidly biodegradable GelMA in engineering bone tissue that is meant to break down and reveal a hollow main channel. hMSCs were induced to differentiate into osteoblasts through osteogenic differentiation by printing bio-ink with varying quantities of VEGF in the outer layers. In contrast, a novel 3D printer was designed for producing bone, cartilage, and skeletal muscle by integrating jetting, extrusion, and laser-induced forward transfer [41]. The integrated 3D printer was fine-tuned to manufacture bio-constructions of varying sizes while maintaining enough structural integrity. Bioprinting was performed with a carefully calibrated mixture of cell-laden hydrogels, acellular hydrogel, and PCL polymer as the scaffold. Ear, mandible bone, and muscle are just few of the organs and tissues that have been successfully 3D printed. This bioprinting demonstrates the potential for printing strong structures out of live tissue. Medical engineers have employed fibrin, a fibrous protein, to repair skin, blood vessels, and bone, and fibrin blends are used in both scaffold and scaffold-free technologies. Because of their exceptional biocompatibility, biodegradability, and benignness, fibrin-based bio-inks have been frequently employed as feedstocks since the advent of bioprinting. Artificial skin, hearts, and neurological structures are all within printing range with the help of bio-inks. The use of 3D printing technologies has greatly increased the study of soft tissues like nerves, blood vessels, and fibrous tissues [42]. Natural polymer (chitosan) and biodegradable biopolymer (PLA) were employed for printing in this investigation. Artificial ducts were mostly made of chitosan, with PLA serving as a support structure. The printed ducts matched the mechanical parameters of the soft tissue that they were meant to replace (tensile strength, Young's modulus, and fracture strain) very well.

Photothermal therapy (PTT) using near-infrared (NIR) light has shown promise for killing cancer cells in humans. The area of interest is illuminated with a near-infrared laser using fibre optics. The use of a photothermal substance for the ablation of cancer cells is less intrusive than traditional treatments like chemotherapy or radiation therapy. PTT's specificity is a major benefit over competing therapies. This quality helps this treatment strategy restrict tissue loss close to the target area. The future of photothermal materials for PTT seems bright when their nanoscale dimensions are combined with good NIR absorption, a high quantity of photostability, minimal photo corrosion, and low cytotoxicity. When exposed to IR light, metal nanoparticles exhibit surface plasmon resonance; however, after being exposed to particularly bright light, most metal nanoparticles lose their photostability. Polyaniline, polypyrrole, and polydioxolene (PDOT) are all examples of conducting polymers with excellent IR absorption capabilities and significantly lower cytotoxicity. This is why in vivo and in vitro cell research makes extensive use of these polymers.

As previously noted, 3D printing allows for the rapid and economical delivery of complicated geometries in the form of individualised consumer goods. Polymers and composites with varying characteristics (strength, stiffness, chemical resistance, etc.) are also available. These benefits make 3D printing a perfect fit for modern robotics applications. Due to limitations in current manufacturing methods, conventional robotics requires the time-consuming assembly of several individual parts. By doing away with the need for specialised moulds and numerous assembly procedures, 3D printing offers solutions to the engineering difficulties inherent in robot fabrication. Robotics employing 3D printing is becoming increasingly popular in the industrial sector. Making robot grippers, for instance, may be quite a costly endeavour due to the extensive personalisation each application requires. Improved performance, such as faster movement and heavier load, can be achieved by printing these parts in lieu of the conventional heavy, non-customised designs. For instance, Haddington Dynamics is creating robot arms with the use of 3D printers, including a gripper for NASA and GoogleX. The ability of soft robots to carry out complex manoeuvres in a dangerous or uncertain setting has garnered them a lot of attention. Soft robots make use of biological forms that have been optimised over millions of years. This is made possible by the development of 3D printing processes and improvements of functional soft materials [43]. The stress at the interface of materials mismatched in compliance is reduced thanks to the stiffness gradient of the robot's body, which allows for smooth transitions from a rigid core to a soft outside. Another feasible application for MJ is the construction of a hexapod robot with hydraulic actuators [44]. The entire hexapod robot was printed in one MJ process. The hexapod robot's fluidic channels were manufactured using a combination of photopolymers and non-curing fluids to allow for hydraulic actuation of the robot's movement. Octobot is a robot octopus that was created utilising cutting-edge 3D printing technology [45]. Moulding PDMS was used to create the octobot's form, and integrated 3D printing was used to create the pneumatic networks. Upon catalytic degradation of fuel, a regulated microfluidic system produces gas and causes downstream fluid flow, allowing the octobot to move. In the same vein, multi-material multi-nozzle 3D printing was used to produce a soft robotic millipede walker. The printed walker included soft silicone (representing the muscles) and hard silicone (representing the legs). The pneumatic actuators in the robot's leg cause it to move when the soft silicone muscle buckles, which displaces the leg laterally and vertically.

The frequency of the actuators controls the robot's speed. Mechanical metamaterials have recently been credited with shifting the focus from form to function in robotics. 3D printing of elastomers has led, for instance, to pneumatic actuators that buckle when internal pressure is lower than external pressure. The core of these actuators does not buckle, yet it is connected to multiple columns that did. The rotation of the central portion is caused by the pillars' misalignment with respect to the central portion. They have numerous desirable properties for use in robots, including portability, safety, and the ability to function as a soft gripper. In addition, when used in conjunction with smart materials like SMPs, robots may be able to move without the need for batteries or on-board electronics if metamaterials are used [46]. The SMP muscle or engineered bistable element can supply the power needed for actuation.

5.4 CONCLUSION

3D printing requires simple digital product design and no hardware changes. It also reduces waste compared to subtractive manufacture. Due to its flexibility in designing difficult geometries and capacity to meet individual desires, AM offers unique views in biology, electrical energy storage, electronics, and robotics. Polymers, ceramics, metals, and composites are customised to each 3D printer. Printing procedures can be tuned for thermoplastics, photopolymers, and other polymers depending on the desired output. Conducting polymers are attractive materials for energy storage, flexible electronics, and bioelectronics due to their polymeric nature, good electrical and mechanical properties, stability, and biocompatibility. CPs are exciting to research due to their unique features as organic materials with electrical conductivity. Rapid prototyping and design flexibility are growing 3D printing polymer uses that can accelerate innovation in numerous sectors. Feedstocks using conductive elements like carbon nanotubes and metal nanoparticles boost polymer composites' electronic component potential. 3D printing can create custom-shaped multipurpose electronics including stretchable electronics with sensors, circuits, and other embedded components and structural electronics. Additionally, 3D printing is widely used in energy storage. Electrodes and super capacitors use electrochemical energy storage technology to build reliable energy infrastructure. Supercapacitors store 1000 times more energy than dielectric capacitors, the main difference. Chalcogenides, CNTs, graphene, metal oxide particles, metal particles, and other polymers are currently of interest as heterogeneous functional components. Conductometry, potentiometry, amperometry, voltammetry, gravimetry, pH-based, and incorporated-receptor-based sensing modes have all contributed to CP sensor development. Chemical sensors detected target analytes more selectively in mixed samples and with significant levels of unforeseen chemicals by changing the applied voltage. Researchers have employed CPs and CP composites for biosensors because of their biological compatibility. 3D printing allows rapid creation of customised tissues and organs utilising biocompatible polymers including synthetic and natural polymers. 3D printing is ideal for current robotics. Conventional robotics require time-consuming assembly of several elements due to manufacturing limitations. 3D printing solves robot fabrication engineering problems by eliminating the requirement for specialised moulds and assembly steps. Since 3D printing technology matured, soft polymer materials have undergone dramatic changes. This means 3D-printed soft polymer materials will remain popular. This review outlines the latest CP research and uses, as well as their electrical and electrochemical properties.

REFERENCES

[1] Yuk H, Lu B, Lin S, Qu K, Xu J, Luo J, Zhao X (2020) 3D printing of conducting polymers. Nat Commun 11:4–11

[2] Park S, Shou W, Makatura L, Matusik W, Fu K (Kelvin) (2022) 3D printing of polymer composites: Materials, processes, and applications. Matter 5:43–76

[3] Zhou LY, Fu J, He Y (2020) A review of 3D printing technologies for soft polymer materials. Adv Funct Mater 30:1–38

[4] Shi Y, Peng L, Ding Y, Zhao Y, Yu G (2015) Nanostructured conductive polymers for advanced energy storage. Chem Soc Rev 44:6684–6696

[5] Someya T, Bao Z, Malliaras GG (2016) The rise of plastic bioelectronics. Nature 540:379–385

[6] Yuk H, Lu B, Zhao X (2019) Hydrogel bioelectronics. Chem Soc Rev 48:1642–1667

[7] Ramanavicius S, Ramanavicius A (2021) Conducting polymers in the design of biosensors and biofuel cells. Polymers (Basel) 13:1–19

[8] Huang WS, Humphrey BD, MacDiarmid AG (1986) Polyaniline, a novel conducting polymer. J Chem SOC, Faraday Trans 1 82:2385–2400

[9] Zhang X, Jiang XN, Sun C (1999) Micro-stereolithography of polymeric and ceramic microstructures. Sens Actuators, A Phys 77:149–156

[10] Kazmer D (2017) Three-Dimensional Printing of Plastics, Second Edi. Elsevier Inc.

[11] Ahn SH, Montero M, Odell D, Roundy S, Wright PK (2002) Anisotropic material properties of fused deposition modeling ABS. Rapid Prototyp J 8:248–257

[12] Turner BN, Strong R, Gold SA (2014) A review of melt extrusion additive manufacturing processes: I. Process design and modeling. Rapid Prototyp J 20:192–204

[13] Kruth JP (1991) Material incress manufacturing by rapid prototyping techniques. CIRP Ann—Manuf Technol 40:603–614

[14] Jin Y, Li H, He Y, Fu J (2015) Quantitative analysis of surface profile in fused deposition modelling. Addit Manuf 8:142–148

[15] Klein S, Avery MP, Richardson RM, Bartlett P, Frei R, Simske SJ (2015) 3D printed glass: Surface finish and bulk properties as a function of the printing process. HP Lab Tech Rep 9398:1–9
[16] Namsheer K, Rout CS (2021) Conducting polymers: A comprehensive review on recent advances in synthesis, properties and applications. RSC Adv. 11:5659–5697
[17] Yang L, Yang L, Wu S, Wei F, Hu Y, Xu X, Zhang L, Sun D (2020) Three-dimensional conductive organic sulfonic acid co-doped bacterial cellulose/polyaniline nanocomposite films for detection of ammonia at room temperature. Sens Actuators, B Chem 323:128689
[18] Muth JT, Vogt DM, Truby RL, Mengüç Y, Kolesky DB, Wood RJ, Lewis JA (2014) Embedded 3D printing of strain sensors within highly stretchable elastomers. Adv Mater 26:6307–6312
[19] Lopes AJ, MacDonald E, Wicker RB (2012) Integrating stereolithography and direct print technologies for 3D structural electronics fabrication. Rapid Prototyp J 18:129–143
[20] Wu S, Ding Y, Wu F, Li R, Hou J, Mao P (2015) Omega-3 fatty acids intake and risks of dementia and Alzheimer's disease: A meta-analysis. Neurosci Biobehav Rev 48:1–9
[21] Shen K, Mei H, Li B, Ding J, Yang S (2018) 3D printing sulfur copolymer-graphene architectures for Li-S batteries. Adv Energy Mater 8:1701527
[22] Roldán S, Barreda D, Granda M, Menéndez R, Santamaría R, Blanco C (2015) An approach to classification and capacitance expressions in electrochemical capacitors technology. Phys Chem Chem Phys 17:1084–1092
[23] Zhao C, Jia X, Shu K, Yu C, Wallace GG, Wang C (2020) Conducting polymer composites for unconventional solid-state supercapacitors. J Mater Chem A 8:4677–4699
[24] Nagamuthu S, Vijayakumar S, Muralidharan G (2013) Biopolymer-assisted synthesis of λ - MnO 2 nanoparticles as an electrode material for aqueous symmetric supercapacitor devices. Ind Eng Chem Res 18262–18268
[25] Singh AK, Sarkar D, Khan GG, Mandal K (2014) Hydrogenated NiO nanoblock architecture for high performance pseudocapacitor. ACS Appl Mater Interfaces 6:4684–4692
[26] Muller GA, Cook JB, Kim HS, Tolbert SH, Dunn B (2015) High performance pseudocapacitor based on 2D layered metal chalcogenide nanocrystals. Nano Lett 15:1911–1917
[27] Zhou J, Fang G, Pan A, Liang S (2016) Oxygen-incorporated MoS2 nanosheets with expanded interlayers for hydrogen evolution reaction and pseudocapacitor applications. ACS Appl Mater Interfaces 8:33681–33689
[28] Sahoo R, Sasmal AK, Ray C, Dutta S, Pal A, Pal T (2016) Suitable morphology makes CoSn(OH)6 nanostructure a superior electrochemical pseudocapacitor. ACS Appl Mater Interfaces 8:17987–17998
[29] Huang J, Virji S, Weiller BH, Kaner RB (2003) Polyaniline nanofibers: Facile synthesis and chemical sensors. J Am Chem Soc 125:314–315
[30] Le TH, Kim Y, Yoon H (2017) Electrical and electrochemical properties of conducting polymers. Polymers (Basel) 9:150
[31] Talaie A (1997) Conducting polymer based pH detector: A new outlook to pH sensing technology. Polymer (Guildf) 38:1145–1150
[32] Chandrasekhar P (2018) Electronic structure and conduction models of graphene. Conduct Polym Fundam Appl 101–106
[33] Forzani ES, Li X, Tao N (2007) Hybrid amperometric and conductometric chemical sensor based on conducting polymer nanojunctions. Anal Chem 79:5217–5224
[34] Athawale AA, Bhagwat SV, Katre PP (2006) Nanocomposite of Pd-polyaniline as a selective methanol sensor. Sen Actuators, B Chem 114:263–267
[35] Kwon OS, Park CS, Park SJ, Noh S, Kim S, Kong HJ, Bae J, Lee CS, Yoon H (2016) Carboxylic acid-functionalized conducting-polymer nanotubes as highly sensitive nerve-agent chemiresistors. Sci Rep 6:1–7

[36] Yoon H, Ko S, Jang J (2008) Field-effect-transistor sensor based on enzyme-functionalized polypyrrole nanotubes for glucose detection. J Phys Chem B 112:9992–9997
[37] Dey A, Kaushik A, Arya SK, Bhansali S (2012) Mediator free highly sensitive polyaniline-gold hybrid nanocomposite based immunosensor for prostate-specific antigen (PSA) detection. J Mater Chem 22:14763–14772
[38] Hui N, Sun X, Niu S, Luo X (2017) PEGylated polyaniline nanofibers: Antifouling and conducting biomaterial for electrochemical DNA sensing. ACS Appl Mater Interfaces 9:2914–2923
[39] Williams D (1991) Solid-state gas sensors: Prospects for selectivity. Anal Proc 28:366–377
[40] Do AV, Khorsand B, Geary SM, Salem AK (2015) 3D printing of scaffolds for tissue regeneration applications. Adv Healthc Mater 4:1742–1762
[41] Kang HW, Lee SJ, Ko IK, Kengla C, Yoo JJ, Atala A (2016) A 3D bioprinting system to produce human-scale tissue constructs with structural integrity. Nat Biotechnol 34:312–319
[42] Zhao CQ, Liu WG, Xu ZY, Li JG, Huang TT, Lu YJ, Huang HG, Lin JX (2020) Chitosan ducts fabricated by extrusion-based 3D printing for soft-tissue engineering. Carbohydr Polym 236
[43] Bartlett NW, Tolley MT, Overvelde JTB, Weaver JC, Mosadegh B, Bertoldi K, Whitesides GM, Wood RJ (2015) Robot powered by combustion. Science (80-) 349:161–165
[44] Yang D, Mosadegh B, Ainla A, Lee B, Khashai F, Suo Z, Bertoldi K, Whitesides GM (2015) Buckling of elastomeric beams enables actuation of soft machines. Adv Mater 27:6323–6327
[45] Wehner M, Truby RL, Fitzgerald DJ, Mosadegh B, Whitesides GM, Lewis JA, Wood RJ (2016) An integrated design and fabrication strategy for entirely soft, autonomous robots. Nature 536:451–455
[46] Chen T, Bilal OR, Shea K, Daraio C (2018) Harnessing bistability for directional propulsion of soft, untethered robots. Proc Natl Acad Sci U S A 115:5698–5702

6 3D-Printed Conducting Polymers for Electrochemical Energy Storage Devices

Alexandra Robinson, Anjali Gupta, Felipe M. de Souza, and Ram K. Gupta

6.1 INTRODUCTION

Electric energy is present in nearly every foundation of society nowadays, as it is applied in virtually every type of modern technology. The increasing energy demands pushed the development of energy storage devices such as supercapacitors and batteries. Such devices are widely employed in electronic products, backup power systems, wearable devices, pacemakers, and electric vehicles, among many others [1, 2]. Through that, the relevance of such devices in the market is noticeable. Supercapacitors, for instance, present high power density accompanied by long cycle stability and fast charge/discharge processes. However, they display low energy densities that are usually below 10 Wh/kg [3, 4]. Batteries, on the other hand, present a considerably higher energy density that can reach around 170 Wh/kg and is accompanied by a much longer charging and discharging time which is translated into low power density [5, 6]. Scientists have been focusing on optimizing the electrochemical performance of such devices. In this sense, such materials should display satisfactory electroactive properties that are strongly related to surface area, conductivity, electrostability, and chemical versatility. Even though there has been considerable progress in this area, the traditional techniques may present some hindrances in terms of obtaining more hierarchical structures in terms of electrode geometry and architecture. Some of the approaches to optimization consist of fabricating 3D structures for electrodes or solid electrolytes, which can lead to a more porous structure that can better utilize the limited space in energy storage devices. These approaches are commonly referred to as additive manufacturing or 3D printing techniques. There are several advantages related to these processes, as they allow better control of the device's design as well as the electrode's dimensions such as thickness, which enables an effective optimization process accompanied by a lower cost when compared to some traditional methods [7]. The progress in 3D printing has led to the development of several approaches, which can be divided into seven categories: (i) extrusion of material (i.e., fused deposition modeling (FDM) and direct ink writing (DIW)), (ii) powder bed fusion (i.e., direct metal laser (DML), selective laser

 DOI: 10.1201/9781003415985-6

sintering (SLS)), (iii) vat photopolymerization stereolithography (SLA)), (iv) material jetting (i.e., inkjet printing (IJP)), (v) binder jetting, (vi) sheet lamination (i.e., laminated object manufacturing), and (vii) direct deposition. A schematic of these 3D printing techniques is presented in Figure 6.1.

Aside from the techniques, there is also the need to explore the available materials that can be suitable to make inks for printing. Some of the main examples of those include carbon-based materials such as graphene, reduced graphene oxide (rGO), graphene oxide (GO), and carbon nanotubes (CNTs), among other carbon structures. Some of their attractive features are related to their high electronic and thermal conductivity, high porosity and surface area, chemical stability, and tunability.

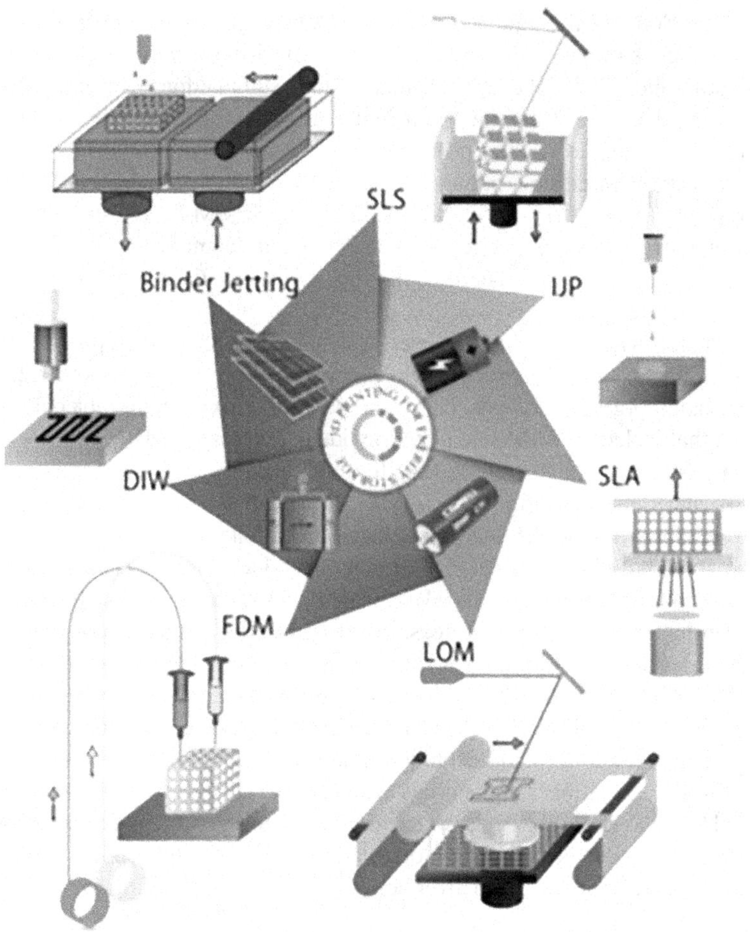

FIGURE 6.1 Schematics displaying the main 3D printing techniques utilized for the fabrication of energy storage devices. Reprinted with permission [9]. Copyright 2019, Royal Society of Chemistry.

Graphene-based materials can be obtained in the form of aerogels that can be used as templates for the electrodeposition of other materials, which can function as current collectors or electrodes. Also, carbon black and activated carbon can be introduced in the formulation of printable inks through blending with polymeric matrices. Similar approaches are also performed on CNTs that, despite presenting most of the desired properties for an energy storage material, tend to agglomerate, which can drastically decrease its active surface area. CNTs are often mixed with solvents, surfactants, and polymers to increase their dispersibility. Another common approach consists of functionalizing a surface with hydrophilic groups such as carboxylic acids or hydroxyls that can improve its interaction with water. Alongside graphene, the MXenes, which consist of 2D nanomaterials based on transition metal carbides and nitrides, have also been implemented in 3D printing techniques. Likewise, carbon-based materials also present high conductivity and porous surfaces along with an inherently high capacitance. Some of the unique aspects of MXenes are based on their high hydrophilicity which, similarly to GO, can facilitate their dispersion in aqueous media for the preparation of colloidal dispersions that can be suitable for 3D printing inks [8].

Following that, aside from the printable inks used for the fabrication of electrodes, there are also strategies used to print solid-state or gel polymer electrolytes (GPEs). GPEs are usually based on a polymeric matrix such as poly(ethylene glycol) (PEG), poly(vinyl alcohol) (PVA), poly(methyl methacrylate) (PMMA), and solvents (organic or aqueous) mixed with inorganic ionomers, which are compounds that can produce ions when dissolved such as neutral salts (i.e., $LiClO_4$, LiCl, Na_2SO_4), strong bases (i.e., KOH, LiOH, NaOH), or acids (i.e., H_2SO_4 and H_3PO_4). It is worth noting that despite the eco-friendly aspects of water as a solvent, it limits the operating voltage window, as it can lead to electrochemical decomposition at voltages above 1.23 V. On the other hand, organic GPE inks can be prepared by mixing high-molecular-weight polymers that include PMMA and poly(vinylidene fluoride) (PVDF), along with salts in organic solvents, which may include tetrahydrofuran (THF), dimethyl sulfoxide (DMSO), and dimethyl formamide (DMF), for example. These solvents allow for voltage windows around 3.5 V, which makes them suitable for Li-ion batteries, for instance. Ionic liquid-based GPEs can display several advantages such as non-volatility or flammability and thermal and electrochemical stability [10, 11]. The importance of 3D printing extends to several areas; in that sense, the application of this technology in the development of energy storage devices can provide a more in-depth level of control, which allows for the fabrication of hierarchical structures that can greatly improve the overall efficiency of electrodes used in supercapacitors and batteries. There is a plethora of different techniques that can be applied for that process, along with a considerable number of materials available that can be converted into printable inks used to make either electrodes or electrolytes in energy storage devices. However, there are some challenges related to formulating inks that can keep the electroactive materials properly dispersed to deliver better electrochemical performance.

6.2 ELECTROCHEMICAL ENERGY DEVICES

Finding new or clean energy sources is not the only challenge facing the ever-evolving modern world as its demand for fuel to drive its industrial advancements rapidly increases; it also requires updated energy storage mechanisms. Currently, research attention has been drawn to converting and storing energy in electrochemical energy devices [12]. This section will predominantly focus on the working principles and mechanisms of supercapacitors, metal-ion batteries, metal-air batteries, and metal-sulfur batteries. There are three main categories of electrochemical energy storage devices: capacitors, batteries, and supercapacitors. Capacitors are efficient in energy distribution, but they are lacking in the ability to maintain large amounts of energy per unit volume about other devices. Meanwhile, batteries do have that ability, but they struggle to distribute this energy effectively as observed in capacitors. Therefore, supercapacitors were created to find a middle ground between the two by mixing their mechanical structures in a way that produces greater capacitance and energy density than capacitors and greater power density than batteries [13]. Accordingly, the characteristics of an electrochemical energy device rely heavily on the chemistry of its electrodes [14].

A standard supercapacitor contains a liquid electrolyte with a membrane separating two solid electrodes of opposing charges, enforced by voltage. Its major characteristics result from the relationship and properties shared between the chosen electrode and electrolyte, including how evenly the electrolytic ions may disperse across the electrodes. Eq. 6.1 demonstrates these relationships, given as

$$C = (\varepsilon_0 \cdot \varepsilon_r \cdot A)/d \quad (1)$$

where ε_0 is the free space's potential to store electrochemical energy, ε_r is the dielectric material's potential to store electrochemical energy in relation to ε_0, A is the electrode's surface area, and d is a measure of the distance between the two polarized electrodes.

Each supercapacitor is categorized based on its mechanical blueprint and energy storage technique. There are two main energy storage principles: electric double-layer capacitance and pseudocapacitance. However, there are three designated branches of supercapacitors: electric double-layer capacitors (EDLCs), pseudocapacitors (PCs), and hybrid supercapacitors (HSCs) [12]. EDLCs store energy electrostatically by using an applied voltage to drive electrolytic ions across the membrane separator into the oppositely charged electrode, where it is absorbed. As charge accumulates, two layers are formed, hence the electric double-layer storage technique. This is a non-faradaic process that yields impressive durability, fast energy absorption, and no material swelling; however, energy density capabilities are restricted. Alternatively, PCs store energy electrochemically. In this faradaic process, oxidation-reduction reactions across an electric double layer are used to create an electrical current to transfer ions between the electrode and electrolyte, thus storing energy through chemical reactions at the electrode [15]. PCs have higher energy density capabilities than EDLCs but lower cyclic durability due to the chemical reactions. HSCs combine both faradaic and non-faradaic processes using carbon-based electrodes as seen in EDLCs with metal/conducting polymer electrodes as seen in PCs [14].

Metal-ion batteries (MIBs) are a different type of energy storage device that converts energy during charges and discharges by the transfer of ions between polar electrodes, termed the "rocking chair battery" [16]. Lithium-ion batteries (LIBs) are the predominant example of MIBs in real-world applications, such as battery-powered tools, cars, and handheld devices. A typical LIB is made with a carbon anode and a metal oxide cathode, and these components intercalate the Li-ions during charge/discharge. Then it has an electrolyte, usually liquid, that conducts the ions in the direction of their respective electrode. Additionally, a membrane separator prevents a possible short circuit between the polar electrodes [17]. Figure 6.2 depicts this structure [18]. Energy density is primarily dependent upon the material used to make up the cathode [19]. Durability increases with the success of the cathode's ability to intercalate the ions. Other types of MIBs include monovalent sodium-ion batteries or potassium-ion batteries and multivalent zinc-ion batteries or calcium-ion batteries [20].

Metal-air batteries (MABs) are open-system energy storage devices that are different from other battery types as they utilize oxygen from the surrounding air as their cathodic material, yielding a lower price and weight. Like MIBs, they also contain two electrodes, an electrolyte (liquid or solid), and a membrane separator. However, the anode is built with metals, such as zinc or magnesium, but the cathode is made of porous that allows space for air to diffuse [21]. Figure 6.3 depicts the general structure of a zinc-air battery (ZAB), one of the most popular MABs [22]. The anode of an MAB is the site where metals are converted to metallic cations, while the cathode is the site where oxygen is converted to hydroxide anions. Furthermore,

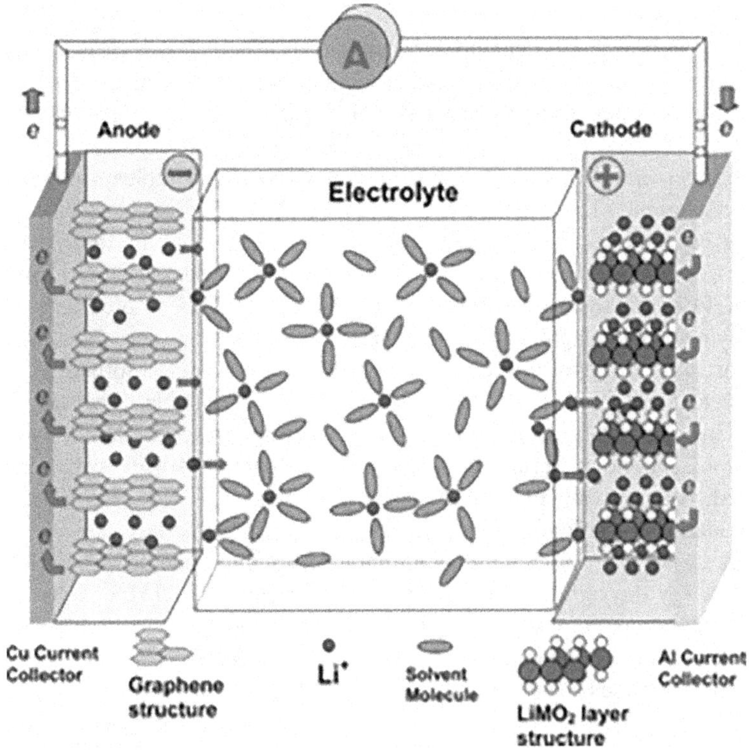

FIGURE 6.2 The general structure of an LIB. Reprinted with permission [18]. Copyright 2004, American Chemical Society.

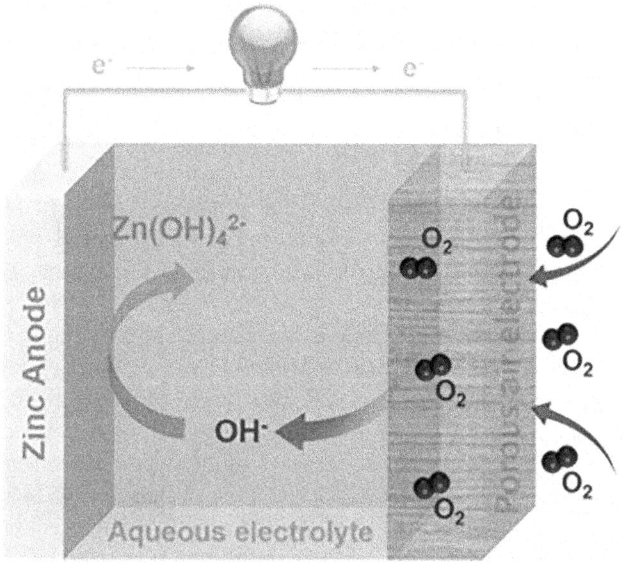

FIGURE 6.3 The general structure of a ZAB. Reprinted with permission [22]. Copyright 2019, American Chemical Society.

the cathode contains three layers vital to its efficiency: the catalyst layer, which serves to speed up the oxygen reduction reaction during discharge and the oxygen evolution reaction during charge; the gas diffusion layer, which holds electrolyte in, keeps external liquid out, allows oxygen to flow through the air to the catalyst layer, and stabilizes the catalyst layer; and a current collector, for the electrons produced from the electrodes [23].

Metal-sulfur batteries (MSBs) are electrochemical energy storage devices that use more complex processes during their charge and discharge mechanisms. These batteries are composed of two electrodes, one of a metal and the other sulfur-based, with a separator saturated with electrolyte in the middle. The metal electrode is the site of oxidation, while the sulfur-based electrode is the site of reduction [24]. The presence of sulfur itself is what drives the 16-electron redox reactions in these batteries since during the discharge process there is the formation of lithium polysulfides. For example, in the popular lithium-sulfur battery (LSB), sulfur is converted to lithium sulfide. While LIBs have an intercalation mechanism, LSBs have a conversion mechanism [25]. Eq. 6.2 illustrates this [26].

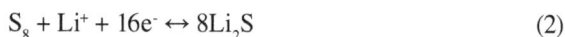

$$S_8 + Li^+ + 16e^- \leftrightarrow 8Li_2S \tag{2}$$

In discharge, solid sulfur at the cathode reacts with metal ions produced by the anode to spur a four-region conversion process of sulfur to polysulfide intermediates to stable metal sulfides. During charge, these metal sulfides are reversely converted back to metal ions and solid sulfur. One of the major challenges of this mechanism is the shuttle effect, caused by the soluble polysulfide intermediates [25]. Additionally, the sulfur reduction reaction (SRR) is kinematically slow, thanks to its many complicated steps [26].

6.3 CONDUCTING POLYMERS

3D printing technology can offer a more versatile approach to the manufacturing of products, as it may remove the need for tooling, polishing, and other processing techniques that can lead to an increase in production time, cost, and waste. Products such as biomedical devices, electronics, and wearable technologies can greatly benefit from this technique. One common approach lies in the compositing of carbon-based materials with nanoparticles that can be printed over a polymeric matrix [27, 28]. The addition of conducting polymers can provide a new array of materials given that such materials usually present low weight, cost, and high synthetical versatility. The conducting polymers consist of repeating units of a monomer that contain conjugated double bonds which enable electronic conductivity. Some examples are polypyrrole (PPy), polyaniline (PANI), and poly(3,4-ethylenedioxythiophene): polystyrene sulfonate (PEDOT:PSS), which are the most utilized ones in 3D printing techniques. Despite their inherent conductivity, this property can be enhanced through the doping of metallic atoms or compositing with fillers such as carbon and graphene, among others [29–31]. PANI is a highly versatile conducting polymer that can be synthesized in a plethora of procedures that can be divided into either chemical or electrochemical approaches. In the chemical route, aniline can be polymerized into PANI using an oxidant agent which can be ammonium persulfate, HCl, H_2SO_4, and $FeCl_3$, among others [32, 33]. In the electrochemical route, aniline can be polymerized and electrodeposited in a substrate through an applied voltage from a potentiostat. Despite the easy synthetic methods for the polymerization of PANI, its processability is often difficult given its low solubility in most solvents and infusibility, which hinders the extrusion printing method of similar techniques that require optimal rheological properties. PANI can be composited with other polymers and functionalized with other chemical moieties that can further improve its conductivity, mechanical, and thermal properties [34–36]. Also, it has been observed that PANI becomes more soluble with the addition of organic acids such as dodecylbenzene sulfonic acid (DBSA) or camphorsulphonic acid (CSA) [37, 38]. There have been some promising approaches that enable the 3D printing of conducting polymeric composites for energy storage devices. One example was presented by Wang *et al.* [39], who fabricated a PANI/rGO composite that could be utilized as an ink for 3D printing. The synthetic approach consisted of obtaining GO through a modified Hummer's method and PANI through oxidative polymerization with HCl and ammonium persulfate. The obtained PANI was later dispersed in N-methyl-2-pyrrolidone (NMP) and mixed with GO to form a gel that was used as an ink through the direct ink writing method. After printing the desired shape, the electrodes were soaked in an HI solution to reduce the GO into rGO yielding the PANI/rGO. The schematic for this synthetic process is presented in Figure 6.4. Through that, a planar solid-state supercapacitor was fabricated that presented satisfactory electrochemical performance by delivering a specific capacitance of around 1329 mF/cm^2.

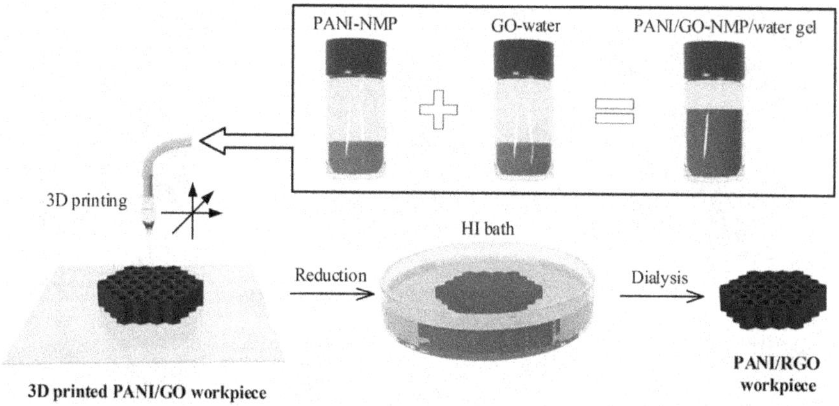

FIGURE 6.4 Schematic for the 3D printing process for a PANI/rGO-based electrode. Adapted with permission [39]. Copyright 2018, American Chemical Society.

In another approach, Yuk *et al.* [40] fabricated a 3D-printable conducting ink based on PEDOT:PSS. For that, an aqueous solution of PEDOT:PSS went through a freeze-drying process that was later redispersed in a solution of water and DMSO. By performing freeze-drying, the water in the solution was evaporated, leading to the formation of a porous gel of conducting polymer. The redispersion of the polymer into a water/DMSO solution led to a stable dispersion of the conducting polymer with the solvent system, allowing it to efficiently printed over a substrate. Then a dry annealing process was performed to evaporate the solvents and increase the contact between the PEDOT:PSS chains to increase the ink's conductivity. The schematic for that process is presented in Figure 6.5a–c. The change in morphology throughout the processes for the fabrication of the polymer conducting-based ink can also be seen through a transmission electron microscope (TEM). It can be seen that the PEDOT:PSS nanofibrils become progressively more compact when comparing the solution (Figure 6.5d), ink (Figure 6.5e), and applied ink after the dry-annealing (Figure 6.5f) process, which can lead to better conductivity, which for this case was around 155 S/cm. Further, the concentration of PEDOT:PSS had a considerable influence on the rheological properties of the ink, as higher concentrations of PEDOT:PSS caused the ink to lose its glossy aspect while increasing its viscosity, as shown in Figure (5g–j). From that, the apparent viscosity of inks with increasing concentrations of PEDOT:PSS could present a range in which they were spreading, printable, or clogging the nozzle, as presented in Figure 6.5k.

6.4 3D-PRINTED CONDUCTING POLYMERS FOR ELECTROCHEMICAL ENERGY

6.4.1 Supercapacitors

The fabrication of electrode materials suitable for supercapacitors through 3D printing is a technology that can provide some advantages when compared to traditional

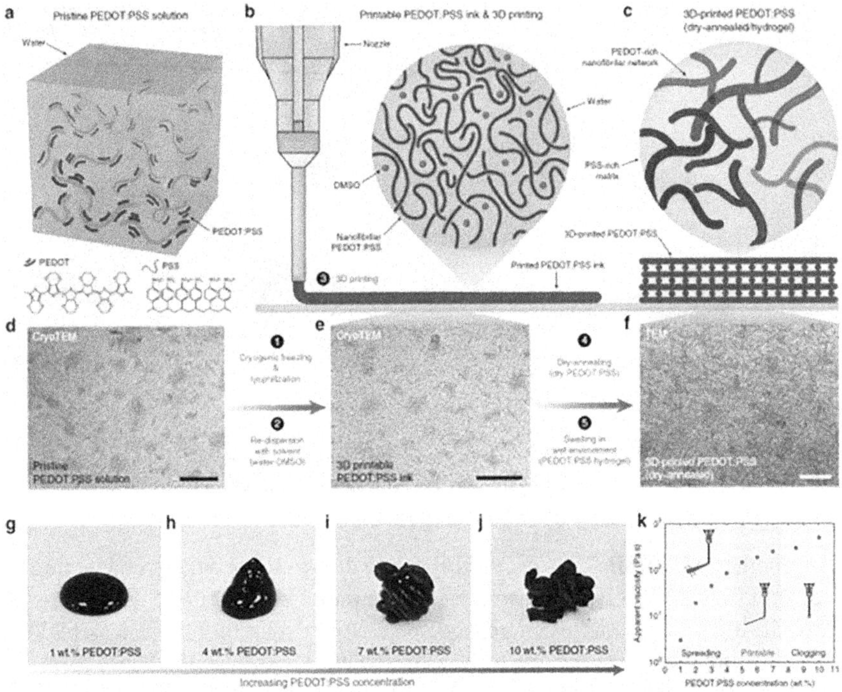

FIGURE 6.5 Schematics for the synthesis and fabrication of a 3D-printable ink based on PEDOT:PSS. (a) Aqueous solution of PEDOT:PSS. (b) 3D-printable ink PEDOT:PSS dispersed in DMSO and water after freeze-drying and redispersing process. (c) 3D-printed PEDOT:PSS after the dry-annealing process. TEM images for (d) pristine PEDOT:PSS aqueous solution, (e) 3D-printable PEDOT:PSS ink, and (f) 3D-printed PEDOT:PSS after the dry annealing process. (g–j) Redispersed inks with increasing concentrations of PEDOT:PSS. (k) Apparent viscosity for 3D print inks with increasing concentrations of PEDOT:PSS. Adapted from reference [40]. Copyright 2020, the authors, published by *Nature*. This article is licensed under a Creative Commons Attribution 4.0 International License.

synthetical approaches such as design flexibility, reduction of material waste, fast prototyping, a wider range of options for customization, facilitating the integration of components, and utilizing a diverse number of materials. Hence, it can facilitate the scaling of production. Aside from these aspects, 3D printing technology can also be an important part of the fabrication of flexible energy storage devices. Based on that, Yang *et al.* [41] proposed the fabrication of an electroactive ink based on PEDOT:PSS that was later composited with CNTs, which yielded a high area capacitance of around 990 mF/cm^2 along with a capacitance retention of 74.7% after 14000 cycles. The geometrical design of the electrode also played an important role in the flexibility of the device. Figure 6.6a–b presents the design of the electrode and the way it can withstand strains in different directions. The analyzed electrochemical properties of the PEDOT:PSS/CNT showed a performance improvement when compared to the neat PEDOT:PSS electrode. The performance improvement could

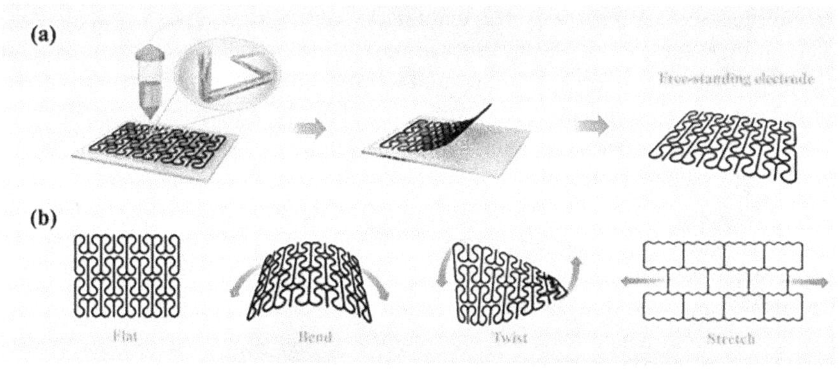

FIGURE 6.6 Schematic displaying the (a) fabrication process of 3D-printed electrodes based on PEDOT:PSS/CNT with (b) a high degree of flexibility. Reprinted with permission [41]. Copyright 2021, Royal Society of Chemistry.

be attributed to the better contact between the aromatic moieties present in both CNT and PEDOT:PSS, which could have improved the conductivity of the electrode. Also, the inherent mechanical properties and electrochemical stability of CNT may have played a role in the improvement of such properties in the device. Furthermore, the electrochemical profile for the assembled device presented a quasi-rectangular shape in the cyclic voltammetry (CV) and a quasi-triangular shape in the galvanostatic charge-discharge (CGD), which implied that most of the energy storage mechanism occurred based on the EDLC process. Hence, the improvement of surface area to allow a better permeation of ions along with the inherent conductivity of both PEDOT:PSS and CNT were likely the major factors that yielded satisfactory performance for the flexible 3D-printed supercapacitor.

6.4.2 Metal-Ion Batteries

Metal-ion batteries are some of the main commercially available energy storage devices due to their satisfactory electrochemical performance. Currently, LIBs represent a larger portion of the market, yet increasing demand and performance optimization led to the need for new approaches to further improve their properties. There is a myriad of materials that can be used in 3D printed LIBs, such as $LiCoO_2$ (LCO), $LiTi_5O_{12}$ (LTO), $LiMnO_4$ (LMO), and $LiFe_xMn_{1-x}PO_4$ (LFMP). These electroactive materials can be added in the formulation of 3D-printable inks, which can be used to fabricate electrodes through different techniques such as DIW and IJP, for instance. However, one of the common issues is related to the low conductivity of these materials. To counter that, conducting additives such as activated carbon, acetylene black, or others can be added in the ink's formulation. Such an addition leads to an improvement in the electrochemical performance. Another aspect related to the 3D printing of electrodes for batteries is that the design plays a major role in the properties and electrochemical stability. Ion migration in the electrode–electrolyte interface can be improved. However, complex structural designs can deteriorate after an excessive number of cycles, potentially leading to a performance decrease. Based on that, some of the factors that influence the overall electrochemical properties of a 3D printed electrode are related to ink additives, structured surface area, electrode geometry/architecture, and structural stability. An important aspect of 3D printing for batteries is that it enables the fabrication of flexible electrodes. Such an approach has been performed by several researchers, such as Kohlmeyer *et al.* [42], who fabricated an ink for 3D printing by adjusting the proportion of LCO, LTO, and LFP, along with carbon nanofibers and PVDF. The employed printing method was DIW, which has several advantages such as low cost, easy operation, a broader range of available materials that can be used, and the capability to fabricate complex 3D structures at room temperature without the need for tooling or complex processing techniques. The DIW consists of an extrusion of a continuous filament composed of a shear-thinning ink. Then the ink is applied on a substrate in a predetermined fashion that follows a layer-by-layer model. The electrodes displayed specific capacitances of 80, 89, and 106 mAh/g for the LCO, LTO, and LFP at 5C rate, respectively.

Interdigitated electrodes are another design that enables closer proximity between electrodes by making an alternating pattern between the anode and cathode with a precision below the millimeter scale. Sun *et al.* [43] fabricated a micro battery that displayed power and energy densities of 9.7 J/cm^2 at 2.7 mW/cm^2 and 1.5 mAh/cm^2 at 1C, respectively. The micro battery consisted of LFP and LTO as positive and negative electrodes, respectively, that were 3D printed through an interdigitated design with the schematics presented in Figure 6.7a–d.

In another work, Cao *et al.* [44] fabricated a micro battery through interdigitated electrodes based on LFP and Li-metal as cathode and anode, respectively. The formulated ink for that process presented hydroxyl functionalized CNFs, which yielded satisfactory rheological properties such as high viscosity, conductivity, water retention, and shear-thinning behavior. The main advantage of this ink was the suppression of Li dendrite growth, which led to higher electrochemical stability. The schematic of the general aspects and design of the interdigitated micro battery is presented in Figure 6.8a–d. Based on those works, is notable that interdigitated and fork-type electrodes can provide some valuable enhancements such as close contact between the electrodes that can reduce the overall system's resistance, attainable electrochemical stability, and reduced space for the device.

6.4.3 Metal-Air Batteries

Traditional batteries function through the conversion of chemical potential energy into electricity based on the chemical reaction of metallic ions from the anode to the cathode. In a MAB, there is a metal-based anode, Zn or Al, and a cathode based on a material that allows the facile diffusion of O_2. In an aqueous electrolyte, O_2 diffuses to receive electrons from the metallic ions that are dissolved in the electrolyte. Then, after receiving the electrons, O_2 is converted into OH^-. In a non-aqueous electrolyte, O_2 is converted to O^{2-} species instead. Such technology can bring a novel way to store energy, as it can provide high energy densities in a relatively small weight when compared to traditional batteries. However, there are still several challenges, such as its life cycle due to relatively low electrochemical stability and constant need for

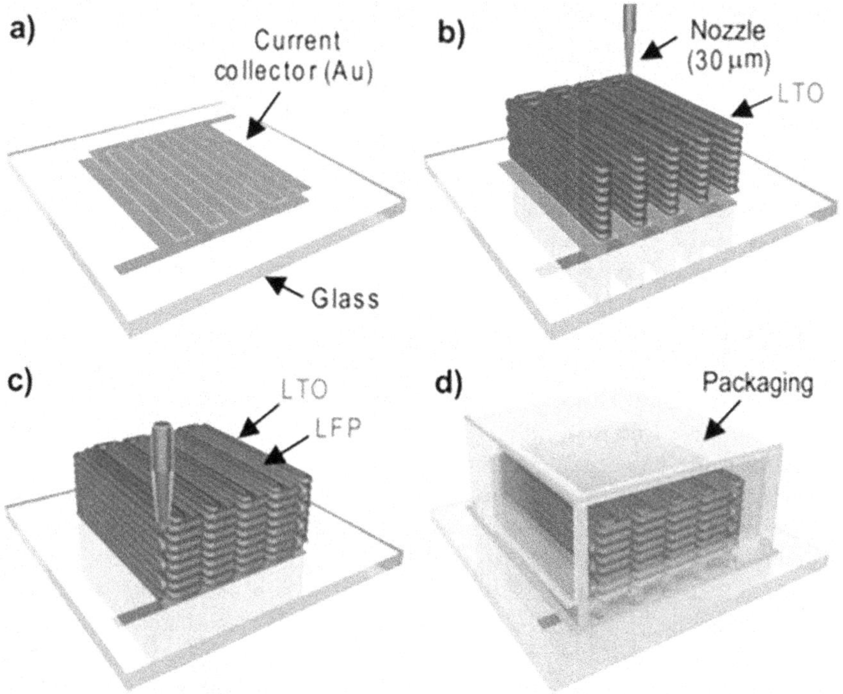

FIGURE 6.7 Illustration for the 3D-printed micro battery with interdigitated architectures over (a) a glass substrate containing an Au as current collector, (b) LTO- and (c) LFP-containing inks that were applied through a 30-μm nozzle, followed by a sintering process, and (d) packing. Reprinted with permission [43]. Copyright 2013, John Wiley and Sons.

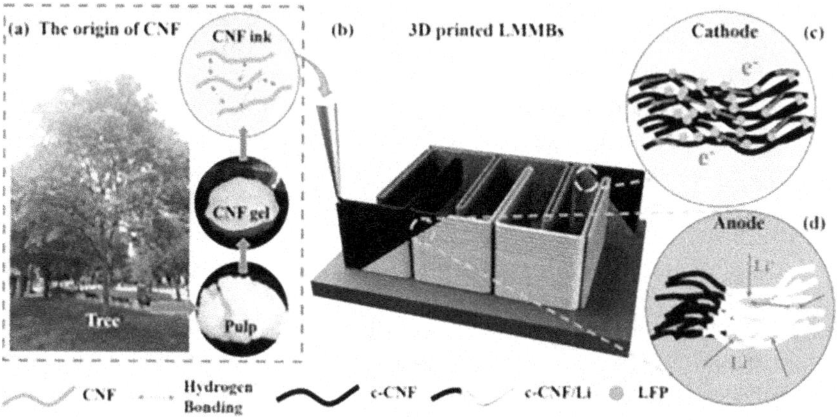

FIGURE 6.8 Schematic for a micro battery based on interdigitated electrodes. (a) Bio-based CNF is used for the fabrication of ink. (b) Design of the micro battery. (c) Cathode based on CNF/LFP and (d) 3D-printed CNF/Li electrode that functioned as a stable scaffold for the intercalation with Li atoms. Reprinted with permission [44]. Copyright 2019, John Wiley and Sons.

O_2 supply from the atmosphere. Since this technology is relatively new, there have been a limited number of studies. However, there has been continuous progress in the elucidation of its challenges. One example was reported by Yu *et al.* [45], who performed laser sintering to fabricate printed anodes for Al-air batteries (AAB). This approach led to better conductivity, as it was proposed that infrared laser sintering could improve the electric contact between the Al nanoparticles, which led to an improvement in electrochemical performance. The device consisted of a Pt/C as an air cathode and a gel containing KOH as an electrolyte. The AAB presented 239 mAh/g of discharge capacity along with 0.95 V of operation voltage. In addition to that, the mass loading of active material could be performed by controlling the number of layers of the 3D-printed electrode, as there was a progressive increase in the areal capacity with the increasing number of layers that were 1.5 (360 μm), 2.8 (560 μm), and 3.23 mAh/cm^2 (680 μm) for one, two, and three layers of electrode, respectively. Based on that, utilizing methods that can lead to a hierarchical and defined structure has been demonstrated to be a core factor for the improvement in the performance of MABs, as properly controlling such aspects can lead to a decrease inside reactions in the anode. Yet gas impurities such as CO_2 and electrolyte instability are still some of the issues related to this technology. The kinetic improvement of ORR and OER is influenced by the material's size, morphology, and crystallographic structure. Along with that, the introduction of carbon-based nanomaterials and doping with heteroatoms are known approaches that can be further explored. In this sense, introducing 3D printing techniques can provide better approaches to systematically combine the optimal aspects of each component of an MAB to lead to an increase in energy density, electrochemical efficiency, and stability.

6.4.4 Metal-Sulfur Batteries

The LSB is a highly regarded technology, as it can potentially surpass the capacity of current LIBs, as in theory, it can provide around 1675 mAh/g. However, the inherently high resistance of S and the low amount of active materials enabled to be introduced into the electrode hinder the optimization of its performance. Part of the solution to address that can be potentially found in the use of 3D printing techniques. One example was presented by Hintennac *et al.* [46], who utilized direct laser printing for the fabrication of a small LSB electrode. The device delivered a specific capacitance of around 1300 Ah/kg with a specific capacitance of around 75% at 1C after 400 cycles. Such performance was attributed to the relatively low amount of active material in the battery. In another approach, Manthiram *et al.* [47] prepared an aqueous-based ink containing aligned multi-walled CNTs, which was used to print a collector that prevented the dissolution of polysulfides. The fabricated cathode displayed an areal discharge capacity of around 7 mAh/cm^2 and 7 mg/cm^2 of S. Similarly, some approaches involving inkjet printing have also been performed to fabricate LSB [48]. In another study Shen *et al.* [49] fabricated a printable ink based on a sulfur copolymer-graphene composite. For that, a dispersion of S and GO was prepared by sublimating the S into a GO solution, which was followed by slight evaporation of solvent. Then 1,3-diisopropenylbenzene (DIB) was mixed into the system. The obtained ink could then be used to 3D print layer-by-layer designs.

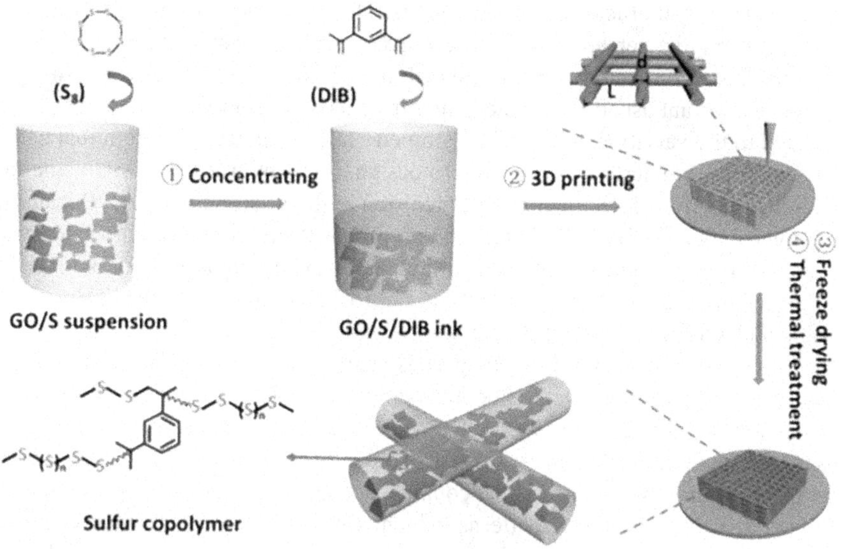

FIGURE 6.9 Schematic for the synthesis of 3D-printed sulfur copolymer graphene. Reprinted with permission [49]. Copyright 2017, John Wiley and Sons.

Following that, a freeze-drying process was performed to remove the solvent. Then a heating process was performed to convert the cyclic S_8 into a linear polysulfide that could react with DIB over the GO's surface [50]. The rheological properties of the composite ink based on S/DIB/GO displayed a shear-thinning behavior, which made it favorable for extrusion 3D printing techniques. The schematic for the synthesis and fabrication of the 3D-printed sulfur copolymer graphene is presented in Figure 6.9. Thermal treatment played an important role in this process, as it set the ink over the substrate but could also reduce the oxygenated groups over the GO's surface while promoting the chemical bond with the polysulfide. Hence, this approach led to the fabrication of an electrode for LSB that could partially suppress the dissolution of polysulfides as they were chemically bonded to the rGO's surface. The 3D-printed sulfur copolymer graphene presented a reversible capacity of around 812.8 mAh/g along with satisfactory cycle stability. Despite being relatively laborious due to the multiple steps required for the fabrication of the electrode, this approach presents a path for the fabrication of LSB electrodes that can diminish some of their inherent issues such as the high resistance of S, the shuttling effect, and poor electrochemical stability. One of the challenges remains in increasing the amount of active material that can be incorporated into an electrode without leading to a drastic decrease in stability and capacitance after several charge-discharge cycles.

6.5 CONCLUSION AND FUTURE REMARKS

3D printing technology has provided a new array of opportunities for the fabrication of electrodes for different types of energy storage devices such as batteries, supercapacitors, and fuel cells, among others. The methods available for 3D printing itself are also quite versatile, as there are several ways in which an electrode can be fabricated based on the type of ink and technique utilized to print it over a substrate. Another important aspect that makes the use of the 3D printing technique attractive is related to the variations in terms of geometry and architectural design that can be explored both in macro as well as micro design. Such an aspect allows for the fabrication of electrodes with a designed precision at the micrometer scale. Because of that, some materials that would present undesirable electrochemical properties when obtained through traditional methods could be revisited with the use of 3D printing techniques that can potentially lead to an improvement in performance. Hence, the design of electrodes with novel structures and morphologies through 3D printing can be a convenient way to optimize the performance of well-known materials. Despite the several techniques available, the most employed ones within the field of energy storage devices are generally FDM, DIW, and IJP. However, despite the myriad of possibilities with this technology, there are some issues. For example, the IJP method may often require the use of ink that contains particles within the nanoscale to avoid clogging and proper printing patterns, which may not always be feasible in certain conditions. Also, a relatively limited number of materials have been utilized in 3D printing technologies, which are often carbon-based, such as GO, rGO, CNTs, and graphene. Hence, other materials with similar properties and dimensions are likely to be utilized in such techniques. Also, it is worth noting that the structural design and morphology of some electroactive materials suitable for electrodes are highly dependent on the 3D printing technique as well as the post-processing procedure, which includes freeze-drying processes and high-temperature annealing, for instance. The extra steps required to reach a functional electrode with optimal performance lead to an increase in cost and time-consuming fabrication, which can make it challenging for a scaling process. Hence, there is a need to improve on these techniques to enable processing that is efficient as well as convenient to enable their proper use in larger-scale production. Based on these aspects, 3D printing can offer a broad range of designs and customization for the fabrication of components for energy storage devices, which is due to the detail in controlling the porous structures that can lead to an improvement in contact between the electrode and electrolyte. Based on that, combining the advantages of 3D printing with traditional manufacturing processes can potentially meet the demand and requirements for the optimization of performance.

REFERENCES

[1] B. Mendoza-Sánchez, Y. Gogotsi, Synthesis of two-dimensional materials for capacitive energy storage, Adv. Mater. 28 (2016) 6104–6135.

[2] A.G. Olabi, C. Onumaegbu, T. Wilberforce, M. Ramadan, M.A. Abdelkareem, A.H. Al—Alami, Critical review of energy storage systems, Energy. 214 (2021) 118987.

[3] X. Li, B. Wei, Supercapacitors based on nanostructured carbon, Nano Energy. 2 (2013) 159–173.

[4] X. Tian, J. Jin, S. Yuan, C.K. Chua, S.B. Tor, K. Zhou, Emerging 3D-printed electrochemical energy storage devices: A critical review, Adv. Energy Mater. 7 (2017) 1–17.

[5] B. Scrosati, J. Garche, Lithium batteries: Status, prospects and future, J. Power Sources. 195 (2010) 2419–2430.
[6] S.W. Running, Ecosystem disturbance, carbon, and climate, Science (80-.). 321 (2008) 652–653.
[7] A. Azhari, E. Marzbanrad, D. Yilman, E. Toyserkani, M.A. Pope, Binder-jet powder-bed additive manufacturing (3D printing) of thick graphene-based electrodes, Carbon N. Y. 119 (2017) 257–266.
[8] T. Nathan-Walleser, I.-M. Lazar, M. Fabritius, F.J. Tölle, Q. Xia, B. Bruchmann, S.S. Venkataraman, M.G. Schwab, R. Mülhaupt, 3D micro-extrusion of graphene-based active electrodes: Towards high-rate AC line filtering performance electrochemical capacitors, Adv. Funct. Mater. 24 (2014) 4706–4716.
[9] P. Chang, H. Mei, S. Zhou, K.G. Dassios, L. Cheng, 3D printed electrochemical energy storage devices, J. Mater. Chem. A. 7 (2019) 4230–4258.
[10] Y. Lin, Y. Gao, F. Fang, Z. Fan, Recent progress on printable power supply devices and systems with nanomaterials, Nano Res. 11 (2018) 3065–3087.
[11] X. Wei, D. Li, W. Jiang, Z. Gu, X. Wang, Z. Zhang, Z. Sun, 3D printable graphene composite. Sci Rep. 5: 11181 (2015).
[12] M. Şahin, F. Blaabjerg, A. Sangwongwanich, A comprehensive review on supercapacitor applications and developments, Energies. 15 (2022) 674.
[13] A. Rajapriya, S. Keerthana, N. Ponpandian, Chapter 4—Fundamental understanding of charge storage mechanism, in: C.M. Hussain, M.B.B.T.-S.S. Ahamed (Eds.), Smart Supercapacitors: Fundamentals, Structures, and Applications, Elsevier, 2023: pp. 65–82.
[14] S. Najib, E. Erdem, Current progress achieved in novel materials for supercapacitor electrodes: Mini review, Nanoscale Adv. 1 (2019) 2817–2827.
[15] Z.S. Iro, C. Subramani, S.S. Dash, A brief review on electrode materials for supercapacitor, Int. J. Electrochem. Sci. 11 (2016) 10628–10643.
[16] Y. Liu, R. Holze, Metal-ion batteries, Encyclopedia. 2 (2022) 1611–1623.
[17] J. Verma, D. Kumar, Metal-ion batteries for electric vehicles: Current state of the technology, issues and future perspectives, Nanoscale Adv. 3 (2021) 3384–3394.
[18] K. Xu, Nonaqueous liquid electrolytes for lithium-based rechargeable batteries, Chem. Rev. 104 (2004) 4303–4418.
[19] X. Lin, Y. Sun, Y. Liu, K. Jiang, A. Cao, Stabilization of high-energy cathode materials of metal-ion batteries: Control strategies and synthesis protocols, Energy & Fuels. 35 (2021) 7511–7527.
[20] M. Kiai, O. Eroglu, N. Aslfattahi, Metal-ion batteries: Achievements, challenges, and prospects, Crystals. 13 (2023).
[21] A.G. Olabi, E.T. Sayed, T. Wilberforce, A. Jamal, A.H. Alami, K. Elsaid, S.M. Rahman, S.K. Shah, M.A. Abdelkareem, Metal-air batteries—A review, Energies. 14 (2021).
[22] J. Yi, X. Liu, P. Liang, K. Wu, J. Xu, Y. Liu, J. Zhang, Non-noble Iron Group (Fe, Co, Ni)-based oxide electrocatalysts for aqueous zinc–air batteries: Recent progress, challenges, and perspectives, Organometallics. 38 (2019) 1186–1199.
[23] C. Wang, Y. Yu, J. Niu, Y. Liu, D. Bridges, X. Liu, J. Pooran, Y. Zhang, A. Hu, Recent progress of metal–air batteries—A mini review, Appl. Sci. 9 (2019).
[24] F. Shi, J. Yu, C. Chen, S.P. Lau, W. Lv, Z.-L. Xu, Advances in understanding and regulation of sulfur conversion processes in metal–sulfur batteries, J. Mater. Chem. A. 10 (2022) 19412–19443.
[25] H. Ye, Y. Li, Room-temperature metal–sulfur batteries: What can we learn from lithium–sulfur? InfoMat. 4 (2022) e12291.
[26] L. Zhou, D.L. Danilov, F. Qiao, J. Wang, H. Li, R.-A. Eichel, P.H.L. Notten, Sulfur reduction reaction in lithium–sulfur batteries: Mechanisms, catalysts, and characterization, Adv. Energy Mater. 12 (2022) 2202094.

[27] K.R. Ryan, M.P. Down, N.J. Hurst, E.M. Keefe, C.E. Banks, Additive manufacturing (3D printing) of electrically conductive polymers and polymer nanocomposites and their applications, EScience. 2 (2022) 365–381.

[28] M.P. Down, E. Martínez-Periñán, C.W. Foster, E. Lorenzo, G.C. Smith, C.E. Banks, Next-generation additive manufacturing of complete standalone sodium-ion energy storage architectures, Adv. Energy Mater. 9 (2019) 1803019.

[29] C.W. Foster, M.P. Down, Y. Zhang, X. Ji, S.J. Rowley-Neale, G.C. Smith, P.J. Kelly, C.E. Banks, 3D printed graphene based energy storage devices, Sci. Rep. 7 (2017) 42233.

[30] Z. Rymansaib, P. Iravani, E. Emslie, M. Medvidović-Kosanović, M. Sak-Bosnar, R. Verdejo, F. Marken, All-polystyrene 3D-printed electrochemical device with embedded carbon nanofiber-graphite-polystyrene composite conductor, Electroanalysis. 28 (2016) 1517–1523.

[31] S.J. Rowley-Neale, D.A.C. Brownson, G.C. Smith, D.A.G. Sawtell, P.J. Kelly, C.E. Banks, 2D nanosheet molybdenum disulphide (MoS2) modified electrodes explored towards the hydrogen evolution reaction, Nanoscale. 7 (2015) 18152–18168.

[32] S. Bhadra, D. Khastgir, N.K. Singha, J.H. Lee, Progress in preparation, processing and applications of polyaniline, Prog. Polym. Sci. 34 (2009) 783–810.

[33] D. Coltevieille, A. Le Méhauté, C. Challioui, P. Mirebeau, J.N. Demay, Industrial applications of polyaniline, Synth. Met. 101 (1999) 703–704.

[34] X. Cheng, V. Kumar, T. Yokozeki, T. Goto, T. Takahashi, J. Koyanagi, L. Wu, R. Wang, Highly conductive graphene oxide/polyaniline hybrid polymer nanocomposites with simultaneously improved mechanical properties, Compos. Part A Appl. Sci. Manuf. 82 (2016) 100–107.

[35] S. Wang, Y. Zhou, Y. Liu, L. Wang, C. Gao, Enhanced thermoelectric properties of polyaniline/polypyrrole/carbon nanotube ternary composites by treatment with a secondary dopant using ferric chloride, J. Mater. Chem. C. 8 (2020) 528–535.

[36] H. Yan, K. Kou, Enhanced thermoelectric properties in polyaniline composites with polyaniline-coated carbon nanotubes, J. Mater. Sci. 49 (2014) 1222–1228.

[37] S.H. Park, K.-H. Shin, J.-Y. Kim, S.J. Yoo, K.J. Lee, J. Shin, J.W. Choi, J. Jang, Y.-E. Sung, The application of camphorsulfonic acid doped polyaniline films prepared on TCO-free glass for counter electrode of bifacial dye-sensitized solar cells, J. Photochem. Photobiol. A Chem. 245 (2012) 1–8.

[38] Y. Xia, J.M. Wiesinger, A.G. MacDiarmid, A.J. Epstein, Camphorsulfonic acid fully doped polyaniline emeraldine salt: Conformations in different solvents studied by an ultraviolet/visible/near-infrared spectroscopic method, Chem. Mater. 7 (1995) 443–445.

[39] Z. Wang, Q. Zhang, S. Long, Y. Luo, P. Yu, Z. Tan, J. Bai, B. Qu, Y. Yang, J. Shi, H. Zhou, Z.Y. Xiao, W. Hong, H. Bai, Three-dimensional printing of polyaniline/reduced graphene oxide composite for high-performance planar supercapacitor, ACS Appl. Mater. Interfaces. 10 (2018) 10437–10444.

[40] H. Yuk, B. Lu, S. Lin, K. Qu, J. Xu, J. Luo, X. Zhao, 3D printing of conducting polymers, Nat. Commun. 11 (2020) 4–11.

[41] J. Yang, Q. Cao, X. Tang, J. Du, T. Yu, X. Xu, D. Cai, C. Guan, W. Huang, 3D-Printed highly stretchable conducting polymer electrodes for flexible supercapacitors, J. Mater. Chem. A. 9 (2021) 19649–19658.

[42] R.R. Kohlmeyer, A.J. Blake, J.O. Hardin, E.A. Carmona, J. Carpena-Núñez, B. Maruyama, J. Daniel Berrigan, H. Huang, M.F. Durstock, Composite batteries: A simple yet universal approach to 3D printable lithium-ion battery electrodes, J. Mater. Chem. A. 4 (2016) 16856–16864.

[43] K. Sun, T.-S. Wei, B.Y. Ahn, J.Y. Seo, S.J. Dillon, J.A. Lewis, 3D Printing of interdigitated Li-ion microbattery architectures, Adv. Mater. 25 (2013) 4539–4543.

[44] D. Cao, Y. Xing, K. Tantratian, X. Wang, Y. Ma, A. Mukhopadhyay, Z. Cheng, Q. Zhang, Y. Jiao, L. Chen, H. Zhu, 3D printed high-performance lithium metal microbatteries enabled by nanocellulose, Adv. Mater. 31 (2019) 1807313.

[45] Y. Yu, M. Chen, S. Wang, C. Hill, P. Joshi, T. Kuruganti, A. Hu, Laser sintering of printed anodes for Al-air batteries, J. Electrochem. Soc. 165 (2018) A584.

[46] S. Rosenberg, A. Hintennach, Laser-printed lithium-sulphur micro-electrodes for Li/S batteries, Russ. J. Electrochem. 50 (2014) 327–335.

[47] C. Milroy, A. Manthiram, Printed microelectrodes for scalable, high-areal-capacity lithium–sulfur batteries, Chem. Commun. 52 (2016) 4282–4285.

[48] C.A. Milroy, S. Jang, T. Fujimori, A. Dodabalapur, A. Manthiram, Inkjet-printed lithium–sulfur microcathodes for all-printed, Integrated Nanomanufacturing, Small. 13 (2017) 1603786.

[49] K. Shen, H. Mei, B. Li, J. Ding, S. Yang, 3D Printing sulfur copolymer-graphene architectures for Li-S batteries, Adv. Energy Mater. 8 (2018) 1701527.

[50] W.J. Chung, J.J. Griebel, E.T. Kim, H. Yoon, A.G. Simmonds, H.J. Ji, P.T. Dirlam, R.S. Glass, J.J. Wie, N.A. Nguyen, B.W. Guralnick, J. Park, Á. Somogyi, P. Theato, M.E. Mackay, Y.-E. Sung, K. Char, J. Pyun, The use of elemental sulfur as an alternative feedstock for polymeric materials, Nat. Chem. 5 (2013) 518–524.

7 3D Printed Conducting Polymers for Perovskite Light-Emitting Diodes and Solar Cells

Esmaeil Sheibani and Bo Xu

7.1 ESSENTIAL MATERIAL REQUIREMENT FOR PRINTING PEROVSKITE OPTOELECTRONIC DEVICES

The future of energy generation is closely intertwined with the global economy's urgent need for environmentally friendly advancements and technologies aimed at reducing emissions. The performance and durability of solar cells depend heavily on material selection and the utilization of innovative technologies. In this regard, the advancements in 3D printing, also known as additive manufacturing, have made significant progress and found practical applications in various fields. 3D printing offers several benefits, including uniform coating across large areas, minimal waste with excellent material utilization, and the flexibility to integrate roll-to-roll (R2R) and sheet-to-sheet systems. This digital, mask-free approach to thin-film deposition holds great potential for scalability. To ensure the scalability and commercial viability of perovskite solar cells (PSCs), various solution-based printing technologies have been developed, such as blade-coating, slot-die coating, spray-coating, flexographic printing, gravure printing, screen-printing, and inkjet printing.

These techniques are complemented by solvent extraction methods like anti-solvent quenching, air-flow quenching, and vacuum quenching, which address key concerns in PSCs manufacturing [1]. Perovskite materials possess exceptional qualities, including high diffusion length, strong optical absorption across the visible spectrum, high carrier mobility, availability of affordable precursor materials, appropriate band gap, and convenient fabrication through low-temperature methods [2]. Flexible PSCs, with their flexibility, lightweight nature, portability, and ability to conform to curved surfaces, are gaining increasing interest for mass production [3].

Recent reports have highlighted the successful use of perovskite precursor ink engineering as an approach to control crystal growth and minimize defects. This involves incorporating various additives such as long-chain polymers, ionic liquids, or low boiling point solvents [4]. In addition to chemical additives and annealing processes, post-growth treatments play a crucial role in the nucleation kinetics and crystal growth of the perovskite layer. The development of perovskite light-emitting diodes (PeLEDs) with unique characteristics and a large specific surface area has

 DOI: 10.1201/9781003415985-7

also been a recent focus. Perovskite quantum dots (QDs) are being explored as efficient absorber materials in applications including photovoltaics, light-emitting diodes (LEDs), and optoelectronics.

However, the rough surface of the perovskite film, including grain boundaries (GBs), defects, uncoordinated sites, and pinholes, can act as traps for holes and electrons, leading to instability and undesired effects on carrier transport, external quantum efficiency (EQE), and blinking of PeLEDs [5–7]. Interfacial engineering and surface modification play a crucial role in addressing these intrinsic challenges. Various methods, including humidity management, additive engineering, interlayer engineering, annealing treatment, compositional engineering, and the recent focus on self-assembly monolayers and crosslinking approaches, have been explored to provide pinhole-free surfaces and adjust perovskite crystal sizes [8]. Among these methods, crosslinking shows special promise for emerging technologies and deserves significant attention.

This chapter emphasizes the impressive progress made in compact, surface solvent-resistant films of charge transporting or perovskite layers, interfacial modification, and ultimately remarkable stability. In comparison to the corrosive and hydrophilic nature of poly(3,4-ethylenedioxythiophene) polystyrene sulfonate (PEDOT:PSS) and the high cost of Poly[bis(4-phenyl) (2,4,6-trimethylphenyl) amine (PTAA), the simple process of p-type cross-linked thermal polymerization not only reduces the overall cost of device fabrication but also demonstrates great potential as an efficient polymeric hole transporting layer (HTL) [9] for both regular/inverted PSCs, PeLEDs, and bulk heterojunction (BHJ) solar cell configurations [10].

A typical solar cell structure comprises a conductive electrode coated on a substrate. The commonly used electrodes in PSC are fluorine-doped tin oxide (FTO) and indium tin oxide (ITO). The other electrode is typically made of aluminum or silver, serving as the cathode in a normal device. However, in inverted devices, the ITO electrode substrate functions as the anode, while the top metal electrode acts as the cathode. Compared to ITO-coated rigid glass substrates, using ITO on flexible polymer substrates has several drawbacks. These include lower conductivity, decreased transmittance, and reduced mechanical durability. Polyethylene terephthalate (PET) and polyethylene 2,6-naphthalate (PEN) (see Figure 7.3) are two typical substrates employed in flexible PSCs on a transparent conductive oxide (TCO). These substrates exhibit minimal deformation until temperatures reach 150 °C. Due to the rigid characteristics of ITO, carbon materials such as graphene and carbon nanotubes (CNTs) and organic materials such as PEDOT:PSS are viable substitutes for TCO due to their exceptional flexibility, optical transparency, and ability to be processed on a large scale using roll-to-roll techniques for cost-effective production [11]. As an electron transfer, TiO_2 films are created through spray pyrolysis or spin-coating methods, both of which necessitate sintering at a high temperature (>450°C) to achieve a dense structure with excellent crystalline quality [12]. This high-temperature process not only adds complexity to the fabrication procedures but is also unsuitable for the polymeric substrates commonly employed in flexible solar cells. Organic small-molecule hole transport materials (HTMs) have garnered significant interest due to their well-defined molecular structures, adjustable photophysical properties, and potential for low-temperature fabrication in flexible solar cells [13–16]. The highest efficiencies in PTAA-based PSCs were achieved by p-doping the PTAA HTM with lithium bis(trifluoromethanesulfonyl)imide (LiTFSI) and *tert*-butylpyridine (TBP) [17]. The effectiveness of PTAA as a HTM in PSCs relies on the molecular weight of the polymer.

From a business standpoint, although there have been significant advancements, there is still ample opportunity for improvement in terms of efficiency, mechanical flexibility, and operational lifetime. Ongoing developments are being made to enhance the 3D printing process and increase the stability of perovskite film with a particular focus on assessing the viability of different coating methods on flexible substrates suitable for low-temperature processes. Achieving a balance between scalability and power conversion efficiency (PCE) is still a difficult challenge, impeding the wider commercialization of PSCs. Although the efficiency and cost of PSCs are well-suited for practical use, the durability of PSCs still needs improvement to align with the longevity standards of conventional PV products. This chapter aims to comprehensively examine the various methods of solar cell fabrication and draws attention to the applications of 3D printing technology used for this purpose.

7.2 VARIOUS 3D PRINTING TECHNIQUES FOR PSCs

In this section, we introduce various methods for manipulating the structure, performance, and morphology of the perovskite active layer in order to successfully fabricate large-area PSCs. This text focuses on the direct comparison between different printing techniques and materials used in solar cell applications. The aim is to provide a better understanding of how processing parameters, as well as solar cell parameters, depend on the specific techniques and printed materials employed.

Spin coating, a popular technique on a laboratory scale for producing consistent thickness, is incompatible with large substrates, and film thinning is difficult, making it unsuitable for mass production [18]. Moreover, using the antisolvent crystallization strategy alongside it is challenging due to the need for careful handling of various processing parameters such as dipping time, volume, and distance to the surface. Ultimately, scaling up this method to large-area printing lines is unlikely due to excessive usage and spilling of washing solvents. In addition, spin coating requires high operating temperatures (450–500°C) during the drying process, leading to film cracking and peeling and also increasing the cost of the final product. PSCs, despite their high efficiency on flexible substrates through lab-scale spin coating, face limitations when it comes to R2R processes due to warping, undulation, and narrow processing windows [3]. The cost of producing materials through vapor-based deposition technologies is likely to increase due to the need for a high vacuum. Moreover, in typical solar cell manufacturing methods, it is difficult to optimize the deposition conditions required [19].

Blade coating is a popular scalable printing technique for perovskite preparation that can easily be used in both lab and industrial settings. While high efficiencies have been achieved using one-step deposition of perovskite solar cells via blade coating, most reported perovskite films were deposited at temperatures greater than 100°C, which is close to or exceeds the boiling point of the solvent used [20]. According to Figure 7.1a, a doctor blade coating involves the placement of a sharp blade at a consistent distance from the surface to be coated. The coating material is then situated in front of the blade. A uniform thickness of coating material can be achieved by maintaining a constant relative movement between the coating material and the substrate. Precise management of solution chemistries is crucial when using doctor blade coating to promote grain development, nucleation temperature, and crystallization temperature [21].

The study found that pre-heating the substrate was helpful in speeding up solvent evaporation and preventing the formation of a needle-like structure. Another method for quickly drying solvents involves the utilization of a low–boiling point solvent system, such as acetonitrile and methyl acetate. This approach led to a certified efficiency of 16.4% with an area of 63.7 cm^2 [22]. Very recently, a perovskite precursor solution that consists of 2-methoxyethanol and 1,3-dimethyl-imidazolidinone has

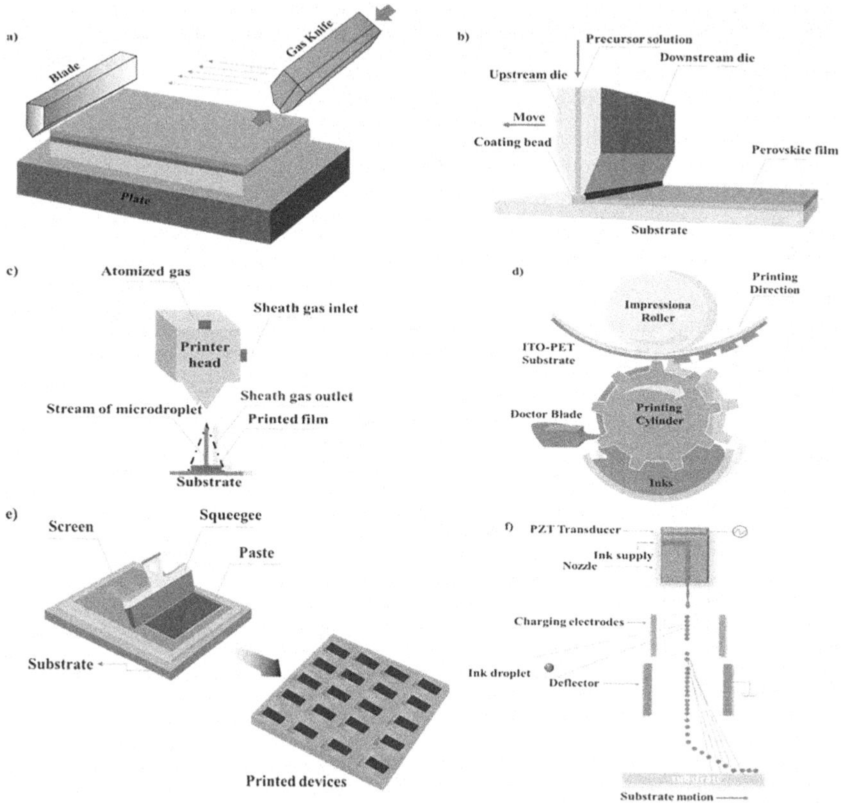

FIGURE 7.1 Schematic illustration of the a) blade-coating technique, b) slot-die printing technique, c) aerosol jet printing slot-die printing technique, d) gravure printing, e) screen printing technique, and f) inkjet printing technique. b, e are adapted with permission [24]. Copyright 2018 J. Phys. Chem. Lett. ACS. c is adapted with permission [26]. Copyright 2011 ACS Appl. Mater. Interfaces, ACS. d is adapted with permission [27]. Copyright 2023 Results in Optics, Elsevier.

been engineered for intermediate-phase stability, enabling scalable production of efficient perovskite solar modules [23]. This solution produces a pinhole-less, uniform perovskite film over an area greater than 100 cm^2, resulting in higher-efficiency PSCs and modules. The n-i-p configuration of the best-performing unit cell and module yields PCE of 23.4% and 20.1%, respectively.

The methods of slot-die coating and blade coating share many similarities in their mechanisms. Compare with blade coating, slot-die coating is a method that offers higher precision and is more efficient in terms of the solution used, as there is minimal wastage of the precursor during the preparation process [24]. The slot-die coating approach is well suited for the R2R process, where ink is continuously fed through a narrow opening to cover the moving substrate (Figure 7.1b). This method is categorized as a pre-metered coating technique, indicating that the thickness of the coating is determined by the mass flow in the coating system.

A recent study conducted by Du et al. yielded impressive outcomes, wherein a slot-die coating technique was utilized, resulting in a solar cell efficiency as high as 22.7%. The study also showed that a 40×40 mm^2 module achieved a stabilized PCE of 19.4%, which stands among the highest made through large-area fabrication processes [25]. Overall, slot-die coating is a favorable option for the high-throughput, scalable manufacture of flexible PSCs due to its superior yield and reproducibility, even though it suffers from utilizing toxic compounds.

Aerosol jet printing (AJP) technique enables accurate deposition of layers, without the need for conventional masks or thin-film apparatuses (Figure 7.1c). The AJP technique can work with any material that can form an aerosol [26]. This process typically involves a combination of high and low-volatility solvents. The high-volatility solvent evaporates quickly after atomization, saturating the carrier gas and ensuring a consistent droplet volume during transportation until it interacts with the sheath gas. The low-volatility solvent then begins to evaporate. Printable ink for (Poly(3-hexylthiophene (P3HT) and PC60BM is prepared by dissolving them in an organic solvent, while PCBTDPP is mixed with PC70BM [26]. The printer carries droplets of the active materials with N_2 gas and deposits them on a substrate coated with a PEDOT:PSS layer. Finally, electrodes such as Ca/Al or LiF/Al are fabricated using thermal deposition, followed by measuring device performance. Moreover, AJP system use is largely limited to nanoparticles, and control of atomization settings, sheath gas flow rate, carrier gas flow rate, and print speed are required to achieve the desired line morphology, including width, thickness, and roughness. The technique shares many advantages with other ink-based methods, such as the flexibility of materials and the ability to deposit nanoparticles without clogging issues associated with inkjet printing.

Gravure printing is a printing technique that achieves high resolution quality with high throughput. The printing apparatus involves a printing cylinder with engraved patterns that is partially immersed in an ink container (Figure 7.1d). The patterns on the cylinder define where the ink will be transferred onto the substrate during the printing process [27]. A blade is used to clean any excess ink from inactive parts of the cylinder and control the film thickness. Applying pressure with the impression roller allows for the ink to transfer from the cylinder to the substrate. The shape and thickness of the designs are determined by the groove depth and patterns on the gravure cylinder. Seo et al. fabricated all layers with gravure printing by using a low-cost R2R process, achieving a PCE of 17.2%. A tin(IV) oxide (SnO_2) ETL was prepared from a SnO_2 NP ink that was dispersed in water and a 2-propanol mixed solvent [28]. After deposition, the ink was dried at 120°C for 10 min, resulting in a uniform and thick film. The $CH_3NH_3PbI_3$ perovskite layer was then gravure-printed over the SnO_2 layer and annealed at 100°C for 10 min. Next, the 2,2′,7,7′-Tetrakis-(*N,N*′-di-4-methoxyphenylamino)-9,9′-spirobifluorene (Spiro-OMeTAD) was printed on top of the perovskite layer in a similar process. Finally, Ag electrodes were deposited on the top using thermal evaporation and a shadow mask. An efficiency of 17.2% along with 15.5% efficiency for maximum power point tracking for 1000 seconds was achieved.

The oldest and most established technique for producing solar cells is screen printing, but its prolonged processing stages significantly increase manufacturing expenses [24]. It comes with limitations such as poor thickness control due to the use of viscous ink, wastage of significant quantities of inks, and a restricted sintering time. These constraints lead to defects induced by solvent interactions within the screen printing ink, ultimately inhibiting the photovoltaic properties of PSCs. According to Figure 7.1e, when screen printing, the paste is pushed through the holes in the emulsion layer and onto the surface of the wafer. This results in the transfer of the paste through a screen in a predetermined pattern that is aligned with the pattern that needs to be transferred.

A nucleation strategy is being developed to produce high-quality perovskite films without the use of antisolvents. However, there are challenges related to the small device size and narrow operating window. To optimize precursor solutions, methylamine ethanol can be volatilized. Large-area (10 × 10 cm) methylammonium lead tri-iodide ($MAPbI_3$) perovskite films can be achieved through phase transition processes by annealing. The average PCE achieved for 0.1 and 1 cm^2 were 19.4% and 17.57%, respectively [29]. The films retain 96.8% of their PCE after 39 days in restricted ambient air with a relative humidity of 5–30%. The PSC or PSC model, however, shows a lower PCE of 13.13% [30].

The inkjet printing technique provides precise control over layer formation and is a fast, contact-free, and digital deposition method. This technique has the potential to create objects with predetermined structures and components. Inkjet printing enables the fabrication of both small- and large-area solar cells, making it a versatile method that can be used on glass or flexible substrates [31]. Figure 7.1f shows a schematic representation of the ink jet printer. Material ink can be applied onto a substrate in the form of droplets. These droplets can be generated either through mechanical compression or by heating the ink.

The presence of N-methyl-2-pyrrolidinone was essential in preventing the quick crystallization of the perovskite film during printing. Additionally, the use of the PbX_2-DMSO (X = Br, I) complex greatly enhanced the quality of the perovskite layer, resulting in significantly larger grain sizes. These improvements led to PCE of 19.6% and 17.9% for device areas of 0.04 and 1.01 cm^2, respectively, in the fabricated PSCs [32]. Zhang and colleagues created a heat-assisted inkjet printing method that can print uniform and compact crystalline perovskites directly onto planar PEDOT:PSS substrates in ambient conditions. Through adjustments to precursor composition, solvent system, printing temperature, and printing parameters, they were able to achieve a PCE of 16.6% [33].

In 2020, the Tajima group utilized their expertise to engineer the composition of a solvent for a multinational perovskite precursor suitable for inkjet printing [34]. They also controlled the deposition process of perovskite wet films in order to enhance the interaction between the ink and the underlying carrier transport layer while regulating nucleation and crystallization. This advancement led to an unprecedentedly high PCE of over 21% with a stabilized power output efficiency of more than 18% with a device architecture of ITO/NiOx/perovskite/C60/bathocuproine (BCP)/Au. In the general area of rapid prototyping, 3D printing technology has the potential to meet the current requirements of photovoltaic technology, such as being lightweight and efficient and having low manufacturing costs.

7.3 CONDUCTIVE POLYMERS FOR PELEDS AND PSCs

Similar to PSCs, the significant drawback hindering the practical implementation of perovskite QDs-based solar cells, LEDs, and other devices is their poor stability. These devices often experience a dramatic decline in performance, lasting only a few hours when exposed to ambient environments. Additionally, the conventional methods used for fabricating and testing these devices involve pre-formed perovskite nanoparticles (PNPs), which necessitate an oxygen/moisture-free atmosphere and high-temperature conditions for synthesis and post-annealing processes. Furthermore, these methods are associated with high energy consumption. A one-step scalable method has been developed to produce freestanding highly-stable luminescent organogels by combining facile photo-polymerization of polyacrylate at room temperature by in-situ perovskite reprecipitation at low energy cost [35]. An exemplary gel called $CH_3NH_3PbBr_3$ NP–embedded poly (butyl acrylate) (PBA) can be formed by following these steps: (1) Dissolve CH_3NH_3Br (MABr) and $PbBr_2$ into dimethylformamide (DMF) solvent at a molar ratio of 3:1. Add oleic acid (OA) and oleylamine (OAm) as ligands to this solution, forming the perovskite precursor (solution A). (2) Prepare the acrylate precursor (solution B) by mixing acrylate monomer with the crosslinker *N,N'*-methylenebisacrylamide (MBA) and the photo-initiator 2-hydroxy-2-methylpropiophenone. (3) Blend solution A and solution B to form the PNP gel precursor. (4) Subject the PNP gel precursor to UV light exposure to initiate photo-polymerization to achieve a photoluminescent gel with evenly distributed PNP that can maintain its photoluminescence in water for over three months

In another approach, researchers have made a groundbreaking discovery that can drastically enhance the stability of perovskite QDs. By encapsulating them within polymers, the stability of these QDs can be greatly improved [36]. For this reason, various thermoplastic polymers are utilized in 3D printing through extrusion as protective encapsulation materials for perovskite nanocrystals. Among them, polycaprolactone (PCL) displayed particularly favorable characteristics and could be effectively blended with perovskite nanocrystals (PNCs) to create PNC–PCL composites. These composites demonstrated deformability and stretchability when heated. The low melting point of PCL allowed the PNCs to retain their optical properties even after the 3D printing process, resulting in fluorescent behavior in the printed objects.

In another exploration siloxane blended poly(N-vinylcarbazole) (PVCz) that using a similar approach siloxane blended PEDOT:PSS film in inverted perovskite solar cells [37]. A 3D siloxane network is created by hydrolysis of tetraethyl orthosilicate (TEOS) in the presence of trace amounts of water that exist in the environment or the film. A strong Si–O–Si covalent bond forms between the silanol groups of the siloxane and the substrate surface. In addition, siloxane chains interlock with PVCz chains, leading to a stiff film on the substrate that is indissoluble in the perovskite precursor solution. The water contact angle of the film was decreased, representative of a suitable bed for growing perovskite crystal associated with reducing the quenching of the excited states in the perovskite emission layer and also hindering direct contact between perovskite and ITO anode layer. In the end, external quantum efficiency (EQE) is boosted from 10.4% to 15.4%.

The perovskite LED configuration includes a glass substrate/ITO anode layer (100 nm)/unblended or siloxane-blended PVCz HTL (20 nm)/Q2D perovskite emission layer (50 nm)/TPBi ETL (50 nm)/LiF EIL (1 nm)/Al cathode layer (100 nm). Recently, a leap forward achieved with NIR-emissive $FAPbI_3$ QDs by surface treatment QDs with ligand conflation 2-phenylethylammonium iodide (PEAI) granted a high photoluminescence quantum up to 61.6%, to some extent compensating for the huge efficiency gap with green and red emission PeLED [38]. Further, the PEDOT:PSS-treated ITO electrode is covered with 9,9-Bis[4-[(4-ethenylphenyl)methoxy]phenyl]-N2,N7-di-1-naphthalenyl-N2,N7-diphenyl-9H-Fluorene-2,7-diamine (VB-FNPD) through a thermally cross-linked strategy as an HTL to provide a suitable place for the adhesion of QD film to the device fabrication process with a smooth and dense surface. In this study, 3',3'',3'''-(1,3,5-triazine-2,4,6-triyl)tris(([1,10-biphenyl]-3-carbonitrile)) (CN-T2T) with good electromobility and a deep LUMO level for better electron injection to the QD layer was employed. Eventually, the champion device obtains the highest EQE 15.4% with a current density of 0.54 mA cm^{-2} and electroluminescent (EL) λ_{max} at 772 nm. To solve the high-phase transition energy barrier of the $CsPbI_3$ black phase, a α-BaF_2 nanoparticle is added to the precursor solution for the formation of strain-free heteroepitaxial growth and hindrance phase transition. Alex K.-Y. Jen et al. synthesized the n-type conjugated molecule HANTA (Figure 7.3) and decorated it with cross-linkable acrylate side chains to achieve a cross-linked polymer via the thiol-ene reaction with PETMP at a partially low temperature of 110°C [39]. The device with all cross-linked charge transporting layers with the arrangement of ITO/c-TCTA/$CH_3NH_3PbI_3$/c-HATNA/bis-C60/Ag had a PCE of 16.08% and 13.42% on rigid and flexible substrates, correspondingly.

The crystallization mechanism of perovskite ink over a module size is still a debated process and parameters such as rapid evaporation of the solvent have a crucial effect on the continuous processing and reproducibility of perovskite devices in module scale. Song and co-workers reported the development details of a printable and stable perovskite nanocapsule ink by 3-aminopropyl tri-ethoxysilane for the high-throughput printing of large-area, highly uniform perovskite films with micron-sized grain structures [40]. The study reveals that the presence of perovskite precursor ink engineering facilitates the release of perovskite nanomaterials, which in turn promote even nucleation through controlled growth processes influenced by diffusion. The printed PSCs and 25 cm^2 modules achieved impressive PCEs of 22.10%

and 16.12%, respectively. The siloxane network at the GBs is crucial, as it functions as a protective layer and prevents ion migration. This leads to an enhancement in the thermal and operational stability of PSCs. Even after continuous operation for over 1000 hours under AM1.5 illuminations, they show minimal loss in efficiency. Additionally, they exhibit excellent thermal stability at 85°C, with over 87% of the initial efficiency retained after aging for 500 hours.

One significant drawback is the increased occurrence of structural defects caused by the expansion of micro-cracks or delamination at GBs with low fracture energy when exposed to mechanical and/or thermal stress during processing, annealing, and operation. Additive engineering offers a straightforward and efficient approach for enhancing crystallization and minimizing defects in PSCs [41]. Polymer additives, in comparison to small molecule additives, possess distinct benefits such as long-range organization and excellent stability, making them extensively utilized in the production of efficient PSCs. Polymers can establish robust and dependable interactions with perovskite grains, potentially linking them together to enhance film stability. Very recently, the crystallization and defect modulation of formamidinium (FA)-based perovskite was cured by introducing the polymer additive polyacrylonitrile (PAN) [42]. The inclusion of nitrile groups in PAN allows for the formation of hydrogen with organic cations and coordination bonds with Pb^{2+}, leading to an improved PCE from 16.80% to 18.33%. Additionally, PAN's hydrophobic properties enhance the device's resistance to moisture. The interaction between the formamidinium iodide (FAI) molecule and added PAN was confirmed by nuclear magnetic resonance (NMR) and X-ray photoelectron spectroscopy (XPS). Figure 7.2a,b depicts the results of these characterizations. NMR analysis revealed the presence of three distinct types of hydrogen atoms in FAI. Upon adding PAN, a C≡N···H–N hydrogen bonding interaction occurred between type a hydrogen in the FA cation and PAN. This interaction resulted in a decrease in the electron cloud density around the hydrogen nucleus, exposing it to a stronger magnetic field. Consequently, the shielding effect of the hydrogen atom was reduced, leading to a downfield chemical shift of the ammonium protons ($^{\delta}$NH signal) from 8.66 to 8.67 ppm. In the high-resolution XPS spectrum of the Pb 4f from the control sample, two main peaks were observed at 138.09 and 142.95 eV, corresponding to Pb $4f_{7/2}$ and Pb $4f_{5/2}$. In contrast, the PAN-interacted sample showed a shift in these peaks to 137.84 and 142.71 eV. This shift indicates that the coordination of C≡N and Pb^{2+} caused the electron cloud density to transition from the ligand to Pb^{2+}, thereby influencing the XPS peaks.

The application of polymer matrix in the active layer of PSCs facilitates a stretchable function, which in turn has shown great application in building integrated photovoltaics and wearable and portable devices. However, the haphazard deposition of polymer chains negatively impacts the photoelectric performance due to the unreleased stress of stretchable PSCs (Figure 7.2c,d). To overcome this issue, a combination of structural bionics and patterned-meniscus coating technology is employed to print polymer chains in an oriented manner on PEDOT:PSS film [43]. The orderly arrangement of the polymer conformation enables the efficient transmission of mechanical stress in a specific direction, safeguarding the stretchable devices' photoelectric performance against damage caused by stress accumulation. Atomic force microscopy (AFM) reveals that the PEDOT:PSS film with a patterned sphere coating

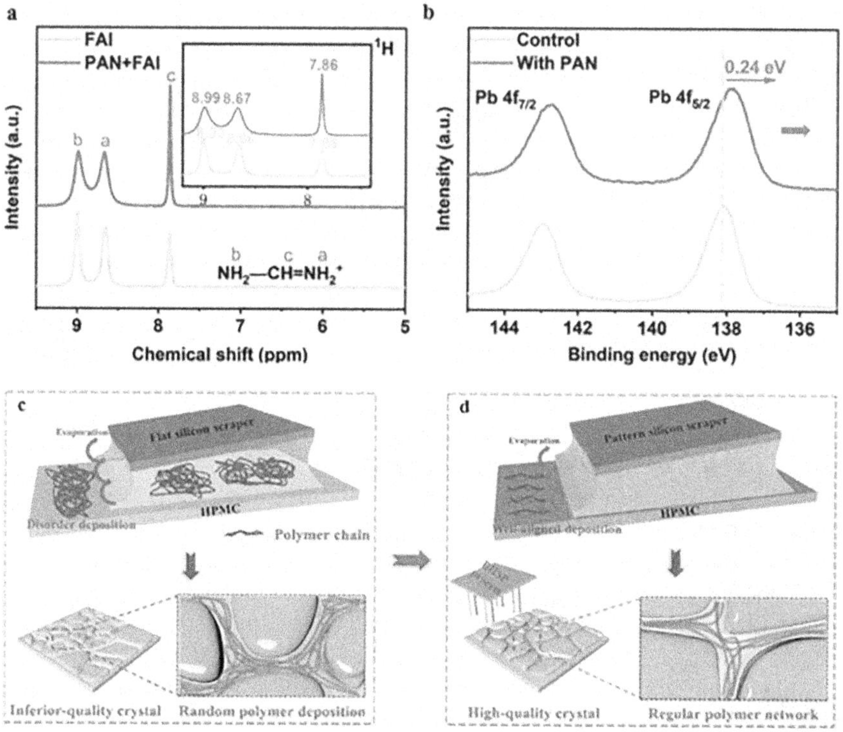

FIGURE 7.2 a) ^{1}H NMR spectra of FAI and PAN+FAI. b) Pb 4f XPS spectra of perovskite films with and without PAN. The deposition form of the polymer chain in meniscus coating process. c) The random deposition of polymer in conventional printing techniques. d) The growth orientation of polymer in the transparency electrode and perovskite grain boundaries by bionic high-speed patterned (BHSP) meniscus coating technique. a, b are adapted with permission [42]. Copyright 2023 Adv. Energy Mater, Wiley. c, d are adapted with permission [43]. Copyright 2023 Adv. Funct. Mater, Wiley.

displays clear evidence of oriented crystal growth. To showcase the advantages of polymer-oriented deposition, the stretchable PSCs with a structure of hydroxypropyl methylcellulose (HPMC)/hc-PEDOT:PSS/SnO_2/perovskite/spiro-OMeTAD/Ag are fabricated. Additionally, the oriented polyurethane with self-healing properties by the sulfur–sulfur bond helps improve the crystal quality of perovskite films and repairs any cracks caused by stress. As a result, the stretchable PSCs achieve a stabilized PCE of 20.04% (1.0 cm^2) and 16.47% (9 cm^2), with only minor discrepancies in efficiency. Notably, even after experiencing 1000 cycles of bending with a stretch ratio of 30%, stretchable PSCs can retain 86% of the original PCE.

In another strategy, elastic organic additives are employed to improve the flexibility, cure defects, and improve the performance of perovskite devices. To achieve this, a cross-linkable multi-functional ionogel (IG)-modified perovskite polycrystalline film was created using a R2R compatible bar-coating method [3]. By manipulating the ionic liquids and polymer network, the mechanical properties and passivation effect of the perovskite devices were effectively controlled. To create a polymer network with self-healing properties at room temperature, the first step involved copolymerizing a soft fluorinated hydrophobic acrylic acid (2,2,2-trifluoroethyl acrylate (TFOL-A)) (Figure 7.3) with strong adhesion properties and a hydrogen-bond-rich stiff acrylamide that was swollen with ionic liquids. Once the initiator is incorporated, the polymerization process can be efficiently carried out using a one-pot method by exposing it to UV irradiation. This resulted in a polymer network containing dynamic non-covalent interactions to engineer the grain boundary structure. The flexible PSCs achieved a remarkable efficiency of 21.76% through a scalable coating process based on a substrate/ITO/(PTAA)/PVK/choline chloride/C60/BCP/Cu device structure. Additionally, these flexible PSCs demonstrated unprecedented enhancements in operational stability (T90 of 1336 hours), mechanical stability (T90 of 25,000 cycles at 5 mm amplitude), and resistance to water damage. Light-based 3D printing has become a focal point of interest for its numerous benefits, such as fast printing speed and excellent resolution. Introducing dynamic bonds into materials used for light-based 3D printing offers numerous advantages such as adaptability, self-healing, and recycling capabilities [44]. These dynamic bonds can reversibly break and reform, facilitating network rearrangement. This strategy not only enables the production of 3D-printed multi-functional materials with unprecedented properties but also aligns with the need for sustainable and nature-inspired design considerations, including adaptability and self-healing. By incorporating dynamic bonds, including imine condensation and boronic ester formation, the potential for expanding the applications of 3D printed materials becomes even greater, meeting a wider range of demands. There is a strong desire to synthesize polymers with rheological properties that make them self-supportive after direct ink writing. The attention of

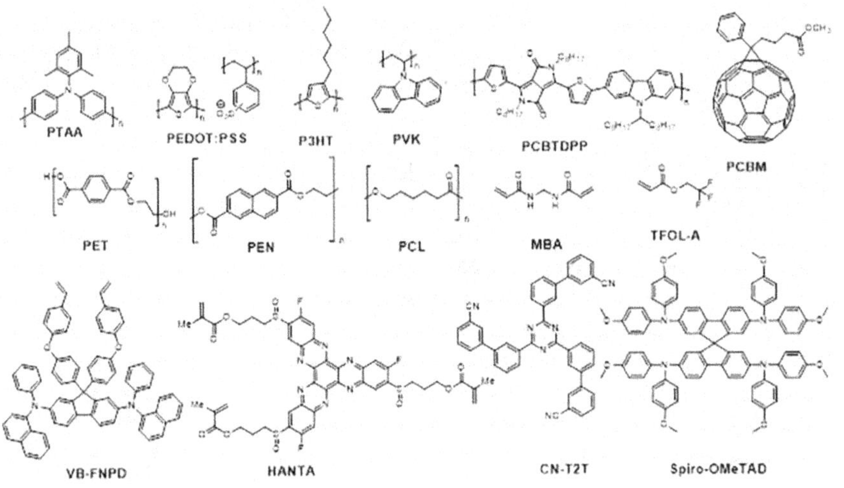

FIGURE 7.3 Chemical structures of organic and polymer abbreviated names mentioned in the text.

chemists has only recently been drawn to bottom-up syntheses of molecular and polymeric systems containing a variety of supramolecular binding for this purpose.

In summary, the development of charge transport polymer materials through cross-linking and dynamic interaction is crucial for PeLED, n-i-p, and p-i-n PSCs as it allows for a cost-effective solution process technique. This cross-linked approach provides a new method to adjust residual strain in perovskite films and simultaneously mitigate defects.

7.4 CONCLUSION AND OUTLOOK

The potential of 3D printing technology to revolutionize solar cell fabrication, improve efficiency, and reduce costs is widely recognized. Unlike conventional manufacturing techniques, 3D printing aligns with the objectives of sustainable development by emphasizing environmental preservation and offering unique advantages in geometric shape design and innovative intelligent manufacturing. Rather than employing subtractive or formative engineering processes, 3D printing focuses on joining materials, making it a promising approach for solar cell production. One significant advantage of 3D printing is its ability to tailor the morphology and composition of perovskite solar cells, leading to improved performance, enhanced stability, and increased efficiency. The solar cell industry faces critical challenges in meeting mass production demands, such as ensuring process scalability and repeatability, developing efficient thin film growth and deposition procedures, enhancing stability and durability, and minimizing toxicity while keeping production costs low. For scalable printing of perovskite solutions, judicious material design is essential, including the use of solvents with specific characteristics. These solvents should have high wettability, low boiling points, and lower toxicity and flammability to enable consistent coating of the precursor solution. Current research focuses on two main areas: exploring conductive ink formulations for printing purposes and developing highly flexible and stretchable substrates. Investigating material composition, printing parameters, and ink formulas will be crucial in addressing these challenges in the future. Despite the remaining hurdles, significant progress has already been made, and the rapid evolution of 3D printing technologies suggests that these issues will be resolved soon. This progress will pave the way for immediate market entrance and the widespread adoption of 3D printing in solar cell manufacturing.

REFERENCES

[1] L. Zeng, S. Chen, K. Forberich, C.J. Brabec, Y. Mai, F. Guo, Controlling the crystallization dynamics of photovoltaic perovskite layers on larger-area coatings, Energy Environ. Sci., 13 (2020) 4666–4690.

[2] J.J. Yoo, G. Seo, M.R. Chua, T.G. Park, Y. Lu, F. Rotermund, Y.-K. Kim, C.S. Moon, N.J. Jeon, J.-P. Correa-Baena, Efficient perovskite solar cells via improved carrier management, Nature, 590 (2021) 587–593.

[3] Y. Kang, R. Li, A. Wang, J. Kang, Z. Wang, W. Bi, Y. Yang, Y. Song, Q. Dong, Ionogel-perovskite matrix enabling highly efficient and stable flexible solar cells towards fully-R2R fabrication, Energy Environ. Sci., 15 (2022) 3439–3448.

[4] Q. Wang, W. Zhang, Z. Zhang, S. Liu, J. Wu, Y. Guan, A. Mei, Y. Rong, Y. Hu, H. Han, Crystallization control of ternary-cation perovskite absorber in triple-mesoscopic layer for efficient solar cells, Adv. Energy Mater., 10 (2020) 1903092.

[5] Y. Zong, Y. Zhou, Y. Zhang, Z. Li, L. Zhang, M.-G. Ju, M. Chen, S. Pang, X.C. Zeng, N.P. Padture, Continuous grain-boundary functionalization for high-efficiency perovskite solar cells with exceptional stability, Chem, 4 (2018) 1404–1415.

[6] Y. Liu, Y. Dong, T. Zhu, D. Ma, A. Proppe, B. Chen, C. Zheng, Y. Hou, S. Lee, B. Sun, Bright and stable light-emitting diodes based on perovskite quantum dots in perovskite matrix, J. Am. Chem. Soc., 143 (2021) 15606–15615.

[7] X. Li, M. Haghshenas, L. Wang, J. Huang, E. Sheibani, S. Yuan, X. Luo, X. Chen, C. Wei, H. Xiang, A multifunctional small-molecule hole-transporting material enables perovskite QLEDs with EQE exceeding 20%, ACS Energy Lett., 8 (2023) 1445–1454.

[8] Z. Dai, S.K. Yadavalli, M. Chen, A. Abbaspourtamijani, Y. Qi, N.P. Padture, Interfacial toughening with self-assembled monolayers enhances perovskite solar cell reliability, Science, 372 (2021) 618–622.

[9] E. Sheibani, L. Yang, J. Zhang, Conjugated polymer for charge transporting applications in solar cells, Organic Electrodes: Fundamental to Advanced Emerging Applications, Springer, 2022, pp. 119–135.

[10] X. Li, S. Fu, S. Liu, Y. Wu, W. Zhang, W. Song, J. Fang, Suppressing the ions-induced degradation for operationally stable perovskite solar cells, Nano Energy, 64 (2019) 103962.

[11] X. Hu, Z. Huang, F. Li, M. Su, Z. Huang, Z. Zhao, Z. Cai, X. Yang, X. Meng, P. Li, Nacre-inspired crystallization and elastic "brick-and-mortar" structure for a wearable perovskite solar module, Energy Environ. Sci., 12 (2019) 979–987.

[12] F. Giordano, A. Abate, J.P. Correa Baena, M. Saliba, T. Matsui, S.H. Im, S.M. Zakeeruddin, M.K. Nazeeruddin, A. Hagfeldt, M. Graetzel, Enhanced electronic properties in mesoporous TiO2 via lithium doping for high-efficiency perovskite solar cells, Nat. Commun., 7 (2016) 10379.

[13] D. Molina, E. Sheibani, B. Yang, H. Mohammadi, M. Ghiasabadi, B. Xu, J. Suo, B. Carlsen, N. Vlachopoulos, S.M. Zakeeruddin, Molecularly engineered low-cost organic hole-transporting materials for perovskite solar cells: The substituent effect on non-fused three-dimensional systems, ACS Appl. Energy Mater., 5 (2022) 3156–3165.

[14] L. Wang, E. Sheibani, Y. Guo, W. Zhang, Y. Li, P. Liu, B. Xu, L. Kloo, L. Sun, Impact of linking topology on the properties of carbazole-based hole-transport materials and their application in solid-state mesoscopic solar cells, Sol. RRL., 3 (2019) 1900196.

[15] E. Sheibani, M. Heydari, H. Ahangar, H. Mohammadi, H.T. Fard, N. Taghavinia, M. Samadpour, F. Tajabadi, 3D asymmetric carbozole hole transporting materials for perovskite solar cells, Sol. Energy., 189 (2019) 404–411.

[16] B. Xu, E. Sheibani, P. Liu, J. Zhang, H. Tian, N. Vlachopoulos, G. Boschloo, L. Kloo, A. Hagfeldt, L. Sun, Carbazole-based hole-transport materials for efficient solid-state dye-sensitized solar cells and perovskite solar cells, Adv. Mater., 26 (2014) 6629–6634.

[17] Z. Li, B. Li, X. Wu, S.A. Sheppard, S. Zhang, D. Gao, N.J. Long, Z. Zhu, Organometallic-functionalized interfaces for highly efficient inverted perovskite solar cells, Science, 376 (2022) 416–420.

[18] N.-G. Park, K. Zhu, Scalable fabrication and coating methods for perovskite solar cells and solar modules, Nat. Rev. Mater., 5 (2020) 333–350.

[19] L. Calió, C. Momblona, L. Gil-Escrig, S. Kazim, M. Sessolo, Á. Sastre-Santos, H.J. Bolink, S. Ahmad, Vacuum deposited perovskite solar cells employing dopant-free triazatruxene as the hole transport material, Sol. Energy Mater Sol. Cells., 163 (2017) 237–241.

[20] B. Chen, J.Y. Zhengshan, S. Manzoor, S. Wang, W. Weigand, Z. Yu, G. Yang, Z. Ni, X. Dai, Z.C. Holman, Blade-coated perovskites on textured silicon for 26%-efficient monolithic perovskite/silicon tandem solar cells, Joule, 4 (2020) 850–864.

[21] Y. Deng, E. Peng, Y. Shao, Z. Xiao, Q. Dong, J. Huang, Scalable fabrication of efficient organolead trihalide perovskite solar cells with doctor-bladed active layers, Energy Environ. Sci., 8 (2015) 1544–1550.

[22] Y. Deng, C.H. Van Brackle, X. Dai, J. Zhao, B. Chen, J. Huang, Tailoring solvent coordination for high-speed, room-temperature blading of perovskite photovoltaic films, Sci. Adv., 5 (2019) eaax7537.

[23] J. Chung, S.W. Kim, Y. Li, T. Mariam, X. Wang, M. Rajakaruna, M.M. Saeed, A. Abudulimu, S.S. Shin, K.N. Guye, Engineering perovskite precursor inks for scalable production of high-efficiency perovskite photovoltaic modules, Adv. Energy Mater. (2023) 2300595.

[24] Y. Rong, Y. Ming, W. Ji, D. Li, A. Mei, Y. Hu, H. Han, Toward industrial-scale production of perovskite solar cells: Screen printing, slot-die coating, and emerging techniques, J. Phys. Chem. Lett., 9 (2018) 2707–2713.

[25] M. Du, X. Zhu, L. Wang, H. Wang, J. Feng, X. Jiang, Y. Cao, Y. Sun, L. Duan, Y. Jiao, High-pressure nitrogen-extraction and effective passivation to attain highest large-area perovskite solar module efficiency, Adv. Mater., 32 (2020) 2004979.

[26] C. Yang, E. Zhou, S. Miyanishi, K. Hashimoto, K. Tajima, Preparation of active layers in polymer solar cells by aerosol jet printing, ACS Appl. Mater. Interfaces, 3 (2011) 4053–4058.

[27] B.R. Hunde, A.D. Woldeyohannes, 3D printing and solar cell fabrication methods: A review of challenges, opportunities, and future prospects, Results Opt. (2023) 100385.

[28] Y.Y. Kim, T.Y. Yang, R. Suhonen, M. Välimäki, T. Maaninen, A. Kemppainen, N.J. Jeon, J. Seo, Gravure-printed flexible perovskite solar cells: Toward roll-to-roll manufacturing, Adv. Sci., 6 (2019) 1802094.

[29] B. Li, Q. Zhang, S. Zhang, Z. Ahmad, T. Chidanguro, A.H. Davis, Y.C. Simon, X. Gu, W. Zheng, N. Pradhan, Spontaneously supersaturated nucleation strategy for high reproducible and efficient perovskite solar cells, J. Chem. Eng., 405 (2021) 126998.

[30] Q. Zhang, Y. Qi, K. Conkle, J. Xiong, D. Box, P. Ray, N.R. Pradhan, T.V. Shahbazyan, Q. Dai, Perovskite films prepared by solvent volatilization via DMSO-based intermediate phase for photovoltaics, Sol. Energy., 218 (2021) 383–391.

[31] B. Parida, A. Singh, A.K. Kalathil Soopy, S. Sangaraju, M. Sundaray, S. Mishra, S. Liu, A. Najar, Recent developments in upscalable printing techniques for perovskite solar cells, Adv. Sci., 9 (2022) 2200308.

[32] Z. Li, P. Li, G. Chen, Y. Cheng, X. Pi, X. Yu, D. Yang, L. Han, Y. Zhang, Y. Song, Ink engineering of inkjet printing perovskite, ACS Appl. Mater. Interfaces, 12 (2020) 39082–39091.

[33] L. Zhang, S. Chen, X. Wang, D. Wang, Y. Li, Q. Ai, X. Sun, J. Chen, Y. Li, X. Jiang, ambient inkjet-printed high-efficiency perovskite solar cells: Manipulating the spreading and crystallization behaviors of picoliter perovskite droplets, Sol. RRL., 5 (2021) 2100106.

[34] H. Eggers, F. Schackmar, T. Abzieher, Q. Sun, U. Lemmer, Y. Vaynzof, B.S. Richards, G. Hernandez-Sosa, U.W. Paetzold, Inkjet-printed micrometer-thick perovskite solar cells with large columnar grains, Adv. Energy Mater., 10 (2020) 1903184.

[35] Y. Zhang, Y. Zhao, D. Wu, J. Xue, Y. Qiu, M. Liao, Q. Pei, M.S. Goorsky, X. He, Homogeneous freestanding luminescent perovskite organogel with superior water stability, Adv. Mater., 31 (2019) 1902928.

[36] C.-L. Tai, W.-L. Hong, Y.-T. Kuo, C.-Y. Chang, M.-C. Niu, M. Karupathevar Ponnusamythevar Ochathevar, C.-L. Hsu, S.-F. Horng, Y.-C. Chao, Ultrastable, deformable, and stretchable luminescent organic–inorganic perovskite nanocrystal–polymer composites for 3D printing and white light-emitting diodes, ACS Appl. Mater. Interfaces, 11 (2019) 30176–30184.

[37] T. Matsushima, R. Nasu, K. Takekuma, T. Ishii, Z. Feng, X. Tang, N. Nakamura, G. Tumen-Ulzii, C. Adachi, Efficient perovskite light-emitting diodes with a siloxane-blended organic hole transport layer, Adv. Photonics., 3 (2022) 2200003.

[38] Z.L. Tseng, L.C. Chen, L.W. Chao, M.J. Tsai, D. Luo, N.R. Al Amin, S.W. Liu, K.T. Wong, Aggregation control, surface passivation, and optimization of device structure toward near-infrared perovskite quantum-dot light-emitting diodes with an EQE up to 15.4%, Adv. Mater., 34 (2022) 2109785.

[39] Z. Zhu, D. Zhao, C.-C. Chueh, X. Shi, Z. Li, A.K.-Y. Jen, Highly efficient and stable perovskite solar cells enabled by all-crosslinked charge-transporting layers, Joule, 2 (2018) 168–183.

[40] Z. Huang, X. Hu, Z. Zhao, X. Meng, M. Su, T. Xue, J. Chi, H. Xie, Z. Cai, Y. Chen, Releasing nanocapsules for high-throughput printing of stable perovskite solar cells, Adv. Energy Mater., 11 (2021) 2101291.

[41] S. Liu, Y. Guan, Y. Sheng, Y. Hu, Y. Rong, A. Mei, H. Han, A review on additives for halide perovskite solar cells, Adv. Energy Mater., 10 (2020) 1902492.

[42] Z. Zheng, M. Xia, X. Chen, X. Xiao, J. Gong, J. Liu, J. Du, Y. Tao, Y. Hu, A. Mei, Enhancing the performance of Fa-based printable mesoscopic perovskite solar cells via the polymer additive, Adv. Energy Mater. (2023) 2204335.

[43] C. Gong, F. Li, X. Hu, C. Wang, S. Shi, T. Hu, N. Zhang, C. Liang, D. Wu, Y. Chen, Printing-induced alignment network design of polymer matrix for stretchable perovskite solar cells with over 20% efficiency, Adv. Funct. Mater. (2023) 2301043.

[44] G. Zhu, H.A. Houck, C.A. Spiegel, C. Selhuber-Unkel, Y. Hou, E. Blasco, Introducing dynamic bonds in light-based 3D printing, Adv. Funct. Mater. (2023) 2300456.

8 3D-Printed Conducting Polymers for Solar Cells

Hamideh Mohammadian Sarcheshmeh and Mohammad Mazloum-Ardakani

8.1 INTRODUCTION

Additive manufacturing (AM), or 3D printing, indicates a group of manufacturing methods that provide fast fabrication of structures with 3D features and a wide range of sizes from sub-microns to several meters. Some advantages of this technique are the reliability and simplicity of using various compatible materials (e.g., metals, polymers, and ceramics). The combination of 3D printing and nanotechnology provides a new opportunity for the fabricating of 3D-engineered materials with optimal and multiple functions. 3D printing has shown many applications in different industries, such as prototyping (fabrication of parts to test designs), aerospace, space, automotive, energy harvesting, energy storage, architecture, dentistry, medicine, nanotechnology, education, food, cultural heritage, consumer goods, art, and jewelry. This chapter summarizes some investigations in the field of AM technologies in solar energy. Different materials are used for 3D printing, such as metals, polymers, ceramics, and composites. While most industrial and scientific studies have concentrated on metal printing due to its advanced technological applications, polymer-based AM methods are the most advanced methods, and their use surpasses all types of materials. Polymers are suitable materials for fast prototyping and indicate an essential effect in the 3D printing of multifunctional and multiphase materials. AM techniques such as vat photopolymerization (stereolithography (SLA), two-photon polymerization (TPP)), powder bed fusion (selective laser sintering (SLS)), material jetting (inkjet and aerosol 3D printing), extrusion (fused deposition modeling (FDM)), and directed energy deposition (DED) are used for polymers. These techniques are different in terms of basic and required material. The choice of 3D printing method depends on the type of conductive polymers utilized and the electrical qualities of the printed components [1]. Unlike conventional manufacturing procedures, AM does not need the use of tooling, which will decrease material waste and costs. Conductive polymers, such as polyaniline (PANI), poly(3,4-ethylenedioxythiophene): polystyrene sulfonate (PEDOT: PSS), and polypyrrole (PPy) include repeated monomer units, where the atomic structures have been altered to increase their conductivity. The applications of conductive polymers and their composites by AM have been highlighted in sensors, bioelectronics, and flexible and stretchable electronics. They are known as a category of organic materials and a new replacement to the usual conductive materials, for example, metals and carbon materials. They indicate low density and good processability and operate as cost-effective conductive materials. They

DOI: 10.1201/9781003415985-8

represent numerous applications, such as light-emitting diodes, tissue engineering, batteries, sensors, supercapacitors, and photovoltaic cells. Although the mentioned polymers are naturally conductive, their conductivity can be enhanced using (p) and (n) dopants. There are different conductive polymers, but the three most important ones utilized in 3D printing are PEDOT: PSS, PANI, and PPy. AM of conductive polymers represents an attractive compound of practical materials and accurate fabricating that is used in progressed applications. Nevertheless, their processing is complicated, which causes problems with their AM. 3D printing or AM techniques for polymers involve specific conditions to obtain suitable printability, such as chemical as well as rheological properties and specific transfer temperatures of polymers. Consequently, conductive polymers are mixed with other polymers to facilitate AM while preserving electrical performance. This results in a reduction in conductivity owing to the presence of a second insulator polymer. To enhance conductivity, some conductive materials were used as filler material for different polymer compounds, including carbon materials (e.g., carbon black; graphene oxide; carbon nanotubes with excellent conductivity, lightweight, and low cost) and nanoparticles (e.g., silver) [2]. Dispersing conductive fillers into polymer matrixes has been utilized to increase conductivity without a severe change in the bulk material; it thus can be utilized in 3D printing by various technologies including SLA, inkjet, and FDM [3].

Recently, 3D printing techniques have shown a revolution in the fabrication and performance of solar cells. The utilization of 3D printing in the field of solar cells includes various benefits: precise control of layer thickness by adjusting the material delivery rate on the substrate. Acceptable thickness uniformity and suitable film quality decrease non-radiative recombination loss. 3D printing methods can control the crystallization rate. They provide opportunities to deposit materials on thin, ultrathin, and flexible substrates to roll-to-roll fabricating devices that are useful for the mass production of solar cells in the industry. The bending durability of these kinds of substrates allows for fabricating solar cells on curved surfaces with different applications such as large-scale roofs, electric chargers, unmanned aerial vehicles, and electronic textiles [4].

The essential purpose of this chapter is to explain the applications of 3D printing in manufacturing solar cells. Numerous types of research were done on progressed materials to address the limitations of solar cells. The advance of solar cells has been hindered by some subjects, including stability, efficiency, scale-up, and module design. AM is a pre-metered technique that creates precise and uniform deposition of materials and simplifies the optimization process. It can achieve some required photovoltaic properties, like efficiency, lightweight, inexpensive fabrication, and decreased waste materials. 3D printing can be incorporated with some methods for fabricating solar cells, such as gravure, slot-die coatings, and doctor blades. Large-area printing of perovskite solar cells (PCSs) was reported using a 3D printer slot-die coating [5]. 3D-printing technologies were applied to manufacturing front electrodes in solar cells. They showed a low-cost process and decreased fragmentation rate owing to their non-contact method. 3D printing can be used to prepare solar cells on glass and flexible substrates. This technique produces high-aspect-ratio parts for solar cells to enhance excellent light trapping and control, increasing efficiency. A concentrator array was prepared by 3D printing to capture external light for thin-film nanocrystal silicon solar cells [6]. This external light trap decreased optical losses and enhanced light harvesting in the entire solar spectrum. It indicated a 15% enhancement for solar cells.

Conversion efficiencies (photothermal and photovoltaic) for a dye-sensitized solar cell (DSSC) module were increased by 3D printing [7]. The 3D-printed concentrator was utilized to improve the photovoltaic efficiency from 5.48% to 7.03%. Also, a 3D-printed microfluidic device was utilized to control the temperature of the solar cells, which was helpful in keeping efficiency. This work showed a photovoltaic and photothermal efficiency of 49% at the optimized conditions.

Some practical 3D printing methods are explained in the following.

8.2 3D PRINTING METHODS

8.2.1 TPP Method

In comparison with conventional methods, 3D printing methods can fabricate micro-scale and nanoscale structures [8]. Thin TiO_2 film with a 3D micro-design was fabricated for DSSCs by the TPP method. It is a 3D micro/nanoscale fabricating procedure for the fast and flexible manufacturing of 3D structures with resolutions < 100 nm. This technique can be used to design light-trapping systems for DSSCs. Generally, the light trapping process in DSSCs includes some difficulties. In high-performance solar cells, the active layer should be nanocrystalline to keep a high surface area for dye adsorption and remain porous for dye regeneration by the redox electrolyte. Polymer stamps were created by TPP and then employed to pattern TiO_2 electrodes. These structures improved the performance by up to ~25% on the photocurrent compared to planar TiO_2. These results will result in the progress of new 3D-printed submicron optoelectronic cells [9].

8.2.2 Electrohydrodynamic 3D Printing (EHDP)

This method utilizes an electrified viscous fluid jet to create 3D micro/nanoscale structures using practical inks. Zhu and coworkers [10] employed this technique to deposit cesium lead halide perovskite ($CsPbX_3$) crystal dots. These $CsPbX_3$ nanocomposites exhibited luminescence similar to the spin coating method when activated by UV light. Generally, transparent electrodes, an essential section of optical devices, are fabricated by the EHDP method. A micro-scale network of silver lines (<10 μm) was produced by this method on a large-area/flexible graphene substrate [11].

8.2.3 Inkjet Printing

It is a rapid, non-contact, and digital deposition method with excellent layer formation control. It is used to deposit perovskite layers with a power conversion efficiency (PCE) of 12.9%. It is utilized to manufacture a top electrode of silver nanowires for semi-transparent and fully solution-processed PSCs [12]. One-step inkjet printing was reported to print PCSs in the air for effective light harvesting [13]. 3D-printed bulk heterojunction solar cells can also be prepared by this technique owning to using proper conductive inks at a lower cost. Inkjet-printed organic solar cells (OSCs) have indicated high efficiencies similar to common methods such as spin coating. Control of the morphology and microstructure of polymer films can improve the performance of these inkjet printer solar cells [14]. Thin-film electrodes of PEDOT: PSS were fabricated by inkjet printing for OSCs [15]. The obtained efficiency was in the range of 2 to 3%. It was 10% lower than the spin-coated solar cells. This is attributed to the poor fill factor in inkjet-printed solar cells due to changes in the thickness of the printed film and the resulting inappropriate morphology. Inappropriate morphologies influence charge mobility and lead to a poor fill factor. In addition, PEDOT: PSS film conductivity is poor. To enhance its conductivity, some additives were used. They were various solvents, such as dimethyl sulfoxide, sorbitol, ethylene glycol, polyethylene glycol, dimethyl sulfate, and triton X-100 (a surfactant) [16]. Increasing PEDOT: PSS conductivity can improve electromagnetic radiation shielding and electrical transport functions. These solvents change the structure of PEDOT from a benzoic structure to a quinoid structure and create a close crystal structure [17]. An enhancement from 7.82×10^{-1} S/cm to 1.52×10^{2} S/cm was provided using glycerol solvent, and it increased more using ethylene glycol butyl ether (EGBE) for the inkjet-printed PEDOT: PSS film. Adding glycerol and ethylene glycol butyl ether increased efficiency from 2.09% to 3.16% for cells based on (poly(3-hexylthiophene) (P3HT): (phenyl-C61-butyric acid methyl ester) (PC60BM). A better fill factor and short circuit current density were obtained after using these solvents due to the improvement of morphology [18]. Aernouts and coworkers [19] fabricated inkjet printing OSC with a polymer: fullerene combination used as an active layer. Due to the coalescing of the ink droplets separately, homogenous and pinhole-free organic thin films were obtained. The adjustment between surface wetting and ink viscosity created smooth PH3T: PCBM layers with lower roughness, leading to a PCE of 1.4%. The results show that inkjet printing is an inexpensive processing method that significantly simplifies the integration of photovoltaic devices into electronic utilizations or fabricating monolithic organic photovoltaic modules. Optimizing the morphology of the active layer by post-production treatments, controlling solvent evaporation, or proper solvent mixing ratios can result in performance comparable to spin-coated devices.

Inkjet printing is a method for the selective deposition of individual droplets provided with an accurate volume. Continuous films can be made by integrating these individually deposited droplets on a substrate. By controlling the inkjet process, hole-free films were obtained that were proper as active layers in OSCs [20]. This method is suitable for controlling the deposition of functional materials in special places on substrates and for the simple and rapid deposition of polymer layers on large surfaces. Therefore, it can be a proper technique for large-scale manufacturing of OSCs. It is also efficient for polymer-based cells. They can be simply made due to their compatibility with different substrates without using additional patterns. Choulis and coworkers investigated the performance of highly effective inkjet printing OSCs. They showed that the morphological and surface properties of inkjet printing P3HT: fullerene composite, as a photoactive layer, can be effectively improved with proper solvent formulations [21]. P3HT is extensively employed as an electron donor material for printing owing to its excellent stability and low cost, and fullerene PC60BM is also the most widely utilized as an electron acceptor material. A micro-fab inkjet printer was utilized first to print the active layer, P3HT: C60, over PEDOT: PSS. The obtained efficiency was poor due to the crystallization of C60 film as a result of its low solubility in ortho-dichlorobenzene (ODCB) and the penetration of the solvent molecules into the crystal lattice [22]. A combination of a high boiling solvent, ODCB, and a low boiling solvent, 1, 3, 5-trimethylbenzene (mesitylene), was utilized to print photoactive layers. The ODCB solvent inhibited the nozzle clogging, and the mesitylene solvent enhanced the spreading of ink. Inkjet printing of P3HT: PCBM dispersed in a solvent mixture (ODCB and mesitylene) showed an efficiency of 3% due to proper morphology [21]. It should be mentioned that layers based on P3HT: PCBM require annealing (150 ^{0}C) after printing to achieve proper morphology. The annealing process is undesirable because it consumes additional energy during the roll-to-roll fabrication of OSCs. Lange et al. [23] introduced inkjet printing for [poly (9, 9-dioctylfluorenyl-2, 7-dyil–CO-(10, 12-bis(thiophen-2-y)-3, 6-dioooctyll-11-thia-9, 13-diaza cyclopentatriphenylene)] (PFDTBTP). It is an amorphous polymer that does not require high-temperature annealing. Also, due to the existence of alkane side-chains in its structure, it is soluble in various solvents. The maximum PCE value of devices manufactured by chlorinated and non-chlorinated solvents was 3.5% and 2.8%, respectively. Poly [N-9′-heptadecanyl-2,7-carbazole-alt-5,5-(4′,7′-di-2-thienyl-2′,1′,3′-benzothiadiazole)] (PCDTBT) is a donor-acceptor copolymer that is employed in the inkjet printing of OSCs. Jung and coworkers [24] represented all inkjet-printed OSCs using PEDOT: PSS/PCDTBT and solvent combination of chlorobenzene: chloroform: mesitylene. A PCE value of 5.05% was obtained.

The effect of different parameters on inkjet printing a PEDOT: PSS layer for OSCs was studied. Various parameters noticeably influence the synthesis of continuous PEDOT: PSS films, such as drop spacing, surface treatment of the substrate, and substrate temperature (printing and annealing). In addition, PC60BM has usually been utilized as an active layer. PCE = 2.64% was acquired using optimized parameters, including drop spacing of 30 mm, ink viscosity of 5.65 cP, and 120 nm thickness, as well as printing and annealing temperatures of 25 ^{0}C and 120 ^{0}C, respectively. Fabricated films indicated good transmittance (89%) [25]. Hermerschmidt and coworkers [26] introduced a combination of low band–gap conjugated polymers of poly [(4, 4′-bis (2-ethylhexyl) dithieno [3, 2-b: 2′, 3′-d] silole)-2, 6-diyl-alt-(4, 7-bis (2-thienyl)-2, 1, 3-benzothiadiazole)-5, 5′-diyl] (Si-PCPDTBT) and PCDTBT as electron donors with [6,6]-phenyl-C71-butyric acid methyl ester (PC70BM) as an electron acceptor for inkjet printing an active layer in OSCs. An enhancement in PCE value (from 3.01 to 3.86%) was achieved by controlling the inkjet printing processes compared to previous work using the same materials. In another study, ink viscosity was changed using polystyrene as a rheological modifier. Different solvent compatibility of the polystyrene/P3HT: PCBM compounds led to phase separation during layer drying, while polystyrene formed a layer–air interface, providing a domain-separated or lattice-like topology [27]. Adding a small amount of polystyrene improves the viscosity with better print quality, indicating its ability to overcome the limited viscosity caused by conventional halogen-free ink fabrications of semiconducting polymers. The inkjet printing technique has been extensively employed to deposit one or more layers in OSCs. It has been known to fabricate all-printed OSCs for their ability to locally deposit low volumes of mask-free functional inks with increased positional precision and low cost. Conductive and semiconducting films with optimal control using inks engineered for inkjet printing provide acceptable efficiencies for printed OSCs between 2 and 5%.

In one study, the work was scaled up from lab to industrial for compatible displays of inkjet-printed OSCs, and all layers were printed using non-chlorinated solvents [28]. Photoactive layers were fabricated by the inkjet printing one of two active materials, P3HT: PC60BM or active ink PV2000. The PCE values for P3HT: PCBM and PV2000 compounds were 1.7% and 4.1%, respectively. Interestingly, inkjet printing can be used to manufacture custom-designed solar cells with various art shapes, for example, a Christmas tree–shaped solar cell. This can significantly develop the solar cell market.

All inkjet-printed multilayer Ag-nanowires and a PEDOT: PSS electrode were represented by Mecerreyes et al. [29]. They reported semi-transparent OSCs with PCE = 4.3%. Another OSC with a total area of 186 cm^2 was fabricated by multilayer printing of PEDOT: PSS, P3HT: PCBM, and Ag. This cell showed PCE = 1.6% with good stability. Eggers et al. [30] fabricated high-quality inkjet-printed triple-cation perovskite films with particular thicknesses (>1 μm), providing PCEs > 21%. These micrometer-thick films exhibited pillar-shaped crystal structures, no horizontal grain boundaries, and a very long charge carrier lifetime, indicating their good optoelectronic properties. Schackmar et al. [31] introduced all inkjet-printed absorber (triple-cation perovskite) and electron-transport PCBM layers in PSCs. A PCE value >17% with low hysteresis and good short-term stability (40 h) was obtained. Pengwei et al. [32] reported inkjet-manipulated uniform large-size perovskite grains for efficient and large-area PSCs. The performance of large-area PCSs is limited due to the non-uniform edge coating and crystalline defects in perovskite films. Thus, homogenous and compact crystallized films are essential for high-performance PCSs. A uniform PbI_2 film led to a compact film with micro-scale crystalline grains and thus PSC with high PCE = 18.64% for small areas (0.04 cm^2) and PCE = 17.74% for large areas (2.02 cm^2). The increased PCE value was attributed to improved perovskite film homogeneity, large grain size, fewer grain boundaries, and desirable crystallite orientation. Inkjet manipulation is a promising method for producing large and high-quality films for future optoelectronics, such as light-emitting diodes and sensors.

8.2.4 Aerosol Jet Printing

It is a non-contact direct writing technique to manufacture fine features on various substrates. It employs AM of thin-film layers on different substrates without conventional masks or equipment [33]. Yang and coworkers [34] utilized this technique to fabricate OSC. Two semiconducting polymers, P3HT and poly-[N-9-heptadecanyl-2,7-carbazole-alt-3,6-bis(thiophene-5-yl)-2,5-diethylhexyl-2,5-dihydropyrrolo-[3,4-]pyrrole-1,4-dione] (PCBTDPP), were utilized as electron donor materials. The fullerene derivatives, PC60BM and PC70BM, were utilized as electron acceptor materials. One of the crucial factors in printing was the selection of suitable mixed solvents to obtain suitable drying speeds for thin liquid films. Without thermal annealing, the optimized device fabricated with P3HT: PCBM showed a PCE of 2.53%. Solvent annealing was useful in increasing the PCE of PCBTDPP-based cells. By solvent annealing, PCE values of 3.92% and 3.14% for a small-area and large-area solar cell, respectively, were obtained. This work showed that aerosol jet printing is an acceptable method for manufacturing large-area OSCs.

8.2.5 FDM Method

It is a famous and practical technique due to its low cost. In this process, the material is extruded after melting at a specific temperature and then deposited layer by layer on a hot substrate. The material solidifies after printing one layer at a time, and the extruder head works coordinately in the x, y, and z directions, allowing the material to deposit in any position. Recently, a study on the effects of critical parameters, for example, layer height, extrusion temperature, and printing speed, on the printing quality of the product in FDM was carried out. Sagil et al. [35] fabricated the electrodes of a DSSC using a FDM 3D printing technique. DSSCs, third-generation solar cells, have drawn much attention due to their low cost of fabrication and material, environment-friendly processes, flexibility, and simple handling. DSSCs could improve more challenging indoor applications due to their better performance under silicon solar cells. The efficiency of DSSCs is lower than that of commercial silicon cells. Furthermore, applying a liquid medium results in liquid leakage, thermal expansion in hot weather, and freezing in cold weather conditions. DSSCs can be fabricated in small quantities by low-cost solution-phase methods, for example, roll-to-roll processing and screen printing methods. Fabrication of DSSCs using 3D printing techniques helps to solve their limitations, such as stopping leakage, supplying more customized design electrodes, and fast commercialization of DSSC technology. FDM 3D printers were employed to print substrates and spacers used in DSSC. The substrate on the photoanode and center spacer were fabricated with this 3D printing technique. In FDM, process parameters, including infill density, layer height, and printing speed, were optimized based on the printing quality of various thicknesses of electrodes. Open-circuit voltage (Voc) of 0.460 and 0.290 V was observed for FTO/glass-based DSSC and 3D-printed DSSCs, respectively.

Sagil and coworkers employed FDM, the most popular technique to print polymers, for affordable construction counter electrodes (CEs) in DSSCs. These CEs were fabricated using nature-inspired fractal designs to enhance their catalytic activity and general performance. The fractal designs utilized in this work were inspired by the veining pattern found in the leaves of the Camellia japonica plant. Figure 8.1 shows a Camellia japonica leaf. CEs require low surface resistance and strong adhesion of the catalyst material to the substrate. These 3D fractal designs provided high surface coverage, a large current density, and low resistance for use in DSSCs. A schematic of the DSSC assembly including the 3D fractal design-based CE can be seen in Figure 8.2. The I-V curve demonstrated a high current due to the high electron collection and transportation of these unique designs [36].

3D printing has been widely investigated for the manufacturing of light-trapping nanostructures and solar concentrators that are utilized to increase the efficiency of solar cells by reducing entropy losses. Regarding macroscale design, 3D printing

FIGURE 8.1 a) Camellia japonica leaf (inset: magnification to show venation pattern); b) 3D fractal design for the CE of the DSSC (inset: sub-branches). Adapted with permission [36]. Copyright (2018), Springer Nature.

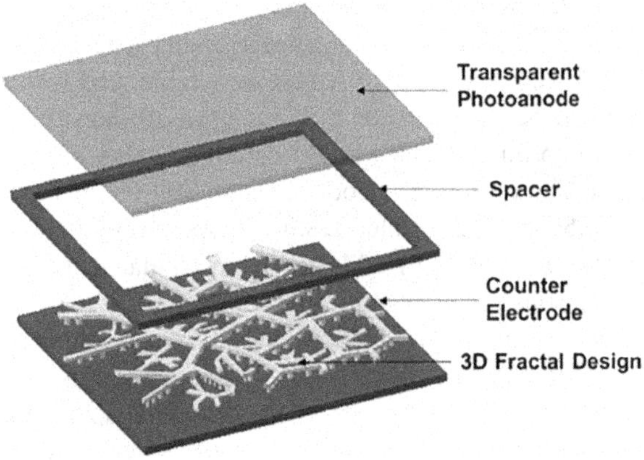

FIGURE 8.2 Schematic of DSSC assembly, including the 3D fractal design-based CE. Adapted with permission [36]. Copyright (2018), Springer Nature.

indicated the possibility of fabricating optimized 3D solar panel frames [37]. Dijk and coworkers [38] showed a 3D-printed external light trap is interesting for all solar cells. It is placed on the sun-facing surface of the solar cells and retro-reflects the lights those are reflected and emitted by the solar cells. The light trap was printed by

FDM and fabricated by smooth thermoplastic with a silver coating. An increment in broadband absorption using a 3D-printed external light trap led to a 15% improvement in photocurrent and PCE value in a thin-film nanocrystalline silicon solar cell.

PSCs are known as inexpensive replacements for conventional solar cells. However, they indicate stability problems when exposed to the environment. Their stability can be affected by water, UV light, oxygen, temperature, and continuous light [39]. Therefore, different procedures have been reported to prolong their stability, such as using different epoxies and desiccants in post-fabrication [40]. In this regard, the perovskite material was encapsulated using a layer-by-layer FDM process of polymer composite filaments. This structure provides the application of different polymer-based materials with special additives. This technique has different advantages: (1) It utilizes composite filaments to align filler particles during printing, improving thermal and electrical properties and thus increasing charge extraction and performance for solar cells [41]. (2) It incorporates a polymer barrier around filler particles, keeping the particles from oxygen and water and thus improving the long-term stability of PSCs. Tai et al. [42] demonstrated some thermoplastic polymers to improve stability in PSCs. Qaid et al. [43] investigated the ability of light enhancement by $CsPbBr_3$-substituted perovskite material during encapsulation in poly(methyl methacrylate) (PMMA) polymer.

The FDM method has remarkably changed the way different products are designed and manufactured. It has shown opportunities to produce new products on-site and on-demand. The blends of manufacturing flexibility and material additives indicate that FDM is a suitable choice for studying PCSs. Weiss et al. [39] introduced a PCS-based polymer using the FDM technique. They increased resistant UV light and stability from other environmental conditions using the remarkable capability of FDM to encapsulate perovskite materials. This work reported printable polymer materials, including the preparation of methylammonium lead iodide ($MAPbI_3$) perovskite microcrystals in the environment, the incorporation of $MAPbI_3$ into a polymer matrix without degradation, and the use of the composite to produce a solar-active material. Results indicated a high transparency with increasing temperature and, as a result, improved conductivity in exposure to simulated sunlight. In addition, increased electron-hole pair generation was reported for FDM-printed composite film with a 45% reduction in resistance.

8.2.6 Direct-Write 3D Printing

This technique is employed for manufacturing 1D to 3D printed structures by deposing concentrated inks with fine nozzles (~ 0.1–250 μm). In contrast to inkjet printing (a drop-based process), direct-write printing contains extruding ink filaments either in or out of the plane. Printed filaments usually match the size of the nozzle. Thus, microscale features can be patterned and formed into larger arrays and multidimensional designs. A variety of inks, such as compounds of ceramic, metallic, polymeric, and sol-gel materials, have been used for direct-write printing. Bok Yeop et al. [44] reported a very concentrated silver ink for 3D printing via this method. The ink was synthesized by dissolving a mixture of poly (acrylic acid) with two kinds of molecular weights in water and diethanolamine. This polymer was used as a capping agent for controlling silver nanoparticles' size. Transparent conductive silver grids were patterned on a flexible polyimide film. They were suitable replacements for transparent conducting oxide materials and utilized in 3D photovoltaics and light-emitting diodes.

8.2.7 SLA Method

SLA is one of the most extensively applied techniques, including consecutive annealing layers of photosensitive inks by UV irradiation within a vat [45]. 3D printing can be used to fabricate luminescent solar concentrators (LSCs). They contain luminescent species with high quantum efficiency in a highly transparent polymer substrate that is utilized to reflect incident light absorbed at a red-shifted wavelength, which can be a low-cost replacement for silicon substrates in integrated solar energy utilization [46]. It is a suitable 3D printing method to manufacture LSCs. Luminescent nanomaterials can be dispersed in methyl methacrylate resins, absorb UV and visible light, and compete with the photoinitiators in the resins. Therefore, the SLA method should employ an infrared laser instead of the UV or visible lasers that are usually utilized to perform photopolymerization. Efficient charge-collecting capacity for the photo-injected electrons is essential for solar cell performance. Using the SLS technique, it was reported that nanoparticle coating on polymeric conductive substrates increases inter-particle physical and electrical contacts [47].

8.3 FUTURE AND CHALLENGES

3D printing technology is advancing to more complicated uses such as complex 3D solar collectors and flexible multilayer organic/inorganic solar cells. It allows for advances in solar cell production and possibly the fascinating revolution of high-performance 3D photovoltaic devices. The progress of multi-material 3D-printed structures capable of combining focused optics, layers, and embedded fluorescent-polymer compounds to manufacture novel generations of developed solar cells based on high-conversion photovoltaics or LSCs is also considerable [48]. In addition, awareness of the progress of environmentally sensitive materials able to actively change configurations over time in response to external stimuli is very attractive because the utilization of shape memory polymers in so-called 4D printing can be significantly attractive for the development of self-configurable, active, and compatible harvesting devices such as solar cells [49].

To complete all the mentioned advances in solar cells, some issues must be overcome. Some of them are significant developments in printable active materials to provide multilayer manufacturing, advancements in high-quality optics, and AM of flexible contact materials [46].

8.4 CONCLUSIONS

3D printing techniques have solved different problems in manufacturing solar cells. They can provide uniformity of coverage over large areas and also proper utilization of materials with low waste. They have remarkable potential for improvement of solar cell use and particularly direct and inexpensive large-area manufacture of OSCs and light traps. These techniques are not only some of the cleanest in the renewable energy supply chain but also decrease costs. They will promote progress in renewable energy applications to reduce the use of fossil fuels.

REFERENCES

[1] I. Jasiuk, D.W. Abueidda, C. Kozuch, S. Pang, F.Y. Su, J. McKittrick, An overview on additive manufacturing of polymers, Jom. 70 (2018) 275–283.

[2] I. Salaoru, S. Maswoud, S. Paul, Inkjet printing of functional electronic memory cells: A step forward to green electronics, Micromachines. 10 (2019) 417–426.

[3] K.R. Ryan, M.P. Down, N.J. Hurst, E.M. Keefe, C.E. Banks, Additive manufacturing (3D printing) of electrically conductive polymers and polymer nanocomposites and their applications, E-Science. 2 (2022) 365–381.

[4] D. Yang, R. Yang, S. Priya, S. Liu, Recent advances in flexible perovskite solar cells: Fabrication and applications, Angew. Chem. Int. Ed. 58 (2019) 4466–4483.

[5] D. Vak, K. Hwang, A. Faulks, Y. Jung, N. Clark, D. Kim, G.J. Wilson, S.E. Watkins, 3D printer based slot-die coater as a lab-to-fab translation tool for solution-processed solar cells, Adv. Energy Mater. 5 (2015) 1401539.

[6] L. van Dijk, E.A.P. Marcus, A.J. Oostra, R.E.I. Schropp, M. Di Vece, 3D-printed concentrator arrays for external light trapping on thin film solar cells, Sol. Energy Mater Sol. Cells. 139 (2015) 19–26.

[7] Q.-Z. Huang, Y.-Q. Zhu, J.-F. Shi, L.-L. Wang, L.-W. Zhong, G. Xu, Dye-sensitized solar cell module realized photovoltaic and photothermal highly efficient conversion via three-dimensional printing technology, Chin. Phys. B. 26 (2017) 38401.

[8] Y. Yan, Y. Jiang, E.L.L. Ng, Y. Zhang, C. Owh, F. Wang, Q. Song, T. Feng, B. Zhang, P. Li, Progress and opportunities in additive manufacturing of electrically conductive polymer composites, Mater. Today Adv. 17 (2023) 100333.

[9] A. Knott, O. Makarovskiy, J. O'Shea, Y. Wu, C. Tuck, Scanning photocurrent microscopy of 3D printed light trapping structures in dye-sensitized solar cells, Sol. Energy Mater Sol. Cells. 180 (2018) 103–109.

[10] M. Zhu, Y. Duan, N. Liu, H. Li, J. Li, P. Du, Z. Tan, G. Niu, L. Gao, Y. Huang, Electrohydrodynamically printed high-resolution full-color hybrid perovskites, Adv. Funct. Mater. 29 (2019) 1903294.

[11] J. Kang, Y. Jang, Y. Kim, S.-H. Cho, J. Suhr, B.H. Hong, J.-B. Choi, D. Byun, An Ag-grid/graphene hybrid structure for large-scale, transparent, flexible heaters, Nanoscale. 7 (2015) 6567–6573.

[12] F. Mathies, H. Eggers, B.S. Richards, G. Hernandez-Sosa, U. Lemmer, U.W. Paetzold, Inkjet-printed triple cation perovskite solar cells, ACS Appl. Energy Mater. 1 (2018) 1834–1839.

[13] C. Liang, P. Li, H. Gu, Y. Zhang, F. Li, Y. Song, G. Shao, N. Mathews, G. Xing, One-step inkjet printed perovskite in air for efficient light harvesting, Sol. RRL. 2 (2018) 1700217.

[14] R.G. Burela, J.N. Kamineni, D. Harursampath, K. Kumar Sadasivuni, K. Deshmukh, M. A. Almaadeed, 3D and 4D printing of polymer nanocomposite materials, multifunctional polymer composites for 3D and 4D printing, Elsevier (2020) 231–257.

[15] K.X. Steirer, J.J. Berry, M.O. Reese, M.F.A.M. van Hest, A. Miedaner, M.W. Liberatore, R.T. Collins, D.S. Ginley, Ultrasonically sprayed and inkjet printed thin film electrodes for organic solar cells, Thin Solid Films. 517 (2009) 2781–2786.

[16] S.-S. Yoon, D.-Y. Khang, Roles of nonionic surfactant additives in PEDOT: PSS thin films, J. Phys. Chem. C. 120 (2016) 29525–29532.

[17] X. Hu, G. Chen, X. Wang, H. Wang, Tuning thermoelectric performance by nanostructure evolution of a conducting polymer, J. Mater. Chem. A. 3 (2015) 20896–20902.
[18] S. Ganesan, S. Mehta, D. Gupta, Fully printed organic solar cells–a review of techniques, challenges and their solutions, Opto-Electron. Rev. 27 (2019) 298–320.
[19] T. Aernouts, T. Aleksandrov, C. Girotto, J. Genoe, J. Poortmans, Polymer based organic solar cells using ink-jet printed active layers, Appl. Phys. Lett. 92 (2008).
[20] B. De Gans, P.C. Duineveld, U.S. Schubert, Inkjet printing of polymers: State of the art and future developments, Adv. Mater. 16 (2004) 203–213.
[21] C.N. Hoth, S.A. Choulis, P. Schilinsky, C.J. Brabec, High photovoltaic performance of inkjet printed polymer: Fullerene blends, Adv. Mater. 19 (2007) 3973–3978.
[22] J. Kim, C. Park, I. Song, M. Lee, H. Kim, H.C. Choi, Unique crystallization of fullerenes: Fullerene flowers, Sci. Rep. 6 (2016) 32205.
[23] A. Lange, M. Wegener, B. Fischer, S. Janietz, A. Wedel, Solar cells with inkjet printed polymer layers, Energy Procedia. 31 (2012) 150–158.
[24] S. Jung, A. Sou, K. Banger, D. Ko, P.C.Y. Chow, C.R. McNeill, H. Sirringhaus, All-ink-jet-printed, all-air-processed solar cells, Adv. Energy Mater. 4 (2014) 1400432.
[25] A. Singh, M. Katiyar, A. Garg, Understanding the formation of PEDOT: PSS films by ink-jet printing for organic solar cell applications, RSC Adv. 5 (2015) 78677–78685.
[26] F. Hermerschmidt, P. Papagiorgis, A. Savva, C. Christodoulou, G. Itskos, S.A. Choulis, Inkjet printing processing conditions for bulk-heterojunction solar cells using two high-performing conjugated polymer donors, Sol. Energy Mater. Sol. Cells. 130 (2014) 474–480.
[27] C.A. Lamont, T.M. Eggenhuisen, M.J.J. Coenen, T.W.L. Slaats, R. Andriessen, P. Groen, Tuning the viscosity of halogen free bulk heterojunction inks for inkjet printed organic solar cells, Org. Electron. 17 (2015) 107–114.
[28] T.M. Eggenhuisen, Y. Galagan, A. Biezemans, T. Slaats, W.P. Voorthuijzen, S. Kommeren, S. Shanmugam, J.P. Teunissen, A. Hadipour, W.J.H. Verhees, High efficiency, fully inkjet printed organic solar cells with freedom of design, J. Mater. Chem. A. 3 (2015) 7255–7262.
[29] M. Criado-Gonzalez, A. Dominguez-Alfaro, N. Lopez-Larrea, N. Alegret, D. Mecerreyes, Additive manufacturing of conducting polymers: Recent advances, challenges, and opportunities, ACS Appl. Polym. Mater. 3 (2021) 2865–2883.
[30] H. Eggers, F. Schackmar, T. Abzieher, Q. Sun, U. Lemmer, Y. Vaynzof, B.S. Richards, G. Hernandez-Sosa, U.W. Paetzold, Inkjet-printed micrometer-thick perovskite solar cells with large columnar grains, Adv. Energy Mater. 10 (2020) 1903184.
[31] F. Schackmar, H. Eggers, M. Frericks, B.S. Richards, U. Lemmer, G. Hernandez-Sosa, U.W. Paetzold, Perovskite solar cells with all-inkjet-printed absorber and charge transport layers, Adv. Mater. Technol. 6 (2021) 2000271.
[32] P. Li, C. Liang, B. Bao, Y. Li, X. Hu, Y. Wang, Y. Zhang, F. Li, G. Shao, Y. Song, Inkjet manipulated homogeneous large size perovskite grains for efficient and large-area perovskite solar cells, Nano Energy. 46 (2018) 203–211.
[33] N.J. Wilkinson, M.A.A. Smith, R.W. Kay, R.A. Harris, A review of aerosol jet printing—a non-traditional hybrid process for micro-manufacturing, Int. J. Adv. Manuf. Technol. 105 (2019) 4599–4619.
[34] C. Yang, E. Zhou, S. Miyanishi, K. Hashimoto, K. Tajima, Preparation of active layers in polymer solar cells by aerosol jet printing, ACS Appl. Mater. Interfaces. 3 (2011) 4053–4058.
[35] S. James, R. Contractor, C. Veyna, G. Jiang, Fabrication of efficient electrodes for dye-sensitized solar cells using additive manufacturing, in: International Manufacturing Science and Engineering Conference, American Society of Mechanical Engineers (2018): p. V001T01A022.

[36] S. James, R. Contractor, Study on nature-inspired fractal design-based flexible counter electrodes for dye-sensitized solar cells fabricated using additive manufacturing, Sci. Rep. 8 (2018) 17032.
[37] A. Zhakeyev, P. Wang, L. Zhang, W. Shu, H. Wang, J. Xuan, Additive manufacturing: Unlocking the evolution of energy materials, Adv. Sci. 4 (2017) 1700187.
[38] L. van Dijk, U.W. Paetzold, G.A. Blab, R.E.I. Schropp, M. Di Vece, 3D-printed external light trap for solar cells, Prog. Photovolt.: Res. Appl. 24 (2016) 623–633.
[39] L. Weiss, T. Sonsalla, Investigations of fused deposition modeling for perovskite active solar cells, Polymers. 14 (2022) 317.
[40] Q. Dong, F. Liu, M.K. Wong, H.W. Tam, A.B. Djurišić, A. Ng, C. Surya, W.K. Chan, A.M.C. Ng, Encapsulation of perovskite solar cells for high humidity conditions, ChemSusChem. 9 (2016) 2597–2603.
[41] T. Sonsalla, A.L. Moore, A.D. Radadia, L. Weiss, Printer orientation effects and performance of novel 3-D printable acrylonitrile butadiene styrene (ABS) composite filaments for thermal enhancement, Polym. Test. 80 (2019) 106125.
[42] C.-L. Tai, W.-L. Hong, Y.-T. Kuo, C.-Y. Chang, M.-C. Niu, M. Karupathevar Ponnusamythevar Ochathevar, C.-L. Hsu, S.-F. Horng, Y.-C. Chao, Ultrastable, deformable, and stretchable luminescent organic–inorganic perovskite nanocrystal–polymer composites for 3D printing and white light-emitting diodes, ACS Appl. Mater. Interfaces. 11 (2019) 30176–30184.
[43] S.M.H. Qaid, H.M. Ghaithan, K.K. AlHarbi, B.A. Al-Asbahi, A.S. Aldwayyan, Enhancement of light amplification of $CsPbBr_3$ perovskite quantum dot films via surface encapsulation by PMMA polymer, Polymers. 13 (2021) 2574.
[44] B.Y. Ahn, S.B. Walker, S.C. Slimmer, A. Russo, A. Gupta, S. Kranz, E.B. Duoss, T.F. Malkowski, J.A. Lewis, Planar and three-dimensional printing of conductive inks, J. Vis. Exp. (2011) e3189.
[45] R. Sheth, E.R. Balesh, Y.S. Zhang, J.A. Hirsch, A. Khademhosseini, R. Oklu, Three-dimensional printing: An enabling technology for IR, J. Vasc. Interv. Radiol. 27 (2016) 859–865.
[46] J.C. Ruiz-Morales, A. Tarancón, J. Canales-Vázquez, J. Méndez-Ramos, L. Hernández-Afonso, P. Acosta-Mora, J.R.M. Rueda, R. Fernández-González, Three dimensional printing of components and functional devices for energy and environmental applications, Energy Environ. Sci. 10 (2017) 846–859.
[47] L. Ming, H. Yang, W. Zhang, X. Zeng, D. Xiong, Z. Xu, H. Wang, W. Chen, X. Xu, M. Wang, Selective laser sintering of TiO_2 nanoparticle film on plastic conductive substrate for highly efficient flexible dye-sensitized solar cell application, J. Mater. Chem. A. 2 (2014) 4566–4573.
[48] M.G. Debije, P.P.C. Verbunt, Thirty years of luminescent solar concentrator research: Solar energy for the built environment, Adv. Energy Mater. 2 (2012) 12–35.
[49] X. Guo, H. Li, B. Yeop Ahn, E.B. Duoss, K.J. Hsia, J.A. Lewis, R.G. Nuzzo, Two-and three-dimensional folding of thin film single-crystalline silicon for photovoltaic power applications, Proc. Natl. Acad. Sci. 106 (2009) 20149–20154.

9 3D Printing for Polymer Electrolyte Membranes in Direct Liquid Fuel Cells

Ajaz A. Wani and Norazuwana Shaari

9.1 INTRODUCTION

An increasing interest in the research and development of sustainable, low-carbon, and clean energy sources and techniques of power production has resulted from the exponential growth in the price of fossil fuels and the requirement to reduce carbon emissions [1–3]. The importance of fuel cells in the future energy infrastructure is well recognized. When hydrogen is used, only clean water is produced as a byproduct of the electrochemical devices' conversion of the chemical energy held in the fuel and oxidant into electricity. Fuel cells' superior efficiency and lower emissions in comparison to traditional combustion processes make them an attractive option for direct energy conversion and power production [4]. Many different kinds of fuel cells have been proved effective in a variety of uses, from very compact portable devices to large-scale fixed power supplies. However, manufacturing processes are a major limitation preventing further development of fuel cells at this time. It is suggested that developing components with complex microstructures might help improve electrochemical performance and reduce losses caused by polarization. Hierarchically organized porous materials, such as those included in the electrodes of DLFCs, may reduce ohmic losses and increase response surface area, hence enhancing mass transfer [3]. The concentration polarization in polymer electrolyte membrane fuel cells (PEMFCs) may be reduced by using flow field designs inspired by natural phenomena, such as leaf veins or homogeneous channels in alveoli. In addition, a hierarchical pattern in the catalyst layer may help remove water from fuel cells and cut down on activation losses [5]. However, traditional subtractive fabrication technologies are either too expensive or impractical to execute these production strategies. This means that producing completely functioning fuel cell components requires a strong manufacturing approach equipped with high precision capabilities. With its ability to produce complex three-dimensional things with varying void percentage, thickness, composition, and homogeneity in a single process, additive manufacturing (AM) has come into its own [6]. This makes AM exceptionally advanced for applications like fuel cells, which necessitate customization, flexibility, and intricate design. Furthermore, AM expedites the implementation of a "fail fast, fail often" strategy, slashing lead times

DOI: 10.1201/9781003415985-9

and curbing costs and waste by producing only necessary parts. Already, AM has displayed substantial potential in crafting unique bespoke designs for energy conversion devices, an area traditionally limited by conventional methods. Integrating AM to propel fuel cell technologies has recently garnered escalating attention, becoming a burgeoning focal point within the fuel cell research community [7].

The current landscape of manufacturing is being reshaped by the advancements in 3D printing techniques, and this transformation is poised to expand into various domains of science and technology in the near future [8]. The process of 3D printing facilitates the digital design and construction of intricate three-dimensional objects, layer by layer, using a diverse range of materials. This method holds numerous advantages over conventional manufacturing procedures, including its capacity for swift prototyping, allowing minor modifications to a core design, and the ability to produce singular and distinct items (Figure 9.1a, b) [9]. Over recent years, 3D printing has gained substantial popularity in both industrial and research settings, demonstrating significant potential for swiftly prototyping intricate designs. With the expiration of 3D printing patents and the widespread utilization of computer-aided design (CAD) models, 3D printing services have become an economically viable route for rapid prototyping, facilitating immediate testing of numerous concepts. The potential for revolutionizing the fuel cell (FC) industry lies within 3D printing, especially as cost and weight reduction are critical factors for the economic viability of fuel cells [10].

The concept of 3D printing has been promoted as a means for rapidly prototyping bipolar plates (BPPs) alongside various other components integral to fuel cells. 3D printing provides the ability to rapidly analyze the performance of BPPs at a minimal time and cost in contrast to other manufacturing processes, since it eliminates the requirement for specialized tools. Employing 3D printing makes it feasible to craft single-piece BPPs without the requirement for welding, even with intricate non-planar geometries such as internal passageways. This innovation could lead to a reduction in the number of components, sealing prerequisites, and the intricacy of assembly, ultimately simplifying the overall manufacturing process. For the successful commercialization of DLFCs, there is a requirement for a fabrication method that accomplishes multiple objectives [11]. This technique should be not only cost effective, automated, and functionally straightforward but also possess attributes like design adaptability, consistency in replication, longevity, and the ability to yield high-performance cells. Among the potential solutions, 3D printing stands out due to its ability to address the limitations associated with the traditional manufacturing processes of DLFCs, all within a single, automated step. The integration of 3D printing processes would serve as a pivotal factor in producing DLFCs that exhibit remarkable durability and reproducibility. Furthermore, adopting 3D printing techniques holds the potential to significantly reduce both the initial setup costs and the overall fabrication expenses [12]. This approach also offers the advantage of enhancing DLFC design flexibility, concurrently curbing energy expenses, and minimizing material wastage. Embracing 3D printing methods, therefore, represents a significant stride toward the eventual commercialization of DLFCs.

This chapter provides an overview of the recent progress in the realm of DLFCs facilitated by AM. We examine and contrast various AM methods with the intention

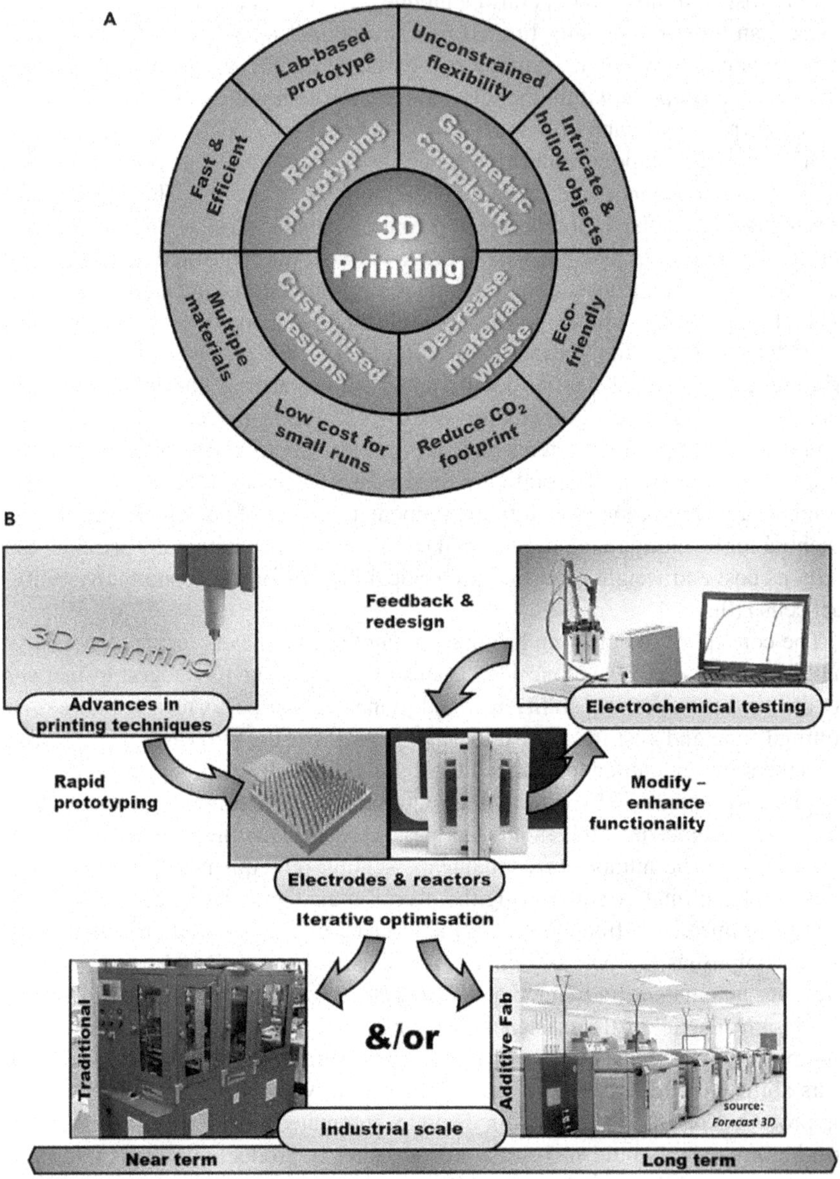

FIGURE 9.1 The advancement in 3D printing and its significance in electrochemical energy conversion and storage applications. (a) Benefits of 3D printing in contrast to conventional manufacturing techniques. (b) Diagram illustrating the integration of 3D printing in electrochemical energy conversion and storage applications, showcasing the gradual inclusion of 3D-printed components into extensive production procedures that will likely take place over an extended period. Adapted with permission from [9]. Copyright Elsevier (2019).

of uncovering the ways in which distinct AM techniques can enhance the production of specific components within fuel cells. Additionally, we pinpoint the challenges encountered when applying AM to fuel cell manufacturing. Drawing from these challenges, we delve into prospective avenues and possibilities for future research and development endeavors.

9.1.1 3D Printing Mechanism

Additive manufacturing techniques like 3D printing work by gradually adding layers of material until the desired object is achieved. The first step is to create a CAD model of the final product. The model is subsequently converted into the stereolithography (STL) file format, which divides the model into layers for the additive manufacturing process. After that, the STL file is sent to the printer, and the final product is built by depositing material layer by layer. In many circumstances, sintering the printed object thereafter is required to give it the requisite strength after being made via 3D printing [13].

9.1.2 3D Printing Process

The process of 3D printing offers a range of advantages over alternative subtractive methods concerning factors like material efficiency, cost, and automation. Unlike techniques such as injection molding and casting, the need for expensive molds, tools, and dies is eliminated [14]. The manufactured component also doesn't need any time-consuming and labor-intensive machining once it's been made. Furthermore, 3D printing operates in an automated manner, removing the potential for human errors. Its capacity to recycle up to 98% of waste materials contributes to its standing as a highly economical technology. Predominantly, 3D printing finds application in generating prototypes, spare parts, and medical/dental implants. Nonetheless, this process does have limitations, including a restricted array of available materials, relatively lower strength, and elevated costs for large-scale production of items.

9.1.3 Types of 3D Printing

The first step in 3D printing is to create a digital model of the final product. This may be accomplished with the use of CAD software, photogrammetry, or a 3D scanner, the latter of which involves constructing the model from a set of images taken from various vantage points [15]. After the 3D model has been created, it must be exported to the STL file format (originally developed for stereolithography), which stores the model's surface data as a set of coordinates for triangulated sections. All 3D printing software can read this format, making it a de facto standard. This information is transformed into a G-code file by a procedure called "slicing." Different 3D printing methods have been created, and they all deposited material in a somewhat different way. The mechanics behind these procedures allow us to classify them into four broad categories: (1) photopolymerization, (2) extrusion, (3) powder-based, and (4) lamination [16].

9.1.3.1 Photopolymerization

In the 1980s, photopolymerization emerged as one of the first methods used for 3D printing. In 1981, Hideo Kodama came up with the idea of employing ultraviolet light to harden a photosensitive polymer, allowing for the creation of three-dimensional objects [17]. To do this, UV light is shined on a polymer that is still in liquid form, solidifying it there. By systematically immersing and solidifying subsequent layers, they can be stacked atop one another with strong adherence, ultimately forming a complete 3D object. This process, depicted in Figure 9.2a, utilizes a photopolymer resin exposed to light of a specific wavelength, enabling complex polymerization. As depicted in Figure 9.2b, utilizing computer-aided manufacturing software, a linked material dispenser places the ink using a process akin to other additive manufacturing methods, allowing for the application of different propelling forces to guide the ink dispensation [18]. Stereolithography (SLA) and digital light processing (DLP) are two leading photopolymerization technologies with specific relevance to fuel cell applications. The light source is where SLA and DLP diverge most significantly; SLA uses a laser, whereas DLP makes use of a projector. The polymer-based resins used in these processes may be any of the three types described previously: transparent, flexible, or castable [19]. These resins include monomers, oligomers, initiators, and absorbers, and they are typically found in a liquid state. To obtain the necessary mechanical qualities or increased biocompatibility, it is common practice in biomedical engineering to include additional fillers into the resin, such as ceramics. A laser may be moved over the polymer surface in accordance with the layer design (direct laser writing), or UV light can be projected from a digital mirror device once for each layer. Both top-down (in a bath arrangement) and bottom-up (via an optically transparent window) polymerization are possible, with the stage lowering between solidified layers. The latter strategy is useful when there is a shortage of resin polymer and the need to use it in smaller amounts [20]. PolyJet technology utilizes inkjet techniques to deposit photocurable materials exactly as instructed by the layer design,

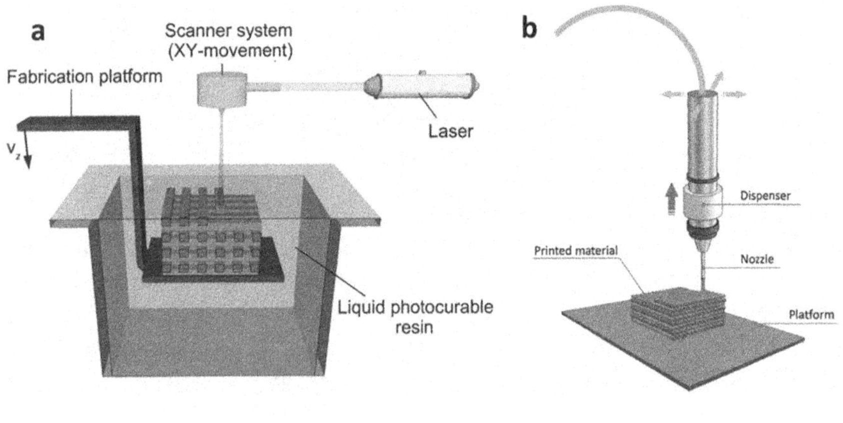

FIGURE 9.2 (a) Diagram illustrating stereolithography. (b) Direct ink writing processes illustration. Adapted with permission from [18]. Copyright Wiley (2017).

similar to how photopolymerization works. When exposed to UV radiation again, the coating quickly becomes rigid. When printing complex patterns, a structurally sound substance is placed at the same time to ensure the print lasts. The resolution is much improved with this Stratasys invention [21].

9.1.3.2 Extrusion

In 3D printing technologies that rely on extrusion, the model material (and potentially the support material) is dispensed straight from a nozzle head after preparation steps like liquefaction. Most often, this method is employed in fused deposition modeling (FDM), which uses thermoplastics to create three-dimensional objects [22]. FDM, developed by Scott Crump in 1989, allows for the creation of three-dimensional structures by successively depositing layers of semi-molten thermoplastic material into a nozzle. Once deposited, the material hardens into a homogeneous layer that stacks perfectly over the layer below it, just as intended in the model. This technique is becoming the standard in 3D printing since it is easy to use and inexpensive equipment is readily available. In filament form (as feeding spools), thermoplastics including polylactic acid (PLA), acrylonitrile butadiene styrene (ABS), polycarbonate (PC), and polyamide (PA) are widely used (Figure 9.3a) [23]. Multiple materials may be utilized in one printing session because of the ability to employ multiple nozzles at once. A material's role may also be that of a sacrificial or supporting structure, with the latter being discarded after the print job is complete. Using a similar premise, robocasting (also known as "direct ink writing") technology enables the precise placement of a wider variety of materials and material combinations than are possible with conventional casting. This technology's adaptability is remarkable; it can be used to deposit everything from ceramics

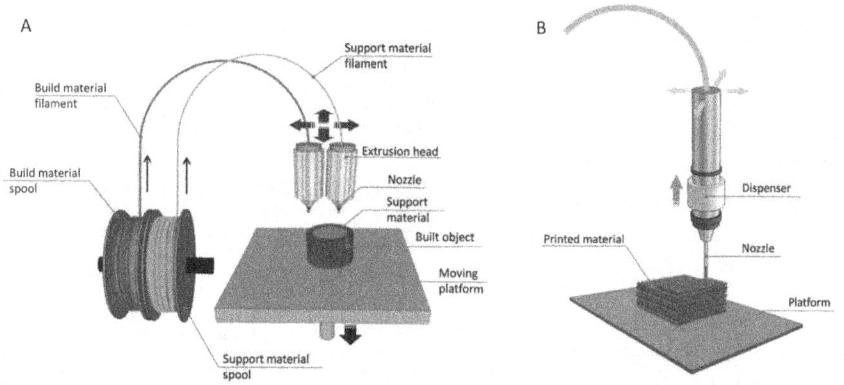

FIGURE 9.3 (a) A simplified diagram of how fused deposition modeling works (FDM). A thermoplastic filament is fed into a nozzle, which is then moved in three dimensions over a construction platform. Polymer "voxels," or molten units, are deposited onto the surface. (b) Diagram of robocasting, commonly known as direct ink writing. Here, a computer-controlled robot moves around a building platform carrying a material dispenser that applies ink in sequential layers. Adapted with permission from [23]. Copyright Royal Society of Chemistry (2016).

and polymers to foods and even live organisms (Figure 9.3b). The ink's characteristics, including material type, viscosity, density, and particle size, determine the suitable dispenser type. Nozzle dimensions and other specific parameters can be fine-tuned for optimal deposition. While the resulting object is usually self-supporting, additional post-fabrication steps might be necessary to enhance mechanical properties and durability through procedures like drying, heating, or sintering.

9.1.4 Powder-Based

Using particles between 50 and 100 mm in size, this additive manufacturing technique uses solid materials as building blocks to produce three-dimensional objects. This method is conceptually similar to SLA, except it uses a powdered photopolymer instead of a liquid one [24]. Rollers or squeegees are used to spread the powder in thin layers over the construction site. Towards this point, the cross-sectional model is aimed at the powder layer, where a binding mechanism (such as liquid glue or laser beams) is used to efficiently connect particles in specific places. Once a solid layer is formed, the stage may be lowered to allow for the deposition of a powder layer on top of it, setting the stage for the succeeding binding phase. Plaster, ceramics, acrylic, wood, marble, and metal powders are all suitable for use with this technique since they may be found in minute particle forms. The unrestricted, non-bonded powder has the potential to act as a support material for the printer, reducing the requirement for co-printed components. Powder that is not utilized during a printing job may be recycled and used again, thus minimizing waste [25]. The most fundamental method of binding makes use of a standard inkjet printing dispenser to precisely distribute liquid glue (or other compatible chemical binder) inks onto the powder bed. The term "binder jetting" describes this procedure. Binding the powder particles using high-energy laser beams is another option. SLS occurs when the beams heat the particles to a point just below their melting point, whereas selective laser melting (SLM) causes the particles to melt [26]. Steel, pure titanium, aluminum, nickel, and a wide variety of precious metal–based alloys are the most common particles bound using laser beam systems. In order to improve the mechanical characteristics of the metal being forged, it has recently been proposed to employ electron beams that can reach greater temperatures (electron beam melting, EBM) [18].

9.1.5 Lamination

After the geometry of a layer has been defined using cutting tools, the lamination 3D printing process involves stacking laminated materials (components accessible in sheet form) on top of each other [27]. Laminate object manufacture (LOM) is a process that Helisys developed in the 1980s [28]. The first step is to lay the material flat on a stage, where it will be cut using a laser or blade. In this step, the cross-sectional shape of the layer is determined, fitting a predetermined CAD model. After the surplus is cut away, another sheet is laid on top of the first, and the shape is traced once again. Layers are joined together using either adhesives or welding processes, and the materials used may range from paper to metal to plastic and beyond. This cycle is performed several times to give the 3D item its final form [29].

9.1.6 3D Printing of Polymer Exchange Membranes

Fuel cells have enormous promise as a means of producing and storing electricity because of their many useful features, including their low working temperatures, scalability, and high power density [30]. These qualities make them attractive for use in transportation and mobility-related fields. The MEA, which includes the membrane, anode and cathode catalyst layers (CLs), and gas diffusion layers, is the central component of DLFCs. Other components of a fuel cell, such as gaskets and BPPs, are required to install the MEA (Figure 9.4) [31]. Microstructure management of the individual parts is crucial to the devices' overall performance. In order to improve cell performance, hierarchically constructed MEAs with functionally graded designs have the potential to increase the available active surface for reactions, ease interfacial transport, and decrease diffusion distances [32]. All of the conductivities—aquatic, thermal, electrical, and ionic—are enhanced by this arrangement. To further facilitate water removal, the plates that make up the cell flow field have been thoughtfully designed to improve mass transport qualities and guarantee uniform reactant distribution to the CL. Because of this, very tiny amounts of catalyst loadings may be used. This evidence demonstrates that the shape of these devices has a significant impact on their efficiency and production and hence their marketability [33]. One of the key obstacles in the widespread adoption of DLFC systems is the development of

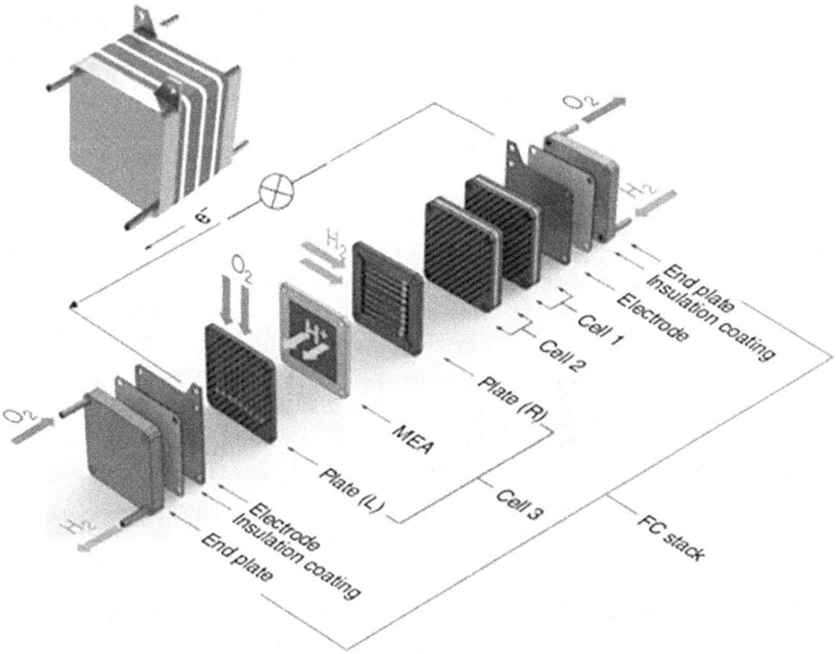

FIGURE 9.4 Illustration of PEM fuel cell stack and components with the CL and GDL housed on plates R and L, respectively. Adapted with permission from [31]. Copyright Wiley (2017).

new fabrication techniques that will allow for the inexpensive mass manufacturing of highly efficient and repeatable components.

9.1.7 Polymeric Exchange Membrane Fuel Cells

With the use of 3D printing technology, PEM components of varying sizes and shapes may be created without sacrificing the active materials' many useful properties. This paves the way for energy devices to be designed in a more efficient and optimum manner. In the following sections, we'll show you how designers have used 3D printing to make different parts of PEM DLFCs. Since the early 1990s, the materials used in PEM cells have stayed relatively the same [34]. Electrolyte membranes made on perfluorosulfonic acid/tetrafluoroethylene copolymer (PFSA) have been widely used. When saturated with water, these thin, flexible sheet membranes transmit electricity and provide strong proton conductivity and chemical stability. In both the anode and cathode, platinum on carbon catalysts are utilized. Platinum electrodes used in commercial applications generally contain between 0.2 and 0.4 mg/cm^2, yielding power densities of 0.5 to 0.7 W/cm^2 [35]. Due to its effect on the whole cost of the device, minimizing Pt consumption per kW is of paramount importance in PEM technology. When compared to air at 60°C, the diffusion coefficient for oxygen in water is 5,700 times lower. Hence proper water management is also critical for PEM system efficiency. The system incorporates a gas diffusion medium to address this concern. In addition, flow plates or bipolar plates facilitate the even distribution of fuel gas and air, provide electrical conductivity between cells, cool the work area, and prevent leakage. High-density graphite is often utilized for bipolar plates, while metals and graphite/polymer composites may also be seen in usage as well.

9.1.8 DLFC Electrolyte

Nafion membranes are widely used in commercial DLFCs, because they have excellent protonic conductivity, low permeability to reactant gases and electrons, and excellent chemical and mechanical stability at temperatures below 100°C. On the other hand, DLFCs' economic viability is negatively impacted by the high manufacturing costs of these membranes, especially for lower production runs [36]. Inkjet printing directly onto the CLs of traditional gas diffusion electrodes is a new revolutionary method. The results of this strategy have been shown to be quite promising. During pure H_2/O_2 operation, the membrane showed negligible resistance and minimized charge transfer, resulting in a remarkable power density (about 4 W/cm^2). This enhancement is often attributed to the thin dimensions (8–25 m) and the decreased interfacial contact resistance between the membrane and ionomer at the CLs. It is noteworthy that the system's power output remains constant at 1 W/cm^2 even while operating in completely dry circumstances. This consistency is due to the thin membrane being humidified from the inside out by back-diffusion of the water it produces. PEM fuel cells with very low platinum loadings (0.102 and 0.029 mg/cm^2) in the gas diffusion layer (GDL) were then fabricated using this technology by Breitwieser et al. As a result of these efforts, an efficiency of 88 kW/g was reached in platinum use [37].

9.1.9 DLFC Catalysts Layer

The widespread usage of platinum as a key component has made the fabrication of catalyst layers crucial to the development of DLFCs as they entered the realm of mass production [38]. The current cutting-edge manufacturing process for commercial catalyst layers offers two distinct approaches [38]. The first technique involves integrating two GDLs into the membrane after applying a catalyst-containing ink to the membrane. The second technique involves hot-pressing the membrane and GDLs together to create a membrane-electrode assembly after applying the catalyst ink to the GDLs. Hot pressing, gravure coating, and screen printing are some of the most conventional methods used for depositing catalyst layers. Catalysts still contribute around 34% of the entire cost of PEM stacks, despite significant advances since the 1960s resulting in a drop in platinum loading by two orders of magnitude [39]. In light of this, researchers have lately investigated the feasibility of using cutting-edge additive printing techniques to further decrease platinum loading without sacrificing performance or longevity [40]. The use of inkjet printing (IJP) has recently emerged as a viable alternative method for producing catalyst layers. Its very high resolution (1,200 dpi), outstanding control over Pt-loading and distribution, and low cost are all major selling points [41]. One of the first applications of this method is the work by Taylor et al., who developed inkjet-printed MEAs with ultra-low catalyst loading (0.020 mg Pt/cm^2) [42]. Maximizing triple phase boundary (TPB) by a reduction in catalyst layer thickness reached via IJP resulted in an outstanding power-to-Pt usage ratio of 17.6 kW/g of Pt. IJP also made it possible to deposit catalysts in varying thicknesses (a method that has shown increased Pt consumption), which outperforms equally dispersed catalysts by a small margin. Saha et al. further compared the efficacy of CLs manufactured using different coating techniques [43]. Inkjet-printed CLs were found to have a more uniform distribution of carbon support and ionomer, and they achieved the same level of performance as a traditionally manufactured catalyst layer, although containing 20% less Pt. Recently, the influence of ionomer loading on ultra-low Pt loading electrodes was studied using IJP as the deposition technique [44]. This research showed that interfacial oxygen transport resistance was more important than bulk thin film transport resistance, indicating a discrepancy between cell performance and Nafion loading. The 47.6 kW/g Pt usage rate is excellent. Wang and Nagao, as an alternative to spraying the catalyst layer, sought to apply the Nafion ionomer as a transport medium onto the CL using IJP [45]. This method exposed a closer membrane/electrode contact, which resulted in enhanced Pt consumption and a higher number of TPBs. Inkjet-printed PEMFCs have been shown to function better than expected in this research.

9.1.10 DLFC Gas Diffusion Layer

The GDL plays a significant role in determining the power output of a DLFC due to its effect on the reactive gas distribution throughout the surface area of the membrane. The macro-porous layer and the micro-porous layer (MPL), are the two main components of a GDL. Each layer is then put on a base made of carbon, such as carbon paper or fabric. The MPL is made when a hydrophobic substance (such as PTFE) is applied to the substrate, and the micro-porous layer is formed when carbon black and more PTFE are coated onto the MPL to help in controlling the water content of the MEA. Multiple initiatives have emerged in recent years to use 3D printing's benefits in GDL production. To combat issues associated with carbon-based materials in GDLs, Jayakumar and Al-Jumaily created a carbon-free gas diffusion substrate using SLS, using a combination of aluminum and polyamide (alumide) [46]. This study demonstrated the potential success of their approach, but more post-processing using graphene-based materials was necessary to enhance the electrical conductivity of the printed GDL. The alumide substrate also experienced electrochemical deterioration under real-world circumstances. In light of the promising outcomes of SLS and its streamlined method, the same group of researchers created a GDL substrate out of titanium and polyamide. Despite the superior microstructural features made feasible by the SLS manufacturing approach and the increased conductivity of titanium powder, GDL's electrical conductivity remained relatively low, resulting in unsatisfactory electrochemical performance and underlining the need for further development. In any case, SLS is a good choice because of its ease of use, low price, short production time, and conservation of raw materials.

9.1.11 DLFC Bipolar Plates and Flow Fields

Water management is a key factor in determining the success or failure of a PEM single cell. For maximum conductivity, keeping the membrane at a constant degree of moisture is crucial. The volume fraction of the conducting phase within the membrane grows as the water concentration in the surrounding environment grows in proportion to an increase in relative humidity (RH). However, if too much water is present, it will flood and clog the pores of the GDL, preventing reactants from reaching the CL. Thus, maintaining a stable balance between water production and removal is critical for maximizing fuel cell efficiency. Designing flow fields is crucial for preventing water buildup and ensuring uniform gas distribution throughout the electrode surface [47]. In addition to enclosing the flow fields, BPPs provide essential roles such as providing mechanical support for the stack, establishing electrical connections between cells, and facilitating heat management. BPPs also help cut down on the PEMFC stack's size, weight, and overall cost. The importance of design of these components has led to considerable usage of 3D printing to study them. Compared to micro-electro-mechanical techniques and CNC machining, FDM shows improvement in accuracy and production time when used to create small fuel cell flow fields [48]. For graphite-based BPPs, several researchers explored SLS, which allowed the creation of intricate flow field patterns more efficiently than conventional methods. Liquid phenolic infiltration post-processing greatly increased electrical conductivity after SLS processing, resulting in plates with high porosity and poor mechanical and electrical characteristics. Surface roughness issues were addressed by employing a laser re-melting process to enhance surface smoothness. In addition, BPPs with intricate patterns employing various carbon-based materials showed steady performance on par with standard BPPs [49]. Recent research into metallic BPPs has focused on different 3D printing processes, with titanium-based alloys and stainless steel being of particular interest. By using direct metal laser sintering (DMLS), designers could create complex flow fields and interior cooling channels without resorting to costly custom equipment and traditional machining procedures. However, due to contact resistance issues, surface polishing and coating were necessary after DMLS. Similarly, SLM was utilized for stainless steel BPPs, revealing heat treatment's capacity to improve durability. The fast prototyping techniques of SLA and polyjet 3D printing were also investigated for use in developing new flow field designs for BPP [50].

9.2 CONCLUSIONS AND OUTLOOK

The incorporation of 3D printing technologies into the manufacturing processes of fuel cells and electrolysis systems holds evident advantages. Encapsulation, interconnector components like manifolds, and current collectors may improve, but core components like electrochemical cells are where the bulk of the change will be seen. High-aspect-ratio compact nanoarchitectures that can be printed in their entirety or even full stack single-step fabrication with integrated functionality are examples of more advanced strategies with the potential to fundamentally disrupt the market. Other benefits, such as higher performance via greater cell active area or improved fuel flow distribution throughout the stack, have also been documented, making 3D printing technology a strong candidate to replace traditional production processes for fuel cells and electrolysis. Particularly in contexts requiring reproducibility and mass customization, mobile applications like drones, or segments with varying requirements, such as the commercial sector, could spearhead this implementation. Notably, the growing interest in developing and standardizing multi-material 3D printing processes at an industrial level opens avenues for fabricating intricate devices like fuel and electrolysis cells. This transition from a primary prototyping technique to a fabrication tool holds the potential to reshape the industry landscape.

REFERENCES

[1] N. K. Singh, A. Goswami, P. K. Sadhu, Energy economics and environmental assessment of hybrid hydel-floating solar photovoltaic systems for cost-effective low-carbon clean energy generation, *Clean Technol. Environ. Policy* 25 (2023) 1339–1360.
[2] D. P. Macedo, A. C. Marques, Is the energy transition ready for declining budgets in RD&D for fossil fuels? Evidence from a panel of European countries, *J. Clean. Prod.* 417 (2023) 138102.
[3] N. Shaari, N. N. R. Ahmad, R. Bahru, C. P. Leo, Ionic liquid-modified materials as polymer electrolyte membrane and electrocatalyst in fuel cell application: An update, *Int. J. Energy Res.* 46 (2022) 2166–2211.
[4] Z. F. Pan, L. An, C. Y. Wen, Recent advances in fuel cells based propulsion systems for unmanned aerial vehicles, *Appl. Energy* 240 (2019) 473–485.
[5] X. Peng, Z. Taie, J. Liu, Y. Zhang, X. Peng, Y. N. Regmi, J. C. Fornaciari, C. Capuano, D. Binny, N. N. Kariuki, D. J. Myers, M. Scott, A. Z. Weber, N. Danilovic, Hierarchical electrode design of highly efficient and stable unitized regenerative fuel cells (URFCs) for long-term energy storage, *Energ. Environ. Sci.* 13 (2020) 4872.
[6] J. Wang, N. He, J. Fei, Z. Ma, Z. Ji, Z. Chen, N. Nie, Y. Huang, Flexible and wearable fuel cells: A review of configurations and applications, *J. Power Sources* 551 (2022) 232190.

[7] Y. N. Yusoff, N. Shaari, S. H. Osman, Metal oxide-based materials as an emerging platform for fuel cell system: A review, *Mate. Sci. Technol.* 39 (2023) 2363.
[8] M. A. S. R. Saadi, A. Maguire, N. Pottackal, S. H. Thakur, M. Md. Ikram, A. J. Hart, P. M. Ajayan, M. M. Rahman, *Adv. Mater.* 34 (2022) 2108855.
[9] C. Lee, A. Taylor, A. Nattestad, S. Beirne, G. G. Wallace, 3D printing for electrocatalytic applications, *Joule* 3 (2019) 1835.
[10] G. Athanasaki, A. Jayakumar, A. M. Kannan, Gas diffusion layers for PEM fuel cells: Materials, properties and manufacturing—A review, *Int. J. Hydrog. Energy* 48 (2023) 2294–2313.
[11] N. Shaari, Polymer electrolyte membranes in fuel cell applications, *Recent Adv. Renew. Energy Technol. 1* (2021) 311–352.
[12] K. K. Maniam, R. Chetty, R. Thimmappa, S. Paul, Progress in the development of electrodeposited catalysts for direct liquid fuel cell applications, *Appl. Sci.* 12 (2022) 501.
[13] M. Garg, R. Rani, V. K. Meena, S. Singh, Significance of 3D printing for a sustainable environment, *Mater. Today Sustain.* 23 (2023) 100419.
[14] Y. Yang, X. Zhu, Q. Wang, D. Ye, R. Chen, Q. Liao, Towards flexible fuel cells: Development, challenge and prospect, *Appl. Therm. Eng.* 203 (2022) 117937.
[15] P. E. Éltes, M. Bartos, B. Hajnal, Á. J. Pokorni, L. Kiss, D. Lacroix, P. P. Varga, Á. Lazáry, Development of a computer-aided design and finite element analysis combined method for affordable spine surgical navigation with 3D-printed customized template, *Front. Surg.* 7 (2021).
[16] A. Ambrosi, A. Bonanni, How 3D printing can boost advances in analytical and bioanalytical chemistry, *Microchim. Acta* 188 (2021) 1–17.
[17] J. Izdebska-Podsiadły, History of the development of additive polymer technologies, *Polym. 3D Print. Methods, Prop. Charact.* (2022) 3–11.
[18] X. Tian, J. Jin, S. Yuan, C. K. Chua, S. B. Tor, K. Zhou, Emerging 3D-printed electrochemical energy storage devices: A critical review, *Adv. Energy Mater.* 7 (2017) 1700127.
[19] J. Zhang, L. Wei, X. Meng, F. Yu, N. Yang, S. Liu, Digital light processing-stereolithography three-dimensional printing of yttria-stabilized zirconia, *Ceram. Int.* 46 (2020) 8745–8753.
[20] J. Yuan, Y. Chen, J. Sun, Y. Wang, M. Lin, M. Wang, H. Wang, J. Bai, 3D printing of robust 8YSZ electrolytes with a hyperfine structure for solid oxide fuel cells, *ACS Appl. Energy Mater.* 6 (2023) 4133–4143.
[21] X. Y. Tai, A. Zhakeyev, H. Wang, K. Jiao, H. Zhang, J. Xuan, Accelerating fuel cell development with additive manufacturing technologies: State of the art, opportunities and challenges, *Fuel Cells*, 19 (2019) 636–650.
[22] L. Y. Zhou, J. Fu, Y. He, A review of 3D printing technologies for soft polymer materials, *Adv. Funct. Mater.* 30 (2020) 2000187.
[23] A. Ambrosi, M. Pumera, 3D-printing technologies for electrochemical applications, *Chem. Soc. Rev.* 45 (2016) 2740–2755.
[24] C. L. Cramer, E. Ionescu, M. Graczyk-Zajac, A. T. Nelson, Y. Katoh, J. Haslam, L. Wondraczek, T. G. Aguirre, S. LeBlanc, H. Wang, M. Masoudi, E. Tegeler, R. Riedel, P. Colombo, M. Minary-Jolandan, Additive manufacturing of ceramic materials for energy applications: Road map and opportunities, *J. Eur. Ceram. Soc.* 42 (2022) 3049–3088.
[25] S. A. Rasaki, C. Liu, C. Lao, H. Zhang, Z. Chen, The innovative contribution of additive manufacturing towards revolutionizing fuel cell fabrication for clean energy generation: A comprehensive review, *Renew. Sustain. Energy Rev.* 148 (2021) 111369.
[26] A. Kothuru, S. Goel, Leveraging 3-D printer with 2.8-W blue laser diode to form laser-induced graphene for microfluidic fuel cell and electrochemical sensor, *IEEE Trans. Electron Devices* 69 (2022) 1333–1340.

[27] M. Cannio, S. Righi, P. E. Santangelo, M. Romagnoli, R. Pedicini, A. Carbone, I. Gatto, Smart catalyst deposition by 3D printing for polymer electrolyte membrane fuel cell manufacturing, *Renew. Energy* 163 (2021) 414–422.
[28] H. Windsheimer, N. Travitzky, A. Hofenauer, P. Greil, Laminated object manufacturing of preceramic-paper-derived Si-SiC composites, *Adv. Mater.* 19 (2007) 4515–4519.
[29] P. Rewatkar, S. Goel, Catalyst-mitigated arrayed aluminum-air origami fuel cell with ink-jet printed custom-porosity cathode, *Energy*, 224 (2021) 120017.
[30] R. M. Nauman Javed, A. Al-Othman, M. Tawalbeh, A. G. Olabi, Recent developments in graphene and graphene oxide materials for polymer electrolyte membrane fuel cells applications, *Renew. Sustain. Energy Rev.* 168 (2022) 112836.
[31] T. Sutharssan, D. Montalvao, Y. K. Chen, W. C. Wang, C. Pisac, H. Elemara, A review on prognostics and health monitoring of proton exchange membrane fuel cell, *Renew. Sustain. Energy Rev.* 75 (2017) 440–450.
[32] N. Shaari, Z. Zakaria, S. K. Kamarudin, The optimization performance of cross-linked sodium alginate polymer electrolyte bio-membranes in passive direct methanol/ethanol fuel cells, *Int. J. Energy Res.* 43 (2019) 8275–8285.
[33] N. F. Raduwan, N. Shaari, S. K. Kamarudin, M. S. Masdar, R. M. Yunus, An overview of nanomaterials in fuel cells: Synthesis method and application, *Int. J. Hydrog. Energy* 47 (2022) 18468–18495.
[34] M. L. Perry, T. F. Fuller, An ECS centennial series article A historical perspective of fuel cell technology in the 20th century, *J. Electrochem. Soc.* 149 (2002) S59.
[35] D. A. Boysen, S. Cha, C. R. I. Chisholm, K. P. Giapis, S. M. Haile, A. B. Papandrew, K. Sasaki, From laboratory breakthrough to technological realization: The development path for solid acid fuel cells, *Electrochem. Soc. Interface*. 18 (2009) 53–59.
[36] B. C. Ong, S. K. Kamarudin, S. Basri, Direct liquid fuel cells: A review, *Int. J. Hydrog. Energy* 42 (2017) 10142–10157.
[37] P. P. Singh, Ambika, M. Verma, Recent advances in alternative sources of energy, *Energy Cris. Challenges Solut.* (2021) 55–71.
[38] N. Shaari, S. K. Kamarudin, R. Bahru, S. H. Osman, N. A. I. Md Ishak, Progress and challenges: Review for direct liquid fuel cell, *Int. J. Energy Res.* 45 (2021) 6644–6688.
[39] V. Mehta, J. S. Cooper, Review and analysis of PEM fuel cell design and manufacturing, *J. Power Sources* 114 (2003) 32–53.
[40] N. Singh, G. Singh, Advances in polymers for bio-additive manufacturing: A state of art review, *J. Manuf. Process.* 72 (2021) 439–457.
[41] X. Qian, M. Ostwal, A. Asatekin, G. M. Geise, Z. P. Smith, W. A. Phillip, R. P. Lively, J. R. McCutcheon, A critical review and commentary on recent progress of additive manufacturing and its impact on membrane technology, *J. Memb. Sci.* 645 (2022) 120041.
[42] Z. Xie, C. Song, David, P. Wilkinson, J. Zhang, Catalyst layers and fabrication. Proton exchange membrane, *Fuel Cells* (2009) 61.
[43] M. S. Saha, D. K. Paul, B. A. Peppley, K. Karan, Fabrication of catalyst-coated membrane by modified decal transfer technique, *Electrochem. Commun.* 12 (2010) 410–413.
[44] I. Fouzaï, S. Gentil, V. C. Bassetto, W. O. Silva, R. Maher, H. H. Girault, Catalytic layer-membrane electrode assembly methods for optimum triple phase boundaries and fuel cell performances, *J. Mater. Chem. A* 9 (2021) 11096–11123.
[45] Z. Wang, Y. Nagao, Effects of Nafion impregnation using inkjet printing for membrane electrode assemblies in polymer electrolyte membrane fuel cells, *Electrochim. Acta* 129 (2014) 343–347.
[46] P. Hu, B. Tan, M. Long, Advanced nanoarchitectures of carbon aerogels for multifunctional environmental applications, *Nanotechnol. Rev.* 5 (2016) 23–39.
[47] C. Y. Ling, M. Han, Y. Chen, E. Birgersson, Mechanistic three-dimensional analytical solutions for a direct liquid fuel cell stack, *J. Fuel Cell Sci. Technol.* 12 (2015) 061003.

[48] X. Sun, Y. Li, C. Xie, M. Hao, M. Li, J. Xue, Activating triple-phase boundary via building oxygen-electrolyte interfaces to construct high-performance pH-disparate direct liquid fuel cells, *J. Chem. Eng.* 418 (2021) 129480.
[49] R. Oliveira, J. Santander, R. Rego, Overview of direct liquid oxidation fuel cells and its application as micro-fuel cells, *Advanced Electrocatalysts for Low-Temperature Fuel Cells.* Cham: Springer International Publishing (2018) 129–174.
[50] X. Fuku, A. Mkhohlakali, N. Xaba, M. Modibedi, The potential role of electrocatalysts in electrofuel generation and fuel cell application, *Electrode Materials for Energy Storage and Conversion.* Boca Raton, FL: CRC Press (2021) 289–308.

10 3D-Printed Conducting Polymers for Fuel Cells

Sunil Kumar Baburao Mane, Naghma Shaishta, and G. Manjunatha

10.1 INTRODUCTION

Conducting polymers (CPs) with intrinsic electrical conductivity have emerged as one of the materials with the greatest potential in a variety of programs, such as battery backup, adaptable electronics, and bioelectronics. CPs are intriguing material options. Nevertheless, the production of CPs has primarily depended on traditional methods like ink-jet printing, printing with screens, and electron-beam lithography, whose constraints have slowed down the development of new CPs and their widespread use. These limitations on low resolution over 100 μm; two-dimensional patterns with low aspect ratio; and complex and expensive processes such as multi-step techniques in clean rooms, including arrangements, façades, drawings, and post-assemblies have slowed down the development of CPs and limited their potential for wide-ranging uses.

Three-dimensional printing (3DP), in contrast to these traditional methods, provides the ability to build microscale objects in a configurable, simple, and adaptable way with no constraints on design in 3D space. For instance, recent advances in 3DP materials, including glass, metals, liquid crystal polymers, hydrogels, and bioinks containing living cells, have significantly increased the range of substances that can be used in 3DP. Despite significant attempts, 3DP of CPs has only so far produced modest assemblies like isolated fibers due to current CP inks' poor 3DP ability [1–3].

In order to benefit from sophisticated 3DP for the creation of CPs, Yuk et al. developed a high-performance 3DP ink based on one of the most often used CPs, poly(3,4-ethylenedioxythiophene):polystyrene sulfonate (PEDOT:PSS) [4]. They created a paste-like CP ink via cryogenic freezing of an aqueous PEDOT:PSS solution, followed by lyophilization and carefully regulated re-dispersion in a solution of water and dimethyl sulfoxide (DMSO). The resulting CP ink displays greater 3DP, enabling the manufacture of CPs with high resolution (over 30 μm), a high aspect ratio (over 20 layers), and extremely accurate results. Insulating elastomers can also be easily integrated with the CPs by using multi-material 3DP.

PEDOT:PSS 3D microstructures are extremely conductive and flexible in the dry state thanks to the dry-annealing of the 3DP CPs. Additionally, by subsequently swelling in wet conditions, dry-annealed 3DP CPs can be easily transformed into a soft yet highly conductive PEDOT:PSS hydrogel. They also show how easy it is to quickly fabricate several functional CP systems using multi-material 3DP, such as a

DOI: 10.1201/9781003415985-10

soft neural probe with in vivo single-unit recording capabilities and a high-density stretchy electronic circuit.

Numerous effective uses of microbial electrochemical technologies (METs) are being proven over the course of the last few decades, including bioenergy production, surveillance of the environment, recuperation of resources, and synthesis of key chemicals [5]. The production and marketing of METs remain very difficult, despite their enormous scope. Typical obstacles might involve costly and laborious manufacturing procedures, protracted launch times, intricate engineering specifications, and a capacity limitation for systems of considerable size, based on the planned uses. A novel and extremely exciting technique for constructing METs to show electricity production and biosensing at bench size involves using 3DP technology.

Particularly sophisticated and miniature MET structures could now be quickly and cheaply manufactured, which is not possible with conventional procedures. Utilizing 3DP demonstrated great promise for assisting in the optimization of useful large-scale METs, especially those necessary for scaling-up objectives. Additionally, a bioanode made using 3DP could offer instantaneous startup in today's creation of METs. To the authors' knowledge, fewer published review papers have specifically emphasized the relevance and value of 3DP in creating METs, notwithstanding the fact that many papers have been produced on various scientific and operational components of METs. Therefore, the purpose of this assessment is to present an up-to-date perspective, position, and forecast of 3DP uses for developing METs.

Methods for additive manufacturing (AM) offer several benefits, including customization possibilities, minimum trash, the ability to produce highly intricate frameworks, reduced manufacturing expenses, and quick prototyping. Numerous industries, such as those in the medical, energy, arts, architecture, aviation, and automobile industries, make extensive use of this innovation. It is possible to create numerous objects employing unique filament and powdered substances during the making of 3D-printed goods using a variety of fabrication procedures.

Three primary phases may be distinguished in the AM of a three-dimensional object: designing, printing, and post-processing (Figure 10.1). Computer-automated design (CAD) software is used to design the necessary 3D form. In a conventional fabrication procedure, slicing software slices the intended architecture into 2D pictures (layers) in STL (Standard Tessellation Language) style. The topological data pertaining to the sliced layers is contained in the resultant document from the slicing program. Prior to printing, parameters like duration of exposure are frequently set in the system.

These settings rely on the method and the procedure. Now, 3DP based on various technologies for manufacturing uses G-codes produced by the software that are consistent with many 3D file types. The printer actions, sliced pictures, exposure time, temperature, and other details are all contained within these G-codes. 2D layers are printed one after another until the finished product appears on the assembly head as a neat stack of slices. The third phase involves removing the built object from the build head and putting it through post-processing, which is also known as the "green body" at this point. Depending on the exact method of AM utilized, the substance employed, and the intended use of the generated components, post-processing might be necessary for eliminating extra initial components or applying the last polishing.

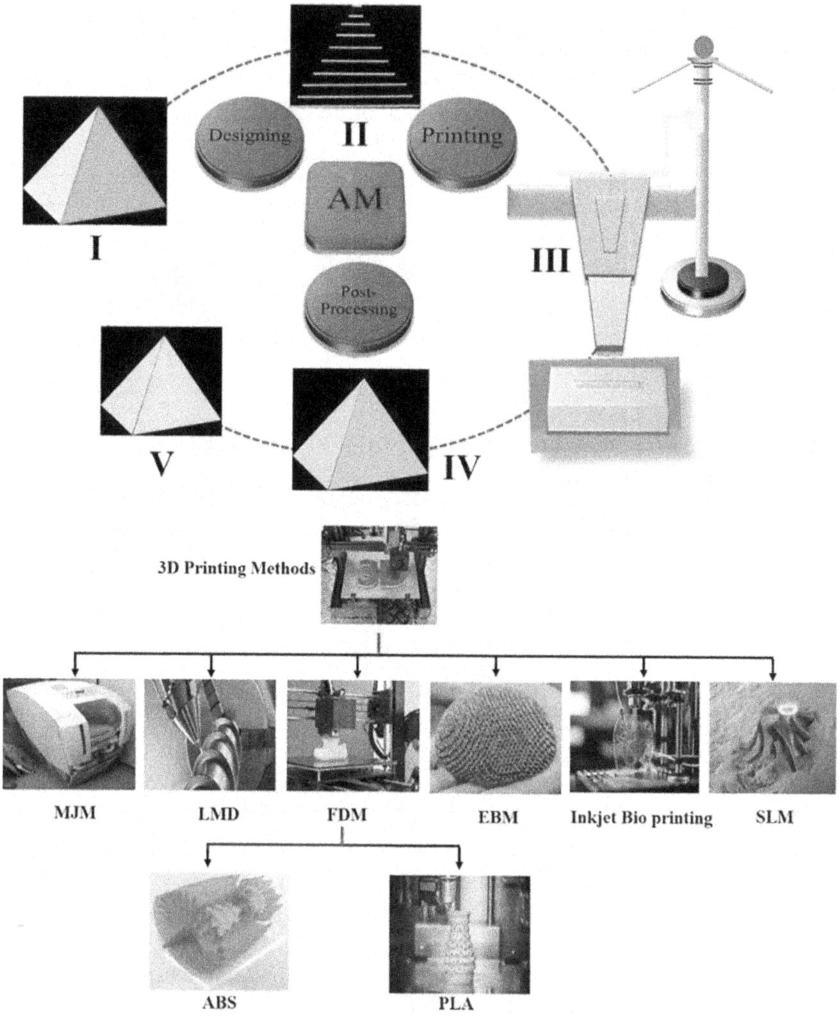

FIGURE 10.1 Detailed development in AM. (I and II) The design step consists of CAD modeling and slicing of a 3D item. (III and IV) In printing management, these slice sheets are printed one following another. The printed body in green form is then subjected to post-processing before developing as a finishing printed body (V).

AM includes several 3DP processes, including stereolithography (SLA), selective laser melting (SLM), inkjet printing, laser metal deposition, fused deposition modeling (FDM), electron beam melting (EBM), and multi-jet modeling (MJM). A detailed summary of the developments in 3DP techniques for electrochemical domains is provided in the current chapter. A thorough analysis of the electrode printing processes used with polymeric and metallic 3DP substances is also provided. Last, this work thoroughly covers the advantages and disadvantages of producing electrodes for energy conversion systems using AM techniques. By layering materials, AM techniques such as polymer 3DP may produce 3D items.

3DP makes use of polymers as its main substance rather than metals or porcelain, which are molecules with long chains built up of units that repeat. To accommodate patterns with a variety of structures, replies, and arrangements that are not feasible with alternatives, polymer printing employs extrusion, resin, and powder 3DP methods. Many readily accessible polymers, such as polylactic acid (PLA), polycarbonate, polyether ester ketone, acrylonitrile butadiene styrene (ABS), polyetherimide, and thermoplastic elastomers, can be printed with this method. Due to its simplicity of usage and low elasticity upon heating and cooling, PLA is the most frequently employed polymer. Although the others have enhanced qualities, they are costly and more challenging to print.

10.2 APPROACHES AND PROCEDURES FOR POLYMER 3DP

The market for AM is dominated by polymer 3DP methods, which are used to create complicated patterns and functioning designs, including final usage products [6–8].

Systems and methods used in polymer 3DP involve the following.

10.2.1 Vat Polymerization with Resin

This is a method of AM that uses light-activated polymerization to precisely cure photopolymer liquid resin to create 3D components. A vat of liquid photopolymer resin is lifted into or elevated out of the manufacturing stage. A laser or light beam is projected over the structure's surface to produce an item. The photopolymer solidifies and adheres to the surface as a result. The structure's foundation is a bit lower or elevated, and an additional photopolymer coating is added after each layer has dried. Until the object is finished, this method is continued [9].

Three generations of resin 3DPs have become accessible. In the first SLA, each layer is created using a laser; in the second, digital light processing (DLP), the whole layer of curing light is projected using a projector chip; and in the third generation of masked mSLA, the entire layer of curing light is projected using an LCD screen. The most recent development of SLA, known as mSLA, substitutes an LED array for a beam of light. To print, the relevant pixels are hidden by the LED array, which only lets the appropriate pixels transmit illumination. The LED array transmits light via an LCD screen. Just the uncovered portion heals as a consequence.

The materials used from pure resin to suspensions includes Oligomers, monomers, and a modest quantity of photoinitiator (PI) are all components of an easier-to-under of photopolymer (resin). When a resin is exposed to a light source, the PI is activated, creating species that are reactive, free radicals (Figure 10.2). These substances interact with oligomers and monomers to create lengthy chains that can then undergo photopolymerization. The production of sufficient reactive species for polymerization cannot be achieved only by monomers and oligomers. As a result, only a minimal amount of PI is required to start the procedure. Increasingly intricate photo-initiating mechanisms, such as co-initiators, inert dyes, photosensitizers, etc., are being created to boost PI yield.

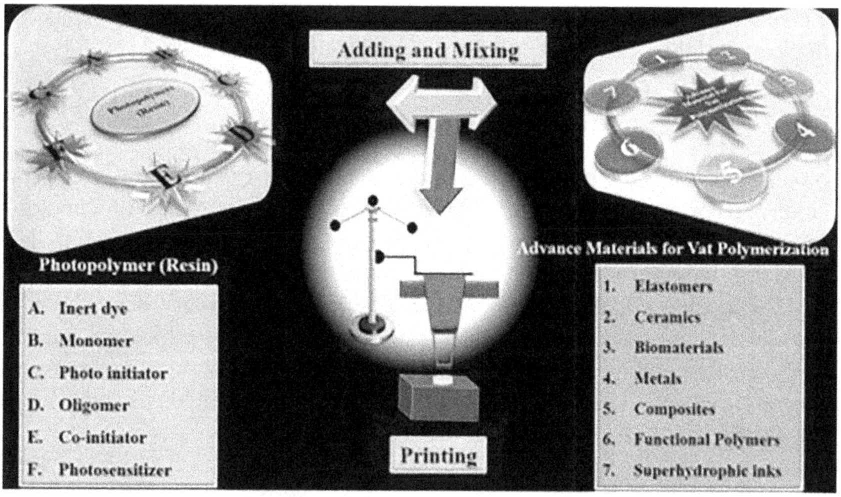

FIGURE 10.2 Representation of material used for vat polymerization method.

Additionally, the base resin composition can have fillers (such as ceramic or metal) added to create a suspension due to the excessive radiation reactivity of resins. Photopolymerized resin serves as a matrix for solid particles in these systems. The organic component is then eliminated in the debinding post-processing stage. After that, the porous solid structure is sintered at a temperature that produces a dense solid portion.

10.2.2 Powder Bed Fusion

In this method, the powdery substance is melted layer by layer with the help of a laser beam. The structure being constructed is covered in a coating of powder that has been uniformly distributed before the laser beam melts the powder to create the required design. An additional coating of powder is applied after lowering the assembled base. SLS is the kind of powder bed fusion that occurs most frequently. The powder undergoes heating beyond its melting point during sintering. It results in the particles bonding together without altering the overall structure of the item. The powder gets raised to a temperature beyond its melting point during melting, allowing it to move and assume the required shape.

10.2.3 Material Extrusion

This is a 3DP technique that produces objects using an uninterrupted line of substance. The initial step of the procedure involves feeding the material, characteristically plastic filament, through a nozzle that is heated. The component is melted by the nozzle and then dumped over the construction surface. A CAD file that specifies the direction of the extruded filament during lamination controls the coating procedure. Subsequently, once the object is finished, it is constructed layer by layer. FDM is a common name for material extrusion, which produces large components from multiple materials with simple geometry.

10.2.4 Material Jetting with Resins

Photopolymer resins are used while printing items. Comparable to how a 2D printer operates, this printing method uses a print head to discharge droplets of a photosensitive polymer substance that hardens when exposed to ultraviolet light. Print heads employed in ordinary inkjet printing are comparable to those utilized in this procedure. Casting for investments uses this kind of equipment to pour viscous liquids to produce wax-like pieces. Parts produced by material jetting are highly accurate in their dimensions and feature a smooth surface. It is feasible to print with numerous substances, and a variety of materials, such as those that resemble ABS, rubber, and entirely transparent substances, are suitable for this process.

10.3 ADVANTAGES OF POLYMER 3DP

Polymers make 3DP more adaptable, effective, and personalized. Any company can use 3DP to produce excellent parts and goods with the appropriate polymer 3DP equipment and a knowledgeable workforce. Comparing this manufacturing technique to more conventional ones, it has several benefits. These benefits involve, among others, those that are time, money, and design related.

10.3.1 Reduced Cast

By doing away with the requirement for equipment and installation, 3DP using polymers can assist firms in lowering their production expenses. Firms can reduce storage costs by simply creating components and goods as required because polymer 3DP is quick and effective. Additionally, 3DP is an additive technique. In subtractive fabrication, components are removed from a block of stuff until the intended form is obtained. Layer-by-layer construction of products with AM reduces waste and the amount of resources needed.

10.3.2 Rapid Prototyping

Employing 3D CAD, rapid prototyping (RP) is the quick creation of a physical part or component. Typically, 3DP is used to create the item. The best-quality prototype is one in which its layout resembles the planned result, compared to a low-quality prototype, which differs significantly from the final outcome. Various techniques for production are used in RP, but layered AM is particularly common. Yet high-speed manufacturing, the casting process, shaping, and extrusion are some more innovations employed in RP. The method of prototyping is sped up by the ability of 3DP to produce parts in a matter of hours. This makes it possible for each step to finish sooner.

In comparison to machining prototypes, 3DP is cheaper and more effective at producing components because the part may be done in a matter of hours. This makes it possible to make every model alteration considerably quicker.

10.3.3 Demand-Side Printing

Another positive aspect of print-on-demand is that, unlike conventional production methods, it doesn't require much space to store goods. As a result of not having a requirement to print in quantity unless necessary, this reduces both space and money. Since the 3D design files are printed employing a 3D model as either a CAD or STL file, they are all kept in an online archive where they may be found and printed as necessary. By altering documents, changes to patterns can be accomplished for relatively little money without wasting outdated stock or spending money on equipment.

10.3.4 Growing Complexity

For intricate patterns, 3DP gives improved printing characteristics. With conventional manufacturing techniques, it might be hard to produce components with complicated patterns. Resin 3DP may create objects with extremely small details and a lesser requirement for post-processing by printing at resolutions as small as 10 microns.

10.3.5 Greater Sustainability

The use of polymer 3DP lessens the requirement for manufacturing plastic parts that might be rapidly discarded and grow outdated. Additionally, it allows for on-demand printing, eliminating the requirement for companies to maintain large stocks of unwanted pieces. Additionally, 3DP can cut transportation and storage expenses, lowering a company's entire environmental impact.

10.3.6 Powerful and Light Components

While various metals may additionally be employed, plastic is the primary material for 3DP. But as plastics are thinner than their metal counterparts; they have benefits. This is crucial in sectors like automobiles and aircraft where lowering weight is a concern and may result in higher mileage. Additionally, pieces may be made from customized substances to offer qualities like heat resistance, greater durability, or being impermeable to water.

10.3.7 Adaptable Structure

More complicated structures may be created and printed using 3DP than using conventional manufacturing techniques. With the aid of 3DP, architectural constraints associated with more conventional procedures are no longer present.

10.3.8 Profitable in Terms of Price

Because 3DP is a one-step production method, it reduces the time and expenses involved with employing different machinery for production. There is also no requirement for workers to be always available when using 3DPs; they can be built up and allowed to complete the task. As already noted, this approach to production may additionally reduce money on materials because it only employs the material needed for the component itself, with little to no waste. Potentially, while purchasing 3DP equipment can be costly, you may reduce this expense by exporting your endeavor to a 3DP solution provider.

10.3.9 Simple Accessibility

As more regional businesses offer industrial service outsourcing, 3DPs are becoming increasingly affordable. Relative to more conventional production methods carried out abroad in various nations, this reduces time and doesn't necessitate high transit expenses.

10.3.10 Reducing Trash

The production of parts only requires the materials needed for the part itself, with little or no wastage as compared to alternative methods, which are cut from large chunks of non-recyclable materials. Not only does the process save on resources, but it also reduces the cost of the materials being used.

10.3.11 Environmentally Responsible

This procedure is, by nature, environmentally benign because it decreases the quantity of material waste required. When you consider things like increased mileage from employing lightweight 3DP components, environmentally friendly advantages are expanded.

10.3.12 Modern Healthcare

By printing human parts like livers, kidneys, and hearts, 3DP has been employed in the field of medicine to save lives. Most of the largest technological advancements are being made in the medical field, where novel applications and advancements continue to be created [10].

10.4 DISADVANTAGES OF POLYMER 3DP

When choosing to employ 3DP technology, it is important to weigh its disadvantages, which are like those of practically any other method.

10.4.1 Minimal Resources

Although 3DP may produce objects from a variety of plastics and metals, the range of initial components is not completely diverse. This is because not every plastic or metal is able to be heated to a temperature that enables 3DP. Additionally, barely any of these printable substances are food safe, and a lot of them cannot be reused.

10.4.2 Limitation of Build Dimensions

The quantity of objects that may be manufactured is presently restricted by the tiny print cartridges of 3DPs. Anything larger needs printing in multiple pieces that are assembled after fabrication. Because the printer requires it to produce more pieces before a laborious process is employed to assemble the components, this might raise prices and production time for bigger goods.

10.4.3 Post-Processing

Most components made from 3DP need a certain amount of tidying up to eliminate material that supports the structure and to smooth the outer layer to obtain the desired result, yet large pieces, as indicated previously, need additional processing. Water jetting, sanding, chemical soaking, washing off air or heat during the drying process, assembling, and other post-processing techniques are utilized. The size of the generated object, its primary use, and the kind of 3DP technique employed during manufacturing all affect how much post-processing is necessary. Therefore, even though 3DP enables quick part output, post-processing has the potential to slow down the manufacturing pace.

10.4.4 Sizeable Volumes

Despite more traditional methods like molding by injection, wherein producing high volumes could prove cheaper, 3DP has a fixed price. While 3DP might require a lower starting price than other means of production, once it is ramped up to mass-produce items in large quantities, the price per unit does not decline as compared with injection molding.

10.4.5 Part Structure

Components are created via 3DP layer by layer. Even while these layers stick together under specific loads or orientations, they may delaminate. While polyjet and multijet components also frequently exhibit increased brittleness, this issue is more serious when products are produced by FDM. Utilizing injection molding in some circumstances might be preferable because it produces uniform pieces that won't separate and disintegrate.

10.4.6 Jobs in Production Are Being Cut

Because printers handle most of the manufacturing automation, a further drawback of 3D technology is the possible loss of human labor. Yet the economies of many developing nations are dependent on low-skill occupations, and this innovation may jeopardize these industrial positions by eliminating the demand for output elsewhere.

10.4.7 Defects in the Design

The kind of equipment or technique utilized is a further potential problem with 3DP because certain printers have poorer tolerances, which means that the result might not match the initial model. This may be rectified in post-processing; however, it needs to be considered that this will extend the production period and expense.

10.4.8 Copyright Concerns

More individuals are going to be able to produce phony and imitation products as 3DP becomes more widespread and affordable, making it nearly impossible to tell what's real. This clearly has problems with assurance of quality and copyright.

10.5 DIFFICULTIES OF 3DP IN POLYMERS

Considering the numerous advantages of 3DP, there remain certain problems that require resolution. The following list highlights various drawbacks.

10.5.1 Quick and Reliable Exchange

It's common in conventional manufacturing and additionally in 3DP with different materials to have to choose between quality and quickness. Yet the greatest polymer 3DPs make it feasible to produce goods quickly without sacrificing the standard of the item.

10.5.2 Controlling Heat Generation

Inadvertent outcomes like bending and breaking might occur as an outcome of the cooling and heating cycles used in polymer 3DP. By employing a 3DP with proactive temperature regulation, these difficulties may be avoided.

10.6 APPLICATIONS OF 3DP

Programmers who are adaptable and accessible are now more in demand. The creation of flexible sources of energy is necessary for satisfying these demands. Fuel cells, solar cells, li-ion batteries, and supercapacitors, for example, are stiff and cannot be used in bendable electronics. As a result, considerable study is being done to develop adaptable energy sources and improve their functionality [11, 12].

Polymer electrolyte membrane fuel cells (PEMFCs), one of the many kinds of energy sources, have possibilities for adaptable uses. PEMFCs have been recognized as viable power sources for cars and other forms of mobility due to their quick responsiveness to changing loads, high energy density, propensity for low operating temperatures, and quick recharge times. Additionally, these benefits may benefit mobile gadgets. PEMFCs might serve as future sources of energy for flexible and lightweight gadgets, including smart watches, implanted medical equipment, and flexible cellphones, if they had bendable properties. [13].

Yoo et al. reported that a flexible PEMFC is realized using 3DP flexible flow-field plates. A membrane electrode assembly (MEA), current collectors, and flow-field plates make up the prepared bendable PEMFC (Figure 10.3). Both the straight and curved orientations of the fuel cell are used to gauge its efficiency. To evaluate

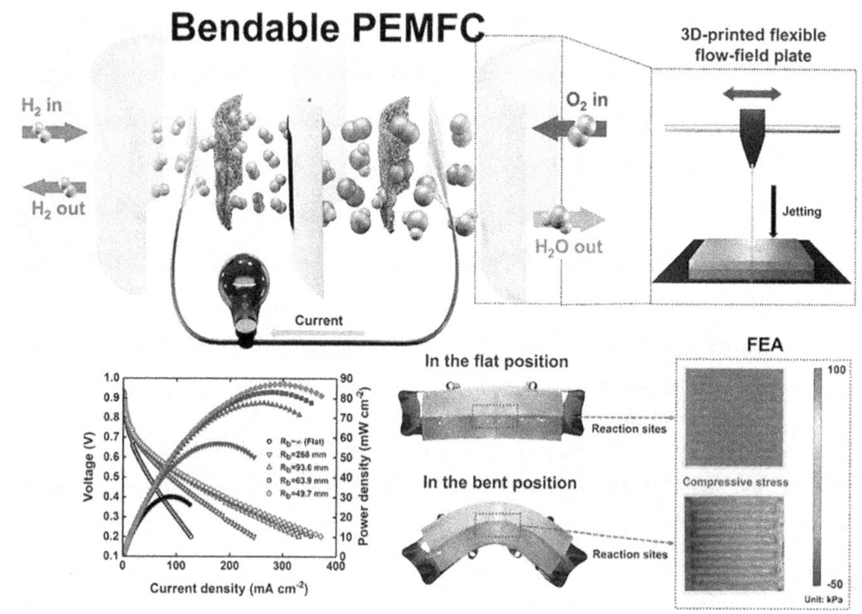

FIGURE 10.3 Graphical abstract of 3DP flexible flow-field plates for bendable polymer electrolyte membrane fuel cells. Adapted with permission [14]. Copyright 2022, Elsevier Ltd.

the effectiveness of the fuel cell, polarization curves, and impedance plots are analyzed [14].

For the fuel cell with the greatest bending, a peak power density of 87.1 mW cm^{-2} is attained, whereas a power density of 30.2 mW cm^{-2} can be achieved in the flat position. The increased efficiency is related to the compressive stress placed on the MEA's locations of reaction throughout the fuel cell's folding. The finite-element analysis (FEA) method is used to compute the compressive stress, and its results show that when fuel cell curvature rises, the compressive stress at the MEA rises, leading to much lower ohmic as well as charge transfer obstacles. The effectively created flexible fuel cell with elastic flow-field plates that were 3DP demonstrates adaptability, respectable efficiency, and a straightforward construction. Due to the utilization of 3DP, it is also affordable from a production standpoint.

Bretosh et al. reported that a femtosecond laser-based method called two-photon polymerization enables the printing of 3D objects in photocurable polymers with submicron precision [15]. For numerous uses in fields like energy, photonics, or multimodal gadgets, making the dielectric 3DP materials conductive might be quite advantageous. The microstructures in question in this work are composed of a silicon-zirconium hybrid organic-inorganic polymer that experiences little shrinking as it develops (Figure 10.4).

The deposition of a gold coating on the outermost layer of the printed microstructures is examined using an easy and effective metallization technique called electroless plating. It investigated how the technique variables affect the quality and

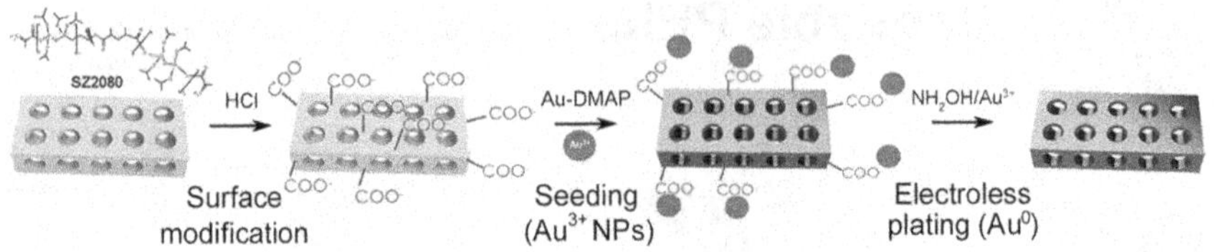

FIGURE 10.4 Graphical abstract of gold metallization of 3D-printed hybrid organic-inorganic polymer microstructures. Adapted with permission [15]. Copyright 2023, Elsevier Ltd.

characteristics of the coated layer. Among these variables, the dosage and step period for the altering agent, in addition to the seeding solution concentration, must be customized for the hybrid microstructures under consideration. The primary variable affecting the shape of the gold layers that are being coated is the amount of metal ions in the plating bath. Greater amounts especially produce sheets that are uniformly smooth and have electrical conductivities that are greater than half those of bulk gold.

Finally, it is demonstrated that the deposited layers cover 3DP microstructures of any shape, demonstrating the conformity of the approach at the micrometric scale. Supplies for fossil fuels are rapidly running out, and burning them harms the ecosystem permanently. As a result, it is now necessary to switch to other forms of energy. PEMFC is a potential technology because of its many positive traits, including excellent performance, high power density, low weight, rapid initialization, low operating temperature, little (or no) pollution output, and silent operation. To be profitable, PEMFCs must be improved even more.

Pashaki et al. reported that the PEMFC is an intriguing method for creating electricity, yet it needs to be improved to be profitable [16]. One of the primary performance-limiting issues for the PEMFC is oxygen delivery to the response center. To increase oxygen transport, many new investigations have concentrated on fuel cell architecture. The current investigation uses mathematics to examine the effects of changing a straight (or planar) PEMFC into an arc-shaped one. The ability of PEMFCs to direct gas flow into the gas diffusion layer (GDL) and apply a centrifugal pull on the conduit, hence improving oxygen movement, gives rise to the concept of PEMFC bending. According to the findings, PEMFC bending can improve efficiency under the tested operating circumstances by roughly 8.33%. Additionally, it has been found that the PEMFC bending influence typically rises with working pressure, stoichiometry ratio, and bending angle however falls with functioning voltage.

Bas et al., reported that ultra-pure hydrogen is created using microbial electrolysis cells from organic waste [17]. Efficiency can be entirely affected by electrodes, making them crucial parts of microbial electrolysis cells. Additionally, these electrodes are expensive and could impair electrolysis efficiency by causing chemical interactions with organic trash (Figure 10.5). According to Pesode et al., utilizing cutting-edge

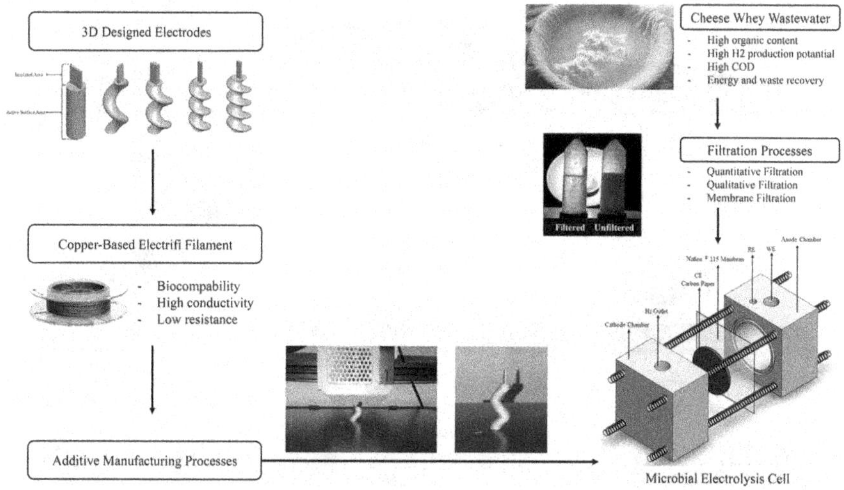

FIGURE 10.5 Graphical abstract of 3D-printed anode electrodes for microbial electrolysis cells. Adapted with permission [17]. Copyright 2022, Elsevier Ltd.

methods helps with biomaterials scientific and technological development as well as the creation of novel sustainable biomaterials [18]. Ecologically sound biomaterials ought to be encouraged. A wide range of biomaterials were developed and created as potential replacements for traditional supplies, and they are being used effectively in a number of biomedical fields (Figure 10.6). The major objective of this chapter is to discuss the sustainability element in the AM of biomaterials. Different metallic biomaterials, such as titanium, stainless steel, magnesium, cobalt-chromium alloy, zinc, and tantalum, are discussed. The implications of various AM approaches on sustainability are studied. Additionally, the characteristics of AM biomaterials and their sustainability are covered in depth.

Bas et al., used a two-chambered microbial electrolysis cell with new, variously shaped anode electrodes that were 3DP as well as cheese whey wastewater as the electrolyte. Employing cupper-based Electrifi filament, electrodes that have been 3D engineered and printed in various geometries (rod, 1-cycled spiral, 2-cycled spiral, 3-cycled spiral, and 4-cycled spiral) are employed to increase mass transfer within the cell. This leads to the observation that the efficacy of microbial electrolysis and the generation of hydrogen are significantly affected by the organic component of waste and electrode shape. In the linear sweep voltammetry examination of the electrochemical analysis, 1-cycled spiral geometry has a current density that is up to 2.6 times higher. Additionally, the 1-cycled spiral shape is five times faster than other electrodes in hydrogen production measurements. It has been found that spiral-shaped electrodes perform better in terms of charge transfer efficiency and the contact zone across the electrode and electrolyte barrier.

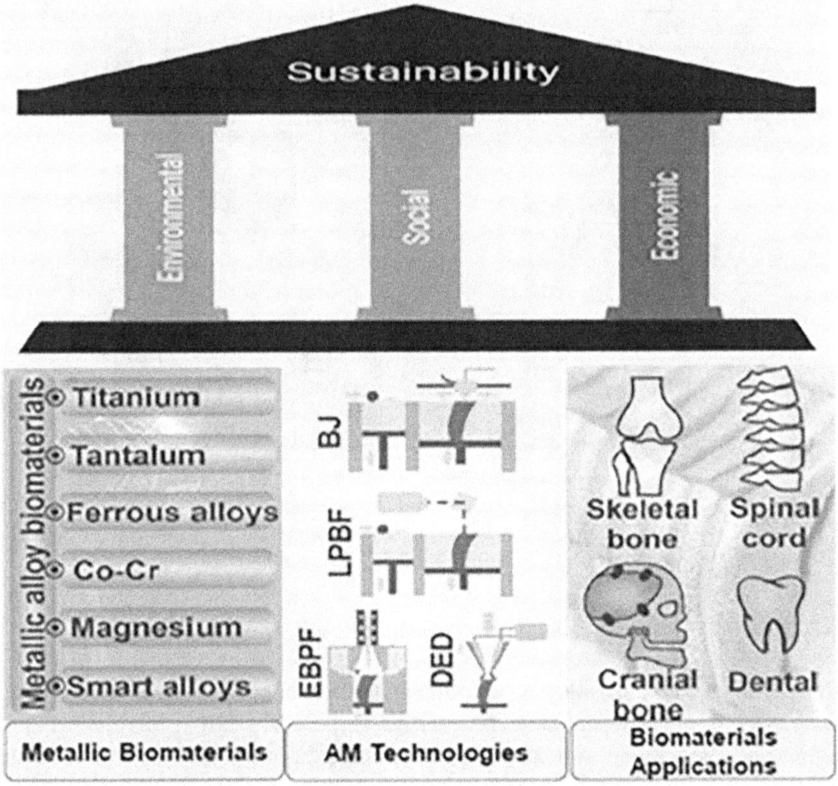

FIGURE 10.6 Graphical abstract of additive manufacturing of metallic biomaterials: sustainability aspect. Adapted with permission [18]. Copyright 2023, Elsevier Ltd.

10.7 FUTURE PERSPECTIVES

Additional expansion will be aided by ongoing developments in 3DP elements and AM technologies, particularly new metal alloys. Watch for new 3DP uses in the aerospace, gadgets, healthcare, power, and transportation sectors in the years ahead. As the number of uses for the method rises and metal additives become more profitable, worldwide demand for AM is anticipated to develop at a rate of 17% year over year through 2023, and from 2020 to 2026, the demand for goods and services related to AM is anticipated to nearly triple. As 3DP has advanced, it has assisted healthcare providers in filling requirements for items like customized prosthetics, immediate diagnostic specimens, and bioengineered organs.

Production, construction, and creative industries are all experiencing the rewards of 3DP, yet the electronics, aerospace, and healthcare sectors are predicted to experience some of the fastest development. Tailored radiators for expensive goods will become a reality in the electronics sector, and 3DP components will become more widely available in the aerospace sector and move from high-end military use to general-purpose aircraft. Personalized treatment will become standard in healthcare as additional substances are tested for therapeutic purposes and financial institutions more widely accept AM.

There is a lot more technology in the works for AM besides what is already available. The direct-ink-writing and thick paste substance extrusion groups are where the intriguing research is currently being performed. For instance, scientists are looking into combining cured photopolymers with sophisticated buildings including thermosets and pottery that may one day be employed for items such as printed circuit boards, less expensive heat transfer systems, and nonbulk porcelain.

When the method is combined with other AM techniques, such as printed circuits built into prosthetics or novel battery dimensions, even more possibilities become apparent. The range of substances that can be printed in 3D is also expanding. Refractory superalloys will support advancements in the aerospace, security, and energy industries. Presently, stronger polymers are anticipated to pass the FDM's needed tests for flame, smoke, and toxicity, which will lower the price of maintenance for corresponding industries.

The development of automated businesses will be accelerated by the incorporation of AM with Industry 4.0 technologies like IoT, AI, and robotics. In order to provide realistic digital prototyping encounters, 3DP, virtual reality (VR), and augmented reality (AR) approaches will be combined. This will enable the development and testing of ideas connected to prototypes before the physical prototypes are produced. The creation of fresh programming has improved 3DP's usability and accessibility, enabling simpler creation of components and assembly. The goal for resources in 3DP is to be more diverse, effective, and sustainable by 2023.

10.8 CONCLUSION

In conclusion, in recent years, 3DP techniques have grown rapidly. This expansion has piqued the curiosity of investigators throughout the world, including those working in fields other than engineering, like materials science, polymer science, and photochemistry. Using AM technology, it is possible to make intricate components or shapes that are challenging to create with conventional manufacturing techniques without the use of a mold. AM, a developing technology, has lately opened the door for the creation of innovative shapes for commercial uses and offers prospective advantages in the electrode production sector.

The AM technologies minimize the amount of recyclables utilized throughout the process of producing the item as well as the amount of energy needed to complete the output. Additionally, the AM process is widely acknowledged to be among the modern approaches to the manufacture of innovative electrodes for use in energy transformation, energy preservation, and electrochemical devices [19]. Because of the cost and intricate nature of the geometric frameworks, it is challenging to manufacture flow pathways used in energy applications like electrodes and bipolar plates using traditional machining techniques. Because it allows for greater architectural liberty, resource funds, and convenience in producing complicated constructions, the AM approach has grown in popularity. With the rapid development of AM technology, a wide range of materials, including metals, ceramics, resins, and esters, can be 3DP using various techniques.

Since the AM technique is still being developed, there aren't numerous investigations on the usefulness of various substances in electrochemical experiments.

Hence, there is still room for advancement and study in the areas of electrochemical energy transfer uses and for broadening the range of substances available for 3DP electrochemical gadget parts. Additionally, AM makes it possible to employ a broad spectrum of printable substances, which will create new possibilities for 3DP methods' architecture and usage fields.

Upcoming electrochemical transformations in various shapes of geometry will be facilitated by the 3DP manufacturing of intricate geometry and electrodes for electrochemical uses employing the AM process. A very significant outcome is that such geometric forms can be created into adaptable devices that work with any animal body in addition to the human body. Adaptable biosensors have made it possible for AM to be used to manufacture robots that interact with humans by offering adaptable frameworks for interaction. They might also be employed in the creation of biocompatible wearable battery devices or methods that make the engineering of fuel cell cars easier.

A high-performance 3DP CP ink based on PEDOT:PSS is able to produce extremely conductive microscale architectures both in the dry and wet states with speed and flexibility. The CP ink demonstrates outstanding 3DP and is readily integrated with other 3DP components in modern multi-material 3DP techniques. This allows us to show how easy, quick, and greatly simplified the 3DP-based manufacture of the soft neural probe and the high-density flexible electrical circuit are. The present study gives a viable manufacturing technique for flexible electronic devices, wearable technology, and bioelectronics based on CPs in addition to addressing the difficulties currently faced in the 3DP of CPs [20].

The prospects for fuel cells appear promising. The fuel cell environment will alter for several reasons in the upcoming years. Fuel cell stack prices will drop, performance is going to improve thanks to technology, and adoption will rise dramatically. There are going to be significant improvements in the integration of green power networks, hydrogen production creativity, and more federal funding. Fuel cells will be the main source of clean energy around the globe in ten years. As a result, this approach will be used more widely in the years to come, especially at the research and development stage and in commercial uses [21–23].

REFERENCES

[1] H. Yuk, B. Lu, X. Zhao. Hydrogel bioelectronics. Chemical Society Review. **2019**, 48, 1642.

[2] V. R. Feig, H. Tran, M. Lee, K. Liu, Z. Huang, L. Beker, D. G. Mackanic, Z. Bao. An electrochemical gelation method for patterning conductive PEDOT: PSS hydrogels. Advance Materials. **2019**, 31, 1902869.

[3] X. Liu, H. Yuk, S. Lin, G. A. Parada, T. C. Tang, E. Tham, C. F. Nunez, T. K. Lu, X. Zhao. 3D printing of living responsive materials and devices. Advance Material. **2018**, 30, 1704821.

[4] H. Yuk, B. Lu, S. Lin, K. Qu, J. Xu, J. Luo, X. Zhao. 3D printing of conducting polymers. Nature Communications. **2020**, 11, 1604.

[5] T. H. Chung, B. R. Dhar. A mini-review on applications of 3D printing for microbial electrochemical technologies. Frontier Energy Research. **2021**, 9, 679061.

[6] Y. He J. Li, L. Zhang, X. Zhu, Y. Pang, Q. Fu, Q. Liao. 3D-printed GA/PPy aerogel biocathode enables efficient methane production in microbial electrosynthesis. Chemical Engineering Journal. **2023,** 459, 141523.

[7] A. Pokprasert, S. Chirachanchai. Tailoring proton transfer species on the membrane surface: An approach to enhance proton conductivity for polymer electrolyte membrane fuel cell. Polymer. **2023**, 265, 125583.
[8] S. Qian, H. Liu, Y. Wang, D. Mei. Structural optimization of 3D printed SiC scaffold with gradient pore size distribution as catalyst support for methanol steam reforming. Fuel. **2023,** 341, 127612.
[9] Garg, R. Rani, V. K. Meena, S. Singh. Significance of 3D printing for a sustainable environment. Materials Today Sustainability. **2023**, 23, 100419.
[10] M. Elbadawi, A. W. Basit, S. Gaisford. Energy consumption and carbon footprint of 3D printing in pharmaceutical manufacture. International Journal of Pharmaceutics. **2023,** 639, 122926.
[11] H. Porthault, C. Calberg, J. Amiran, S. Martin, C. Páez, N. Job, B. Heinrichs, D. Liquet, R. Salot. Development of a thin flexible Li battery design with a new gel polymer electrolyte operating at room temperature. Journal of Power Sources. **2021**, 482, 229055.
[12] S. Luo, Y. Wang, T. C. Kong, W. Pan, X. Zhao, D. Y. C. Leung. Flexible direct format paper fuel cells with high performance and great durability. Journal of Power Sources. **2021**, 490, 229526.
[13] M. Cannio, S. Righi, P. E. Santangelo, M. Romagnoli, R. Pedicini, A. Carbone, I. Gatto. Smart catalyst deposition by 3D printing for Polymer Electrolyte Membrane Fuel Cell manufacturing. Renewable Energy. **2021,** 163, 414.
[14] H. Yoo, O. Kwon, J. Kim, H. Cha, H. Kim, H. Choi, S. Jeong, Y. J. Lee, B. Kim, G. E. Jang, J.-S. Koh, G. Y. Cho, T. Park. 3D-printed flexible flow-field plates for bendable polymer electrolyte membrane fuel cells. Journal of Power Sources. **2022**, 532, 231273.
[15] K. Bretosh, S. Hallais, C. C. Cesar, G. Zucchi, L. Bodelot. Gold metallization of hybrid organic-inorganic polymer microstructures 3D printed by two-photon polymerization. Surfaces and Interfaces. **2023,** 39, 102895.
[16] M. K. Pashaki, J. Mahmoudimehr. Performance superiority of an arc-shaped polymer electrolyte membrane fuel cell over a straight one. International Journal of Hydrogen Energy. **2023**, 48, 13633.
[17] F. Baş, M. F. Kaya. 3D printed anode electrodes for microbial electrolysis cells. Fuel. **2022**, 317, 123560.
[18] P. Pesode, S. Barve. Additive manufacturing of metallic biomaterials: Sustainability aspect, opportunity, and challenges. Journal of Industrial and Production Engineering. **2023,** 40, 464–505. https://doi.org/10.1080/21681015.2023.2229341.
[19] B. Huner, M. Kıstı, S. Uysal, L. N. Uzgoren, E. Ozdogan, Y O. Suzen, N. Demir, M. F. Kaya. An overview of various additive manufacturing technologies and materials for electrochemical energy conversion applications. ACS Omega. **2022**, 7, 45, 40638.
[20] S. Ghaderi, H. Hosseini, S. A. Haddadi, M. Kamkar, M. Arjmand. 3D printing of solvent-treated PEDOT:PSS inks for electromagnetic interference shielding. Journal of Material Chemistry. A, **2023**, 11, 16027–16038. https://doi.org/10.1039/D3TA01021J.
[21] H. Ye, Y. Ma, C. Wang, J. Liu, Y. Liu, Y. Liu, X. Xu, Z. Chen, X. Zhao, T. Tao, Y. Yao, S. Lu, H. Yang, B. Liang. A 3D printed redox-stable interconnector for bamboo-like tubular solid oxide fuel cells. International Journal of Hydrogen Energy. **2023,** 48, 34979–34986. https://doi.org/10.1016/j.ijhydene.2023.05.324.
[22] H. Ashassi-Sorkhabi, A. Kazempour, S. Moradi-Alavian, E. Asghari, J. J. Lamb. 3D nanostructured nickel film supported to a conducting polymer as an electrocatalyst with exceptional properties for hydrogen evolution reaction. International Journal of Hydrogen Energy. **2023**, 48, 29865–29876. https://doi.org/10.1016/j.ijhydene.2023.04.139.
[23] N. A. M. Barakat, S. Gamal, M. M. A. Hameed, O. A. Fadali, O. H. Abdelraheem, R. A. Hefny, H. M. Moustafa. Graphitized corncob 3D biomass–driven anode for high performance batch and continuous modes air–cathode microbial fuel cells working by domestic wastewater. International Journal of Hydrogen Energy. **2023**, 48, 38854–38869. https://doi.org/10.1016/j.ijhydene.2023.06.231.

11 3D-Printed Conducting Polymers for Microbial Fuel Cells

Mehmet E. Pasaoglu, Vahid Vatanpour, and Ismail Koyuncu

11.1 INTRODUCTION

The continuing global energy crisis, caused by rising energy needs and depleting fossil fuel supplies, is a source of concern, as are the environmental and health effects [1, 2]. Primary energy sources include fossil fuels, nuclear energy, and renewable energy [3], although there is rising interest in sustainable alternatives [4, 5]. Along with the energy problem, urbanization, growing populations, and industry exacerbate environmental degradation, needing simultaneous action on both fronts [6]. Microbial fuel cells (MFCs) have the potential to be bioelectrochemical devices that transform the chemical energy of organic matter into electricity via the metabolic activities of electroactive microorganisms. MFC technology is still in its early phases of research, despite having promising applications in biosensors, wastewater treatment, and remote power supply [6].

Conducting polymers are possible materials to be applied in various uses, including bioelectronics, flexible electronics, energy storage, and fuel cells [7, 8]. Nevertheless, for the preparation of conducting polymers, mostly conventional techniques such as screen printing, electron-beam lithography, and ink-jet printing have been used, whose restrictions have hindered quick inventions and extensive applications for conducting polymers [9]. The 3D approach allows for the simple building of conducting polymers into high aspect ratio and high resolution microstructures, which may be combined with other materials such as insulating elastomers using multi-material 3D printing. In addition, 3D-printed conducting polymers can be renewed into greatly conductive and soft hydrogel microstructures [10].

The utilization of conductive polymers can enhance the effectiveness of MFCs. Conductive polymers are well suited for enzyme immobilization and can facilitate charge transfer. Furthermore, biocompatibility is an important issue during the development of implantable MFCs, and the biocompatibility-related aspects of conducting polymers with microorganisms should be considered [11].

11.2 PAST AND PRESENT OF MFCs

Potter [12] recreated a century-old precedent by revealing the electrical effects of microbial decomposition of organic molecules. Despite this breakthrough, practical

 DOI: 10.1201/9781003415985-11

applications were restricted in the following decades. Schröder [13] analyzes the history of microbial electrochemical systems, highlighting major advances and explaining the technology's early lack of interest. The field's rebirth in the 21st century is discussed, as well as its current relevance and prospective future possibilities [14].

In-depth analyses are offered by Santoro et al. [15] and Logan et al. [16] for the fundamental concepts underlying microbial electrochemical systems. These technologies show promise for the treatment of wastewater and the production of green energy. Research has focused on some key areas to make these technologies practical: lowering electrochemical losses and operational costs, improving efficiency, and customizing systems for specific applications (Figure 11.1) [14].

Research on the Clarivate Analytics Web of Science (WOS) platform using the keyword 'microbial fuel cells' found a consistent rise in the number of research papers about MFCs published in credible, peer-reviewed scientific journals between 2004 and 2020 (Figure 11.2). However, it is crucial to note that this depiction just

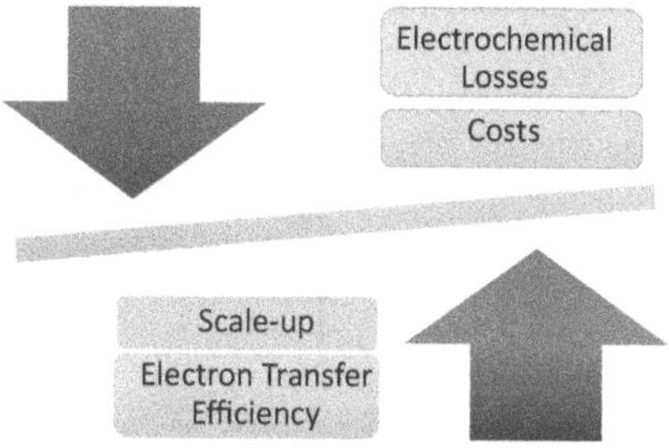

FIGURE 11.1 Four different focus areas of MFC research [14].

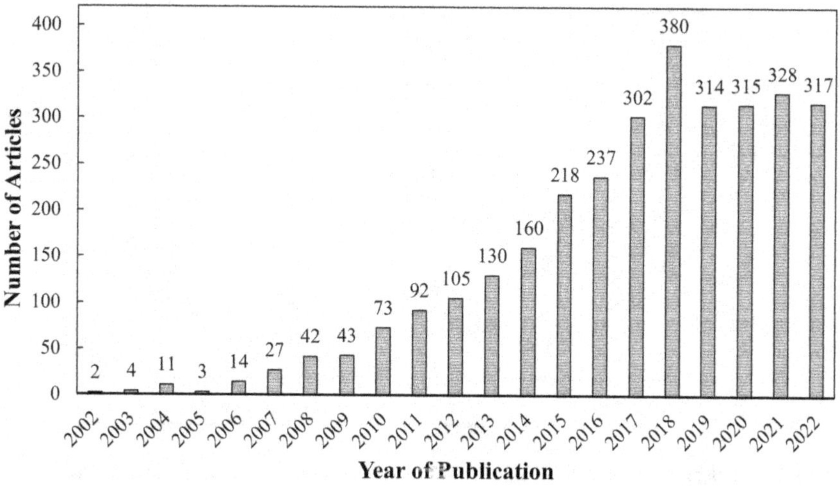

FIGURE 11.2 Year base publication number from Web of Science portal for the keyword 'microbial fuel cell' (2002–2022).

emphasizes the rising trend; if the same search were run on multiple search engines, the results would surely change due to differences in indexed webpages and databases required by different algorithms. Approximately 748 review papers and book chapters have been acquired during a 20-year period by searching the Web of Science for the phrase 'microbial fuel cell'. These review articles and book chapters have been classified as wastewater treatment and recovery, bioelectricity generation, electricity generation in conjunction with wastewater treatment, electrode materials, membranes, microorganisms, design and configuration, biohydrogen production, biosensor, MFC types, and other combinations based on their scientific novelty. However, review papers and book chapters combining three-dimensional (3D) printing of electrodes and membranes for MFC, which are the system's major parts, are still limited.

The anode and cathode chambers of most MFCs, as shown in Figure 11.3, are separated by a proton exchange membrane (PEM) [17]. The anode's active biocatalyst oxidizes organic substrates, generating electrons and protons [18], while the PEM delivers protons to the cathode chamber and external circuits transfer electrons

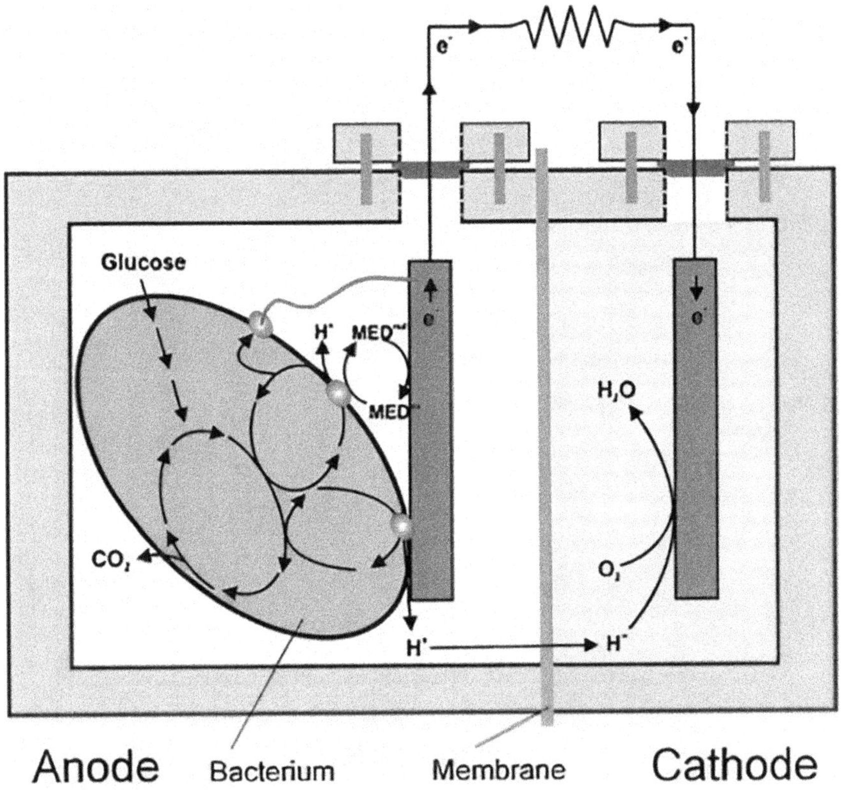

FIGURE 11.3 Anode and cathode compartments of the MFC system. Adapted with permission [16]. Copyright (2006), American Chemical Society.

[19]. In the cathode chamber, protons and electrons combine to reduce oxygen to water [20]. The anode's bio-catalyst oxidizes carbon sources, creating electrons and protons. Figure 11.3 shows the schematic of an MFC system with the reactions in the anode and cathode.

The anodic reaction of acetic acid is depicted in Eq. (1). The anode reaction requires anaerobic conditions because oxygen in the anode chamber limits power output. In a MFC, a membrane between different chambers secures the biocatalyst from reacting with oxygen. This membrane transfers charge between the anode chamber, which contains the bacteria, and the cathode chamber, where electrons interact with oxygen [21]. MFCs are divided into two groups according to the means by which they transmit electrons: those with mediators and those without, with the former requiring the use of exogenous chemicals for electron transport [22].

$$C_2H_4O_2 + 2H_2O \rightarrow 2CO_2 + 8e^- + 8H^+ \tag{1}$$

$$2O_2 + 8H^+ + 8e\text{-} \rightarrow 4H_2O \tag{2}$$

MFCs can be described by a number of electrochemical and biological parameters, including power density and substrate loading rate [23]. However, the performance of MFCs hinges on key elements, including oxygen levels in the cathode chamber, anode chamber substrate oxidation, electron transfer from the anode compartment to the surface, and PEM permeability [20].

In MFCs, the conductivity of the electrode material affects the rate of electron transfer. A more conductive material will permit quicker electron transport, potentially enhancing the overall efficiency of MFCs. The combination of great conductivity and a large surface area in the electrode can considerably improve the performance of MFCs [24]. To promote the power produced by MFCs, large-surface-area anode materials with low cost and great conductivity should be established. In addition, a highly effective and cost-effective catalyst must be developed to enhance cathode electrode efficiency [25].

MFC technology has advanced significantly in recent years; however, practical applications face problems like turbulence and membrane resistance [26]. The creation of power in MFCs is hampered by two major issues: the close relationship between power production and substrate concentrations, which becomes limiting beyond a certain value, and high internal resistance, which consumes a large percentage of power output [20]. The PEM between anode and cathode chambers is principally responsible for this internal resistance. Biocathodes were analyzed to improve O_2 oxidation on the cathode [26]. Novel MFC designs, such as single-chamber MFC, stacked, and upflow MFC, attempt to increase power generation by reducing internal resistance by removing PEM [20].

11.2.1 Effect of Anodes in MFCs

Microbes have a crucial duty in the anode chamber, where they generate electrons. When these electrons move through an external circuit, they are used to decrease electron acceptors in the cathode. Similarly, for the circuit to be complete, protons need to pass through the anode to the cathode by the PEM. As a consequence, this technique creates both electrical power and organic waste disposal at the same time [27].

As previously stated, the anaerobic anode compartment is an important component of MFCs. This section provides all of the criteria required for biomass decomposition. The chamber comprises a substrate, microorganisms, and the electrode of the anode, which serves as an electron acceptor, with the optional addition of a mediator. Eq. (3) depicts the entire reaction that occurs within the anode chamber.

Active microorganism

$$\text{Biodegradable organics} \rightarrow CO_2 + H^+ + e\text{-} \tag{3}$$

Anaerobic environment

It is critical to emphasize that catalysts are necessary to reduce the activation energy required for anodic reactions. These catalysts are often microorganisms found in the anode chamber. As is commonly acknowledged, various factors impact MFC efficiency, including electrode material and equipment configuration. As a result, adjusting these elements might considerably improve MFC performance [2].

Anodic microbial electron transfer is one of the most effective elements that determines the performance of an MFC, boosting the rate of microbial electron transfer by numerous approaches, including modifying electrode and cell design besides adding electron mediators [28]. It needs to be mentioned in this respect that electrodes are crucial components of an MFC that contribute significantly to its efficacy.

Furthermore, optimal electrode materials should have the following characteristics: (i) high electrical conductivity and low resistance, (ii) high biocompatibility, (iii) chemical stability and anti-corrosion, (iv) a wide surface area, and (v) enough mechanical strength and toughness [16].

Carbon compounds are commonly used in anode materials such as graphite fiber brushes, carbon cloth, graphite rods, carbon paper, reticulated vitreous carbon, and carbon felt. These materials were chosen for their microbial stability, good electrical conductivity, and large surface area, all of which improve their performance in MFCs [16]. Materials such as graphite granules (GGs) and granular activated carbon are also used. These have a high amount of microporosity and catalytic activity. Notably, GGs are a low-cost solution with high conductivity [29].

11.2.2 Effect of Cathodes in MFCs

The proton exchange membrane transports protons generated in the anode chamber to the cathode, closing the circuit's electrical function. Meanwhile, the electrons created at the anode by Eq. (5) go to the cathode chamber and mix with oxygen.

The reactive oxygen species and positive ions produced in the anode area then engage in a subsequent process that results in the production of water. Water is subsequently carried to the cathode via the ion-permeable membrane, aided by catalysts, as shown:

$$H_2 \rightarrow 2H^+ + 2e^- \quad (4)$$

$$O_2 + 4H^+ + 4e^- \rightarrow 2H_2O \quad (5)$$

The process generates a constant current via a wire connecting the anode and cathode [2]. Factors like oxidant concentration, proton availability, catalyst performance, electrode structure, and catalytic ability all have an impact on cathode reaction yield [30, 31]. Both anodic and cathodic processes need catalysis. The presence of a suitable catalyst reduces activation energy while increasing the reaction rate. Because of its accessibility, strong oxidation potential, and ecologically favorable end product (water), oxygen is widely utilized as an electron acceptor at the cathode [32].

In order to accelerate the process of oxygen's sluggish reduction kinetics on bare graphite, an appropriate catalyst is often required, despite the fact that this method results in high potential levels. This constraint is problematic in MFCs [33]. Potassium ferricyanide ($K_3[Fe(CN)_6]$) has been proposed as a solution to this problem [34].

The disadvantage is that oxygen does not efficiently oxidize $K_3[Fe(CN)_6]$, necessitating repeated refilling. Furthermore, the presence of $K_3[Fe(CN)_6]$ impacts the anaerobic conditions in the anodic chamber since it enters via the PEM. Despite this, ferricyanide has been shown to be beneficial due to its low overpotential on simple carbon electrodes [35, 36].

Platinum is often employed as an abiotic catalyst in cathodic processes, but its sensitivity to specific chemicals in the substrate solution prevents it from being utilized in MFCs [37]. To improve MFC performance, researchers frequently implement alternate oxidants or artificial electron redox mediators, such as potassium permanganate, into the cathode compartment [38, 39]. Low concentrations of potassium permanganate have been demonstrated in studies to greatly boost current, power, and voltage in MFCs [21]. The cathode is subjected to dissolution within the compartment as well as exposure to external air. While the inclusion of a cobalt (Co) catalyst on the air-facing side of the cathode might boost MFC performance similarly to a platinum cathode, wet-proofing may limit proton availability [40]. By adjusting the air pressure on the cathode, one may boost power production by controlling oxygen transition from the membrane to the anode [41].

Researchers have investigated the use of biocathodes to meet the demand for oxygen oxidation catalysis on the cathode [42]. Microorganisms are used to catalyze cathodic reactions in these biocathodes, resulting in enhanced power production in MFCs [43]. This method provides an alternative to the use of artificial mediators or catalysts. In biocathodes, electron acceptors such as nitrate, sulfate, tetrachloroethene, fumarate, perchlorate, trichloroethene, CO_2, H^+, Fe(III), Cr(VI), U(VI), and Mn(IV) have been used to improve cathode performance without the need of external agents [21, 26, 42, 44].

Sonawane et al. reported the fabrication of a cathode catalytic electrode based on the direct synthesis of a polyaniline-copper composite on carbon paper (CP/PANICu). The fabricating procedure of CP/PANI-Cu was based on galvanostatic polymerization, initiating with aniline polymerization to create PANI and next depositing Cu on carbon paper to attain CP/PANI-Cu. The CP/PANI-Cu electrode obtained a power density of 1010 mW/m^2 in MFC systems [45].

Cathode performance is a significant restriction in MFCs [26]. Surprisingly, cathode surface area has minimal impact on power generation, and increasing cathode efficiency may be accomplished by utilizing large-surface-area or granular materials like graphite [46]. On the contrary, finding materials that improve power output and columbic efficiency while keeping prices low is a big difficulty in MFC setup. Carbon paper; carbon felt; carbon brush; carbon fiber; various forms of graphite; and metals such as Pt, Cu, Cu-Au, and tungsten carbide have all been used in cathodes. Alternative polymer binders, such as perfluorosulfonic acid (Nafion), have also been investigated [26, 47].

11.2.3 Effect of Membranes in MFCs

The ion exchange membrane in MFCs electrically separates the cathode and anode electrodes while enabling the proton transport required for the redox reaction that creates the potential of cells. Additionally, the exchange membrane maximizes the coulombic efficacy of the MFC by avoiding oxygen diffusion to the anode electrode [48]. The following factors are considered vital for the well-organized functioning of a MFC when designing its membrane: permeability, ion conductivity, internal resistance, chemical, mechanical, and thermal stabilities, water uptake, biofouling, and cost [49]. A general, commercially accessible membrane that is extensively applied is a sulfonated tetrafluoroethylene copolymer Nafion membrane. Its fluorocarbon hydrophobic backbone ($-CF_2-CF_2-$) and the linked sulphonate hydrophilic group (SO_3^-) confer outstanding proton conductivity. Most MFC technology is based on the use of expensive commercial membranes such as cation exchange membranes (CEMs) and Nafion (Dupont), which are designed for chemical fuel cells [50]. Some other commercially available membranes are CMI-7000, Zifron, Hyflon, Ultrex, and AMI-7000 [51]. The two-dimensional forms of these membranes offer low geometric versatility in the MFC architecture design. It is shown that for a constant anode chamber volume and electrode size, the power density of MFCs balances with the ion exchange surface area [52]. The increased specific surface area of the ion exchange facilitates larger ionic conductivity and lowers internal resistance, causing a greater power density in MFC [53]. The membrane areas smaller than those of the electrodes meaningfully increase MFC internal resistance [54].

Accordingly, 3D-printed and cast, ionically conductive, monolithic structures can present an expected solution to the challenges of enhancing the specific area of ion exchange membranes and decreasing the total dimensions of the MFCs [55]. New ion exchange membrane materials with fast fabrication processes such as 3D printing meaningfully enlarge the range of probable MFC geometries and increase the ease of manufacture and assembly of multiple units.

11.3 MATERIALS USED IN THE 3D PRINTING METHOD

11.3.1 Polymer-Based Materials

Polymer materials are commonly used in additive manufacturing because they are simple to manufacture and less expensive than conventional construction materials. Figure 11.4 depicts the overall pattern of polymer consumption for the AM technique in 2014. Plastics dominate the industry, accounting for 99% of utilization, and they play a role in the creation of structural engineered components similar to metals.

Polymers in additive manufacturing have a greater range of applications than metals, including energy, sustainability, health, and biology. The fused deposition modeling (FDM) approach is widely utilized and dominates the industry. Various polymeric materials are available for AM (Figure 11.5), with printability and mechanical properties varying depending on the process used.

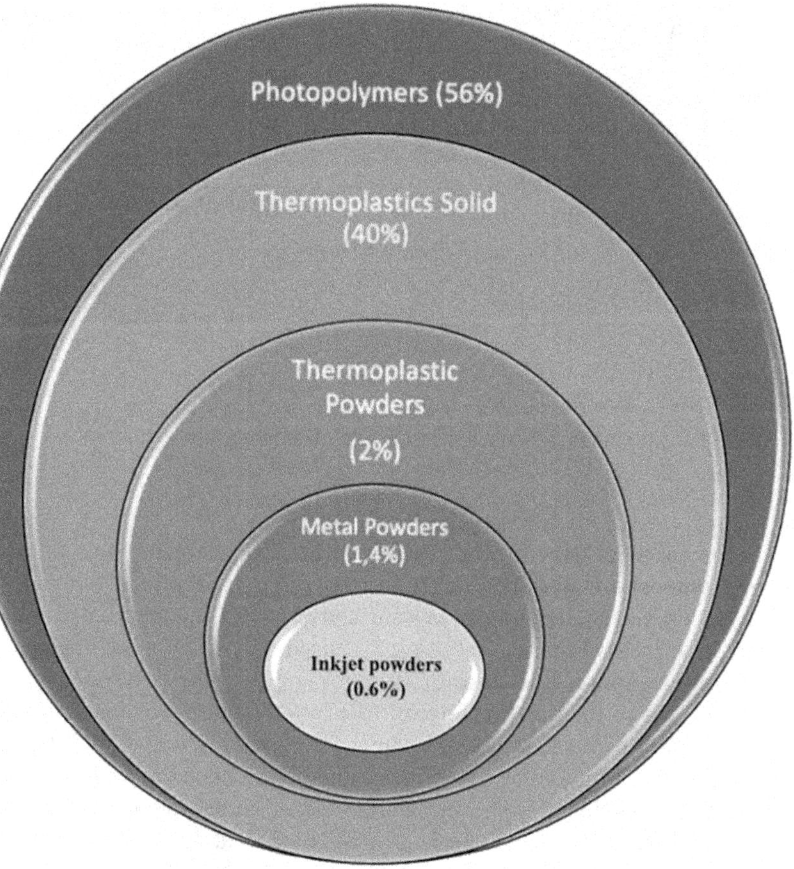

FIGURE 11.4 Materials for 3D printing related to the weight-based use of material in 2014 [56].

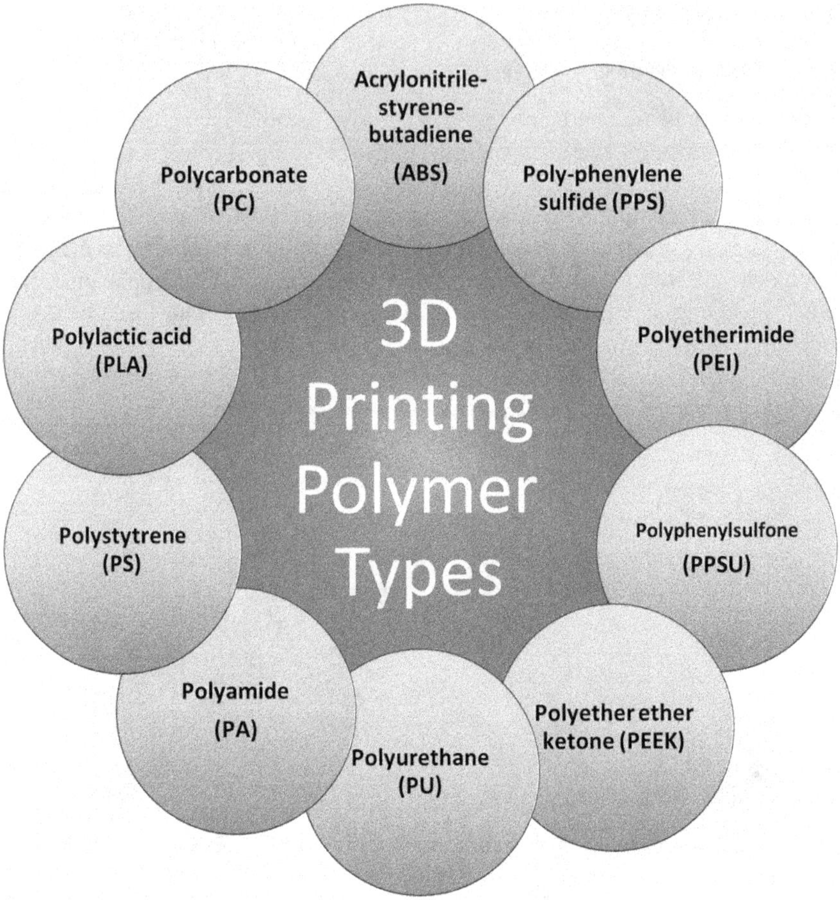

FIGURE 11.5 Commonly used polymer types in 3D printing applications [57].

The potential of AM to improve the structural characteristics of composite/nanocomposite materials as well as to provide conventional polymers with additional properties, including thermal and electrical conductivity, is the reason for the increased interest in AM [58]. Fused deposition modeling stands out among AM techniques because of its affordability. The widespread usage of FDM machines in the manufacture of 3D-printed objects was made possible by the 2009 expiry of Stratasys' FDM patent. The growing use of AM may hasten the development of manufacturing techniques using smart materials, nanocomposites, and biomaterials. The thermoplastics of choice for energy conversion applications are filaments made from polylactic acid (PLA) and acrylonitrile butadiene styrene (ABS) [57].

11.3.2 3D Printing of Conductive Polymer-Based Materials for Electrochemical Applications

The use of various conductive materials can improve the electrical conductivity of PLA and ABS polymer thermoplastics. Metal, carbon, and polymer composites are frequently used as conductivity sources in 3D printing. Composite materials gain conductive properties by including diverse conductive carbon materials such as graphene, carbon black, nanofibers, and carbon nanotubes [59]. A good example of this is the Conductive Graphene PLA Filament that Black Magic 3D (BM) sells, which is made by mixing graphene with PLA filament [60].

A PLA filament with carbon added is available from Proto-Pasta under the name Carbon Black Conductive PLA Filament. Manufacturing touch-sensitive pens, low-voltage circuit uses, and touch-sensor placements are all possible with the help of this conductive PLA filament [61].

Using both metal and polymer-based materials, 3D printing has enabled the manufacturing of electrodes for electrochemical energy conversion operations. This method provides a low-cost solution to material manufacturing in a variety of applications. For example, Baş et al. created 3D-printed anode electrodes for microbial electrolysis cells utilizing conductive PLA filament (copper-based electrifi filament). To improve mass transfer within the cell, they created and printed electrodes with various geometries (rod, 1-cycled spiral, 2-cycled spiral, 3-cycled spiral, and 4-cycled spiral). In a two-chamber microbial electrolysis cell with irregularly shaped 3D-printed electrodes, the researchers conducted electrochemical tests using cheese whey wastewater as an electrolyte. Their findings suggested that the electrode shape and waste's organic content both contributed to the efficiency of microbial electrolysis and the creation of hydrogen [57, 62].

Another study was completed by You et al. (2017) about the applicability of widely available 3D-printed polymer materials as the membrane and anode in MFCs. Gel-Lay performed the best of the tested membrane materials, with a maximum power production of nearly 240 W, which was 1.4 times more than the highest power output of the control cation exchange membrane (CEM), which was around 177 W. During fed-batch cycles, Gel-Lay also displayed greater peak power levels (133.8–184.6 W) than the control (133.4–160.5 W) [63].

A unique method was used in this work to manufacture hierarchical porous carbon foams (HPCFs) as 3D self-supporting anodes for MFCs. These HPCFs were created by pyrolyzing a nanoscale Fe-MIL-88B-NH_2-modified seitan composite. Because of its outstanding biocompatibility, the unusual structure of HPCFs with both macro- and mesoporous properties proved helpful for bacterial adhesion. Furthermore, the N-doped carbon framework and the presence of iron ions in the structure improved the efficiency of charge transfer between bacteria and the electrode. As a result of these characteristics, researchers found out that MFCs outfitted with HPCF anodes had an impressive maximum power density of 11.21 W/m^3 and a high current density of 23.11 W/m^3 [64]. Other studies regarding 3D-printed conducting polymers are summarized in Table 11.1.

TABLE 11.1
Summary of Studies Using 3D Printing for the Fabrication of Key Components in Microbial Fuel Cells [65]

Component	Material	Application	Major Findings	Ref.
Anode	UV curable resin (post-treated via carbonization)	Power generation	The usage of a 3D-printed porous carbon anode resulted in a significant enhancement in maximum voltage generation, approximately 2.4 times higher compared to a carbon cloth anode.	[66]
Anode	UV curable resin (post-treated with copper for surface modification)	Power generation	The 3D-printed electrode outperformed the carbon cloth electrode in terms of maximum voltage production, providing a voltage nearly three times greater.	[67]
Anode	Graphene oxide, ferric ions, and magnetite nanoparticles	Power generation	The volumetric current density of the 3D-printed electrode was significantly greater than that of the carbon felt electrode, providing a current density around 7.9 times higher.	[68]
Anode, cathode, and chassis	PLA (post-treated with graphite or nickel powder coating)	Power generation	A carbon-coated 3D-printed PLA anode's power generation was found to be comparable to that of a modified carbon veil. However, even with the coating, the performance of 3D-printed PLA cathodes was poor.	[69]
Anode (dual- and single-chamber) cathode (dual-chamber only)	PLA	Power generation	3D printing allowed for the precise fabrication of electrodes (measured 9.7 cm^2) at a lower cost.	[70]
Anode	PLA	Power generation	In comparison to the usual plain carbon veil anode, the 3D-printed anode produced much less power, approximately 4.1 times less.	[63]
Membrane	Tangoplus and natural rubber latex	Power generation	The 3D-printed latex membrane had a power density equivalent to that of a commercial CEM; however, the 3D-printed tangoplus membrane was more biofouling resistant. Furthermore, both of these membranes have lower production costs than commercial CEMs.	[55]
Membrane electrode assembly (MEA)	Fimo air-dry clay and terracotta air-dry clay	Power generation	The use of air-dry clay-based 3D-printed MEAs resulted in roughly 50% greater power output than the control samples (commercial CEM and kilned red terracotta). Furthermore, as compared to MEAs based on commercial CEM, the manufacturing cost of 3D-printed MEAs was cheaper.	[71]

11.4 CONCLUSIONS

MFCs are a bio-electrochemical process that proposes to yield electricity by using the electrons derived from biochemical reactions catalyzed by bacteria. Three parts are necessary for electricity production: microorganisms, electrodes, and exchange membranes. The design, surface area, and conductivity of electrodes are important for achieving high-density current and maximum voltage production. 3D printing techniques permit the design and printing of more complex designs than traditional fabricating processes. Using both metal and polymer-based materials, 3D printing has enabled the manufacturing of electrodes for high-efficiency electrochemical energy conversion operations. Notably, low-cost and rapid fabrication of complex and miniaturized designs of METs was achieved, which is not feasible using traditional methods. Remarkably, low-cost and speedy construction of complex and miniaturized designs of electrodes can be achieved, which is not possible using the old procedures. The spacing between electrodes is critical for diminishing internal energy losses in MFCs. Hence, membrane electrode assembly prepared by 3D printing has been used to decrease internal energy losses and so enhance MFC performance. Finally, it can be said that the 3D-printable polymer anode can show noteworthy potential as an easy-to-fabricate and low-cost MFC anode, creating a stable level of power output.

REFERENCES

[1] I. Dincer, Renewable energy and sustainable development: A crucial review, Renew. Sustain. Energy Rev. 4 (2000) 157–175.
[2] M. Rahimnejad, A. Adhami, S. Darvari, A. Zirepour, S.E. Oh, Microbial fuel cell as new technology for bioelectricity generation: A review, Alex. Eng. J. 54 (2015) 745–756.
[3] F. Akdeniz, A. Çağlar, D. Güllü, Recent energy investigations on fossil and alternative nonfossil resources in Turkey, Energy Convers. Manag. 43 (2002) 575–589.
[4] S. Aziz, A.R. Memon, S.F. Shah, S.A. Soomro, A. Parkash, Prototype designing and operational aspects of microbial fuel cell-review paper, Sci. Int. 25 (2013) 49–56.
[5] R. Choudhury, U.S.P. Uday, N. Mahata, O.N. Tiwari, R.N. Ray, T.K. Bandyopadhyay, B. Bhunia, Performance improvement of microbial fuel cells for waste water treatment along with value addition: A review on past achievements and recent perspectives, Renew. Sustain. Energy Rev. 79 (2017) 372–389.
[6] G. Palanisamy, H.Y. Jung, T. Sadhasivam, M.D. Kurkuri, S.C. Kim, S.H. Roh, A comprehensive review on microbial fuel cell technologies: Processes, utilization, and advanced developments in electrodes and membranes, J. Clean. Prod. 221 (2019) 598–621.
[7] T. Nezakati, A. Seifalian, A. Tan, A.M. Seifalian, Conductive polymers: Opportunities and challenges in biomedical applications, Chem. Rev. 118 (2018) 6766–6843.
[8] H. Pang, L. Xu, D.-X. Yan, Z.-M. Li, Conductive polymer composites with segregated structures, Prog. Polym. Sci. 39 (2014) 1908–1933.
[9] H. Yuk, B. Lu, S. Lin, K. Qu, J. Xu, J. Luo, X. Zhao, 3D printing of conducting polymers, Nat. Commun. 11 (2020) 1604.
[10] Y. Zhao, B. Liu, L. Pan, G. Yu, 3D nanostructured conductive polymer hydrogels for high-performance electrochemical devices, Energy Environ. Sci. 6 (2013) 2856–2870.
[11] K. Kižys, A. Zinovicius, B. Jakštys, I. Bružaite, E. Balciunas, M. Petruleviciene, A. Ramanavicius, I. Morkvenaite-Vilkonciene, Microbial biofuel cells: Fundamental principles, development and recent Obstacles, Biosensors 13 (2023) 221.
[12] M.C. Potter, Electrical effects accompanying the decomposition of organic compounds, Proc R Soc B Biol Sci. 84 (1911) 260–276.

[13] U. Schröder, Discover the possibilities: Microbial bioelectro-chemical systems and the revival of a 100-year–old discovery, J Solid State Electrochem. 15 (2011) 1481–1486.
[14] A.S. Vishwanathan, Microbial fuel cells: A comprehensive review for beginners, 3 Biotech. 11(5) (2021) 248.
[15] C. Santoro, M.J.S. Garcia, X.A. Walter, J. You, P. Theodosiu, I. Gajda, O. Obata, J. Winfield, J. Greenman, I. Ieropoulos, Urine in bioelectro-chemical systems: An overall review, Chem Electro Chem. 7 (2020) 1312–1331.
[16] B.E. Logan, B. Hamelers, R. Rozendal, U. Schröder, J. Keller, S. Freguia, P. Aelterman, W. Verstraete, K. Rabaey, Microbial fuel cells: Methodology and technology, Environ Sci Technol. 40 (2006) 5181– 5192.
[17] M. Ghasemi, W.R. Wan, W.R. Ismail, M. Rahimnejad, A.F. Ismail, J.X. Leong, M. Miskan, K. Ben Liew, Effect of pre-treatment and biofouling of proton exchange membrane on microbial fuel cell performance, Int. J. Hydrog. Energy. 38 (2012) 5480–5484.
[18] G. Antonopoulou, K. Stamatelatou, S. Bebelis, G. Lyberatos, Electricity generation from synthetic substrates and cheese whey using a two chamber microbial fuel cell, Biochem. Eng. J. 50 (2010) 10–15.
[19] M. Rahimnejad, G. Najafpour, A.A. Ghoreyshi, Effect of mass transfer on performance of microbial fuel cell, Intech. 5 (2011) 233–250.
[20] Y. Sharma, B. Li, The variation of power generation with organic substrates in single-chamber microbial fuel cells (SCMFCs), Bioresource. Technol. 101 (2010) 1844–1850.
[21] G. Najafpour, M. Rahimnejad, A. Ghoreshi, The enhancement of a microbial fuel cell for electrical output using mediators and oxidizing agents, Energy Sourc. 33 (2011) 2239–2248.
[22] L. Huang, R.J. Zeng, I. Angelidaki, Electricity production from xylose using a mediator-less microbial fuel cell, Bioresource. Technol. 99 (2008) 4178–4184.
[23] M. Rahimnejad, A.A. Ghoreyshi, G. Najafpour, T. Jafary, Power generation from organic substrate in batch and continuous flow microbial fuel cell operations, Appl. Energy. 88 (2011) 3999–4004.
[24] M. Al-Badani, P.L. Chong, H.S. Lim, A mini review of the effect of modified carbon paper, carbon felt, and carbon cloth electrodes on the performance of microbial fuel cell, *International Journal of Green Energy* (2024), 21 (1) 170–186. DOI: 10.1080/15435075.2023.2194979
[25] H.M.A. Sharif, M. Farooq, I. Hussain, M. Ali, M.A. Mujtaba, M. Sultan, B. Yang, Recent innovations for scaling up microbial fuel cell systems: Significance of physicochemical factors for electrodes and membranes materials, J. Taiwan Inst. Chem. E. 129 (2021) 207–226.
[26] G.W. Chen, S.J. Choi, T.H. Lee, G.Y. Lee, J.H. Cha, C.W. Kim, Application of biocathode in microbial fuel cells: Cell performance and microbial community, Appl. Microbiol. Biot. 79 (2008) 379–388.
[27] I.H. Park, M. Christy, P. Kim, K.S. Nahma, Enhanced electrical contact of microbes using Fe_3O_4/CNT nanocomposite anode in mediator-less microbial fuel cell, Biosens. Bioelectron. 58 (2014) 75–80.
[28] U. Schröder, Anodic electron transfer mechanisms in microbial fuel cells and their energy efficiency, Phys. Chem. Chem. Phys. 9 (2007) 2619–2629.
[29] J. Wei, P. Liang, X. Huang, Recent progress in electrodes for microbial fuel cells, Bioresource Technol. 102 (2011) 9335–9344.
[30] M. Zhou, T. Jin, Z. Wu, M. Chi, T. Gu, Microbial fuel cells for bioenergy and bioproducts, in: Sustainable Bioenergy and Bioproducts, Springer (2012) 131–171.
[31] M. Zhou, J. Yang, H. Wang, T. Jin, D.J. Hassett, T. Gu, Chapter 9-Bioelectrochemistry of Microbial fuel Cells and Their Potential Applications in Bioenergy, Editor(s): Vijai K. Gupta, Maria G. Tuohy, Christian P. Kubicek, Jack Saddler, Feng Xu, *Bioenergy Research: Adv. Appl.* (2014), Elsevier, 131–153.
[32] K. Watanabe, J. Biosci, Recent developments in microbial fuel cell technologies for sustainable bioenergy, J. Biosci. Bioeng. 106 (2008) 528–536.

[33] G.C. Gil, I.S. Chang, B.H. Kim, M. Kim, J.K. Jang, H.S. Park, H.J. Kim, Operational parameters affecting the performance of a mediator-less microbial fuel cell, Biosens. Bioelectron. 18 (2003) 327–334.
[34] K.P. Nevin, H. Richter, S.F. Covalla, J.P. Johnson, T.L. Woodard, A.L. Orloff, H. Jia, M. Zhang, D.R. Lovley, Power output and columbic efficiencies from biofilms of Geobacter sulfurreducens comparable to mixed community microbial fuel cells, Environ. Microbiol. 10 (2008) 2505–2514.
[35] A.E. Franks, K. Nevin, Microbial fuel cells, a current review, Energies. 3 (2010) 899–919.
[36] B.E. Logan, J.M. Regan, Microbial fuel cells-challenges and applications, Environ. Sci. Technol. 40 (2006) 5172–5180.
[37] D.H. Park, J.D. Zeikus, Improved fuel cell and electrode designs for producing electricity from microbial degradation, Biotechnol. Bioeng. 81 (2002) 348–355.
[38] S.E. Oh, B. Min, B.E. Logan, Cathode performance as a factor in electricity generation in microbial fuel cells, Environ. Sci. Technol. 38 (2004) 4900–4904.
[39] I.S. Chang, H. Moon, J.K. Jang, B.H. Kim, Improvement of a microbial fuel cell performance as a BOD sensor using respiratory inhibitors, Biosens. Bioelectron. 20 (2005) 1856–1859.
[40] O. Lefebvre, W.K. Ooi, Z. Tang, M. Abdullah-Al-Mamun, D.H.C. Chua, H.Y. Ng, Optimization of a Pt-free cathode suitable for practical applications of microbial fuel cells, Bioresource. Technol. 100 (2009) 4907–4910.
[41] J.J. Fornero, M. Rosenbaum, M.A. Cotta, L.T. Angenent, Microbial fuel cell performance with a pressurized cathode chamber, Environ. Sci. Technol. 42 (2008) 8578–8584.
[42] L. Huang, J.M. Regan, X. Quan, Electron transfer mechanisms, new applications, and performance of biocathode microbial fuel cells, Bioresource. Technol. 102 (2011) 316–323.
[43] M. Rahimnejad, G.D. Najafpour, A. Ghoreyshi, F. Talebnia, G. Premie, G.H. Bakeri, J. Kim, S. Oh, Thionine increases electricity generation from microbial fuel cell using Saccharomyces cerevisiae and exoelectrogenic mixed culture, J. Microbio. 50 (2012) 575–580.
[44] A. Tardast, M. Rahimnejad, G. Najafpour, A.A. Ghoreyshi, H. Zare, Fabrication and operation of a novel membrane-less microbial fuel cell as a bioelectricity generator, Int. J. Environ. Eng. 3 (2012) 1–5.
[45] J.M. Sonawane, D. Pant, P.C. Ghosh, S.B. Adeloju. Fabrication of a carbon paper/polyaniline-copper hybrid and its utilization as an air cathode for microbial fuel cells, ACS Appl. Energy Mater. 2 (2019) 1891–1902.
[46] Q. Deng, X. Li, J. Zuo, A. Ling, B.E. Logan, Power generation using an activated carbon fiber felt cathode in an upflow microbial fuel cell, J. Power Sources. 195 (2009) 1130–1135.
[47] D. Park, J.D. Zeikus, Impact of electrode composition on electricity generation in a single-compartment fuel cell using Shewanella putrefaciens, Appl. Microbiol. Biot. 59 (2002) 58–61.
[48] S. Zinadini, A.A. Zinatizadeh, M. Rahimi, V. Vatanpour, Z. Rahimi, High power generation and COD removal in a microbial fuel cell operated by a novel sulfonated PES/PES blend proton exchange membrane, Energy 125 (2017) 427–438.
[49] J.V. Boas, V.B. Oliveira, M. Simões, A.M.F.R. Pinto, Review on microbial fuel cells applications, developments and costs, J. Environ. Manag. 307 (2022) 114525.
[50] S. Zinadini, A.A. Zinatizadeh, M. Rahimi, V. Vatanpour, K. Bahrami, Energy recovery and hygienic water production from wastewater using an innovative integrated microbial fuel cellemembrane separation process, Energy 141 (2017) 1350–1362.
[51] A. James, P.V. Chellam, Recent advances in the development of sustainable composite materials used as membranes in microbial fuel cells, Chem. Records (2023) e20230.
[52] Y. Zuo, S. Cheng, D. F Call, B. Logan, Scalable tubular membrane cathodes for microbial fuel cell applications, ACS Natl. Meet. Book Abstr. 41 (2007) 3347–3353.

[53] M.M. Ghangrekar, P.S. Jana, M. Behera, Effect of organic loading rates and proton exchange membrane surface area on the performance of an up-flow cylindrical microbial fuel cell, J. *Environ. Sci. Eng.* 54 (2012) 1–9.
[54] S.-e. Oh, B.E. Logan, Proton exchange membrane and electrode surface areas as factors that affect power generation in microbial fuel cells, Appl. Microb. Biotechnol. 70 (2006) 162–169.
[55] H. Philamore, J. Rossiter, P. Walters, J. Winfield, I. Ieropoulos, Cast and 3D Printed ion exchange membranes for monolithic microbial fuel cell fabrication, J. Power Sourc. 289 (2015) 91–99.
[56] A. Forster, Materials Testing Standards for Additive Manufacturing of Polymer Materials: State of the Art and Standards Applicability, NIST Interagency/Internal Report (NISTIR), National Institute of Standards and Technology, Gaithersburg, MD (2015) [online], https://doi.org/10.6028/NIST.IR.8059 (Accessed August 15, 2023).
[57] B. Hüner, M. Kıstı, S. Uysal, İ.N. Uzgören, E. Özdoğan, Y.O. Süzen, N. Demir, M.F. Kaya, An overview of various additive manufacturing technologies and materials for electrochemical energy conversion applications, ACS Omega 7 (2022) 40638–40658.
[58] A.C. de Leon, Q. Chen, N.B. Palaganas, J.O. Palaganas, J. Manapat, R.C. Advincula, High performance polymer nano-composites for additive manufacturing applications, React. Funct. Polym. 103 (2016) 141–155.
[59] E. Redondo, J. Muñoz, M. Pumera, Green activation using reducing agents of carbon-based 3D printed electrodes: Turning good electrodes to great, Carbon. 175 (2021) 413–419.
[60] BlackMagic3D, Black Magic 3D Conductive Graphene Composite (2022) [online], https://3dcompare.com/materials/product/black-magic-3d-conductive-graphene-composite-1–75-mm/ (Accessed 2023).
[61] Proto-Pasta, Proto-Pasta Composite PLA Electrically Conductive Filament (2022) [online], https://www.proto-pasta.com/pages/conductive-pla (Accessed 2023).
[62] F. Bas, M.F. Kaya, 3D printed anode electrodes for microbial electrolysis cells, Fuel. 317 (2022) 123560.
[63] J. You, R.J. Preen, J. Greenman, I. Ieropoulos, 3D printed components of microbial fuel cells: Towards monolithic microbial fuel cell fabrication using additive layer manufacturing, Sustain. Energy. Technol. Assess. 19 (2017) 94–101.
[64] H. Li, L. Zhang, R. Wang, J. Sun, Y. Qiu, S. Liu, 3D hierarchical porous carbon foams as high-performance free-standing anodes for microbial fuel cells, Ecomat. 5 (2023) e12273.
[65] T.H. Chung, B.R. Dhar, A mini-review on applications of 3D printing for microbial electrochemical technologies, Front. Energy Res. 9 (2021) 679061.
[66] B. Bian, D. Shi, X. Cai, M. Hu, Q. Guo, C. Zhang, Q. Wang, A.X. Sun, J. Yang, 3D printed porous carbon anode for enhanced power generation in microbial fuel cell, Nano Energy. 44 (2018) 174–180.
[67] B. Bian, C. Wang, M. Hu, Z. Yang, X. Cai, D. Shi, J. Yang, Application of 3D printed porous copper anode in microbial fuel cells, Front. Energy Res. 6 (2018) 50.
[68] Y.T. He, Q. Fu, Y. Pang, Q. Li, J. Li, X. Zhu, R.H. Lu, W. Sun, Q. Liao, U. Schröder, Customizable design strategies for high-performance bioanodes in bioelectrochemical systems, Iscience. 24 (2021) 102163.
[69] J. You, H. Fan, J. Winfield, I.A. Ieropoulos, Complete microbial fuel cell fabrication using additive layer manufacturing, Molecules 25 (2020) 3051.
[70] E. Jannelli, P. Di Trolio, F. Flagiello, M. Minutillo, Development and performance analysis of biowaste based microbial fuel cells fabricated employing additive manufacturing technologies, Energy Proced. 148 (2018) 1135–1142.
[71] P. Theodosiou, J. Greenman, I. Ieropoulos, Towards monolithically printed Mfcs: Development of a 3d-printable membrane electrode assembly (Mea), Int. J. Hydrog. Energy 44 (2019) 4450–4462.

12 3D-Printed Conducting Polymers for Solid Oxide Fuel Cells

Ahmad Hussain, Nawishta Jabeen, Aasma Tabassum, and Jazib Ali

12.1 INTRODUCTION

Conducting polymers belong to the class of materials possessing suitable electrical conductivity, which are capable of being utilized in energy storage applications owing to their flexibility, electrical and mechanical properties, and stability. Despite their recent progress, the fabrication techniques of such materials with conventional approaches like aerosol printing or ink-jetting have lost their charm recently [1]. Unlike these conventional fabrication techniques, 3D printing is an additive manufacturing technique for the preparation of 3D objects in which the materials are deposited, joined, or solidified under computer-aided design (CAD) models that require softening and melting of constituent materials to yield the layer-by-layer architectures [2]. Traditionally, 3D printing focuses on the construction and treatment of polymer-based materials. Nonetheless, 3D printing techniques have rapidly achieved importance not only for printing numerous polymer-based materials but also for ceramics and metals, which has made 3D printing a versatile technique [3]. The fabrication process is the most crucial aspect, which regulates the material's working and strength. 3D printing has emerged as a relevant manufacturing process, with the potential to revolutionize the micro-fabrication industry [4]. Therefore, 3D printing technology has received substantial importance in the previous two decades owing to its capability to produce delicate and highly complex structures along with geometrical simplicity in a short period. Moreover, the 3D printing technique is most suitable for rapid manufacturing and for several engineering fields like aerospace, electronics, mechanical engineering, civil engineering, and biomedicine [5].

In this chapter, various aspects of the most conventional and extensively adopted 3D printing techniques for the synthesis and design of conducting polymers are discussed. With growing requirements of renewable energy sources, fuel cells represent a sustainable, clean, and highly efficient energy conversion source [6]. Fuel cell technology is an auspicious alternative power production technology, which converts chemical energy into electrical energy by electro-chemical reaction [7]. The first fuel cell was constructed by Sir Humphrey Davy in 1802, with the ability to deliver a weak electric shock [8]. Sir William Grove discovered the functional fuel cell in 1839, which revealed the electrolytic water-splitting reaction [9]. It has been reported that a higher

DOI: 10.1201/9781003415985-12

surface area of supportive material can cause the reduction of agglomeration for metallic particles; hence, it can demonstrate good catalytic morphology for excellent fuel cell performance [10]. Solid oxide fuel cells (SOFCs) and proton exchange membranes (PEMs) are regarded among the most explored fuel cells to date. PEM cells demonstrate lower efficiencies along with the presence of minute impurities, and their operating procedure takes place at low-temperature range (80–120 °C) [11], while the molten carbonate fuel cell (MCFC), which is a renowned type of SOFC, operates at higher temperatures and demonstrates more tolerance to impurities [12]. MCFCs have the ability to exhibit higher efficiency, but the presence of molten alkaline salts in them causes creep and corrosion, which are regarded as disadvantages [13]. Generally, fuel cells are divided into six classes depending on the fuels and electrolytes, each type shows its individual advantages and disadvantages. SOFCs are among those types of fuel cells that have achieved considerable endorsement owing to their low-cost effectiveness, higher efficiencies, and the ability to use a variety of fuels other than hydrogen-based fuels like coal gas, hydrocarbons, etc., [14]. SOFCs are also famous due to their possible utilization as efficient and clean power production abilities. Additional progress in SOFCs is the introduction of 3D printing of conducting polymers as components, which has caused better commercialization and cost reduction for this technology. This chapter presents a detailed review of the current advancements in conducting polymers for 3D printing technologies to be employed for SOFCs. First, various methodologies of 3D techniques will be analyzed capable of enabling conducting polymer materials with their applications for fuel cells and SOFCs.

12.2 3D PRINTING TECHNIQUE

Nowadays, the technique of 3D printing is given great attention from researchers owing to its ability to synthesize complex/hybrid structures and morphologies through simplified fabrication methods with high resolution. Various methodologies are employed for 3D printing technology, including electrohydrodynamic printing (EHDP), extrusion-based printing (EBP), inkjet printing, light-based printing (LBP), and electron beam melting (EBM), which are used with controlled electric and temperature parameters. Figure 12.1 is a representation of various 3D printing techniques followed for the synthesis of conducting polymers.

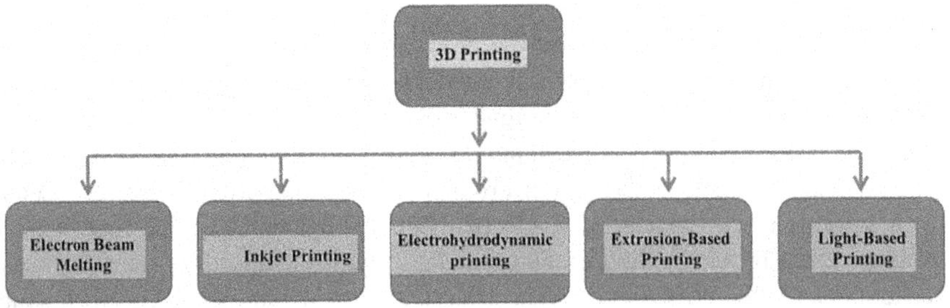

FIGURE 12.1 Various 3D printing techniques followed to develop 3D conducting polymers.

12.2.1 Electrohydrodynamic Printing

Electrohydrodynamic printing (EHDP) is a 3D printing technique in which deposition of the material is made by dissolving it in polarizable liquid. Experimentation is performed by ion transfer under the influence of an applied electric field placed between a grounded substrate and nozzle. The EHDP process is dependent on the material's characteristics, including electrical conductivity, dipole moment, surface tension, and viscosity, as well as the nozzle's parameters like flow rate, pressure, and applied electric field. Moreover, EHDP possesses the ability to demonstrate pulsating or continuous mode that is made for fabricating dots, dashes, or continuous fibers. Hence, in the EDHP process, the deposition method dictates the resolution up to micro or nano levels. For more focused designs, electrodes are placed to control the trajectory of the deposited materials to acquire the micro/nanometer resolution of printing [15].

Wearable electronic devices composed of conducting polymer components are fabricated by the EHDP method. Poly(3,4-ethylenedioxythiophene) (PEDOT) is among such conducting polymers employed to function for such devices as PEDOT-based ink. The most famous PEDOT-based material is its combination with poly (styrenesulfonate) in the form of PEDOT:PSS, which can also be combined with poly(ethylene oxide) (PEO) to achieve PEO/PEDOT:PSS to employ as an EHDP ink solution for 3D printing. Charkhesht et al. synthesized a conductive PEDOT:PSS polymer and utilized it as a carrier of a TiO_2/CB-based electrospun anode. The presence of PEDOT:PSS assisted the anode performance of areal capacity (1.67 mAh.cm^{-2}) and gravimetric capacity (300 mAh.g^{-1}), which is promising for next-generation electrospun lithium-ion battery electrodes [16]. Vijayavenkataraman et al. fabricated a conductive co-polymer, polypyrrole-b-poly(ε-caprolactone) (PPy-b-PCL), which was used in printable ink to synthesize porous 3D scaffolds by EHDP. Various contents of PPy-b-PCL were implemented to fabricate scaffold fiber of diameters from ~33 (0.5%) to ~44 μm (2%) with an average pore size of ~125 μm [17].

12.2.2 Extrusion-Based Printing

Extrusion-Based Printing (EBP) uses a movable nozzle to perform layer-by-layer deposition of the material; such printing follows the specific shape programmed in software connected to a programmable computer. EPB is further divided into two extrusion printing techniques, which are different for manufacturing the polymers. First is fused deposition modeling (FDM), it uses the polymer material in a filament that moves with a gear mechanism into a hot end to get melted. A second technique is direct ink writing (DIW), which employs polymers in semi-melted condition as solutions/pastes, which are slowed down by current variation, screws, pistons, or air. In both methodologies, the utilized polymers must contain specific rheological characteristics in which viscosity values should be 10^1 Pa·s for high shear rates of 10^2 s^{-1} and 10^4 Pa·s for low shear rates of 10^{-1} s^{-1} to work as printable and to retain the required shape [15].

PEDOT nanomaterials and nanocomposites are prepared via extrusion 3D printing techniques. Ou et al. integrated Sb_2Te_3 nanoflakes and multi-walled carbon nanotubes (MWCNTs) into a PEDOT:PSS scaffold to improve the properties of hybrid structures [18]. In a similar way, Yoo et al. fabricated flexible and transparent films by integrating Ag-NWs into a PEDOT:PSS scaffold through screen-printing and roll-to-roll technologies [19]. Gu et al. reported the fabrication of wearable electrodes made by alternating layers of PPy nanotube ink and poly(vinyl alcohol) gel ink following the DIW printing technique. It was observed that conducting polymers constructed via 3D printing demonstrated good mechanical strength; dispersed PPy nanotubes presented a core channel for the transportation of ions. Samples have maintained a retention capacitance of ~93% even at a 120° bending angle [20].

12.2.2.1 Fused Deposition Modeling

Fused deposition modeling (FDM) is a 3D printing extrusion technology, which is also called fused filament fabrication (FFF) and creates 3D components by the continuous melting of conducting polymer in filament to form metal threads or wire. An extruder feeds the plastic filament by a hot extrusion nozzle, which melts the material and later precisely deposits the material layer by layer atop the build platform in an arranged auto-directional path. The vertical and horizontal movement of the nozzle is controlled by a programmable computer-aided software model. The FDM technique is renowned due to its several reported benefits, including reliability, less cost, and simplicity of the synthesis process. The FDM process has exhibited the ability to perform high-resolution, printing parts of ~40 μm [21]. Less mechanical strength is the only disadvantage of this manufacturing technique, which happens because of the irregular cooling, heating, and cycles during the process [22].

12.2.2.2 Direct Ink Writing

Direct ink writing (DIW) is an extrusion-based 3D printing methodology employed to promote the synthesis of 3D morphologies with novel hierarchical structures and composites at micro- and nanoscales. During this fabrication technique, a high-viscosity ink is extruded and deposited layer-by-layer fashion as scaffolds or other 3D geometrical structures with the help of a deposition nozzle by a programmable computer at the translational stage. After the deposition, the solidification process of the 3D-constructed structure is performed to generate the desired featured structure and its properties [23]. Generally, robo-casting is also another name used for DIW, which is in fact related to the two classes of the deposition processes: continuous and droplet ink extrusion [24].

Li et al. employed the DIW 3D printing methodology for the fabrication of micro-supercapacitors of a flexible nature, which were constructed by pre-planned designed PEDOT:PSS/MXene composite gels working as inks, with no extra tedious methods or the addition of any other toxic organics [23].

12.2.3 Inkjet Printing

The inkjet 3D printing technique is a hugely versatile process for certain solid suspensions and liquid materials, which coats and deposits at lower pressure and lower temperature. The jetting head for this processing technique requires three nozzles; thermoplastic material jetting is performed by the first nozzle, while the other two nozzles provide/deposit wax support. The print head raster examines the surface; numerous layers are created by a layer-by-layer development to attain the defined thickness of the layer. The inkjet printing technique involves the exploitation of drop-on-demand ability along with highly precise milling competence. Conductive nanoparticles, dielectric nanoparticles, and polymers may be coated/deposited by this technique, proving it easy to use for a wider range of suitable materials. Low cost, reproducibility, and the ability to coat/deposit materials with various physical and chemical properties are the best merits of this technique. Longer fabrication duration due to less speed is the big disadvantage of this technique, causing roughness and uneven finishing of the surface, and it becomes tough for the jet to operate precisely at high temperatures [15].

Alshammari et al. have reported the utilization of inkjet printing techniques to synthesize conductive composite polymer of PEDOT:PSS with the incorporation of multi-walled carbon nanotubes, in which the orientation of the MWCNT has led to the improvement of conductivity in the printed samples. Aligned MWCNT samples demonstrated 53% varied electrical conductance compared to randomly oriented MWCNT samples [25]. Pan et al. utilized a mixture formed by aniline and phytic acid (0.64–16.0 cP) and ammonium persulfate (1.05 cP) as sequential deposition to fabricate 3D hydrogel patterns by aerosol and inkjet printing technologies. The achieved printed material exhibited an excellent novel hierarchical structure and high electrical conductance, along with a ~480 F g^{-1} capacitance and capacitance retentions of 91% and 83% stability even after 5000 and 10,000 cycles, respectively [26].

12.2.4 Light-Based Printing

Light-based printing (LBP) is a 3D printing methodology, which is dependent on the liquid state photo-polymerization process of monomers or polymers placed in a vat by a well-controlled solidification process with a specific shape and 3D structure [27]. There are three basic sub-techniques that are categorized under LBP: (1) stereolithography (SLA), (2) selective laser sintering (SLS), and (3) digital light processing (DLP). In SLA, the resin is photo-cured under the employment of an applied laser beam that is controlled by deflected mirrors; later liquid solidifies at the surface, where light spots are scanned. A high-power laser is used for the SLS technique for the sintering of tiny particles of polymer powder to transform into a solid structure. For the DLP process, millions of mirrors are employed to directly design 2D images on photo-sensitive material for the fabrication of a digital-micro-mirror device [15].

12.2.4.1 Stereolithography

Nowadays, SLA is implemented as an industrial 3D printing technique to design the concept models, complex components with convoluted geometrical structures, and cosmetic models with a rapid duration of just one day. For example, SLA possesses the ability to perform high-resolution printing of ~10 μm, though the lowest resolution of the layer for the SLS printed part is ~20 μm [28]. SLA parts have the highest resolution, isotropic nature, accuracy, and water-tight prototypes, and end-use parts are refined advanced materials with specific functions. The surface finishing remains smooth and even compared to all other 3D printing techniques. This 3D additive manufacturing method is based on laying down a photo-polymer liquid ink enclosed inside a container, which can be solidified by an ultraviolet argon or helium-cadmium laser. The entire object can be "sliced" in an ordered way with thin layers, and then each layer is deposited onto the support and solidified by a scanned laser. After numerous iterations, the entire object can be fully fabricated [28].

Scordo et al. reported an SLA 3D printing technique of a resin-based composite of conducting polymers (PEDOT:PSS filler) and studied the cumulative volatile organic compound's adsorption characteristics [29].

12.2.4.2 Selective Laser Sintering

SLS is a 3D printing technique, Dr. Carl Decker was the first to introduce and present this technique in 1980 during his stay at the University of Texas. This technique utilizes a highly precise and high-power laser for sintering tiny nanoparticles of polymer powders to convert them to a solid structure. SLS has remained a famous technique for manufacturers and engineers for decades [30]. In this process, material powders such as plastic, metal, or ceramics are selectively sintered by scanning a CO_2 laser to form each layer, followed by cylindrical rolling spreads to achieve a new layer of polymer powder on the fused layer by the fusing process via CO_2 laser. This method is continuously repeated until the required morphological object is prepared. The SLS fabrication method is employed to construct composites of polyamides. For the indirect technique, a polymer is sintered first to prepare a green part, and then a second heat treatment cycle is employed in the furnace and is filtered with bronze/copper to construct the highly dense metallic part [31]. Lupone et al. manufactured composites of conducting polymer of polyamide-12, which were supported by carbon fiber and graphite. These hybrid composites exhibited significant improvement in conductivity [32].

12.2.4.3 Digital Light Processing

DLP is a 3D printing technique in which photo-curable resins are used, as they can be employed to synthesize a 3D single-layer object by a controlled solidification process under a projector light (either white light or UV light). This is a rapid technique for the development of layers. Moreover, it is possible to alter the characteristics and properties of the achieved final products of as-printed materials by varying the photo-curable resin compositions. Hence, in this way, it becomes easy to attain a broad collection of models to prepare structures with efficient/smart features and properties [33].

Goretti et al. employed a DLP-based 3D printing method to analyze the impact of PANI-HCl on the chemical/physical characteristics of acrylic composites based on ethylene-glycol-phenyl-ether acrylate (EGPEA) and diphenyl(2,4,6-trimethyl benzoyl) phosphine oxide (TPO), a photo-initiator, to work as a framework of a percolation prototype [34].

12.2.5 Electron Beam Melting

Electron beam melting (EBM), or electron beam powder bed fusion (EB PBF), uses a high-powered electron beam to melt the conducting metallic powders layer by layer; the following process is repeated until the required morphological objects are achieved. Fabricated parts are highly dense and mechanically strong, finding use in turbine blades and hip implants. The EBM technique permits the synthesis of high-quality parts as they are fabricated under highly controlled environmental parameters with smooth circulation of temperature. This technique is enormously fast, and therefore its production rate is higher [35]. EBM was first presented by a Swedish company known as Arcam AB in 2005. The technique uses the idea of electron generation by a gun, which can be fixed and sped up by employing an electromagnetic lens and then electro-magnetically scanned by a CAD model [36].

Chen et al. studied vaporized hydrogen peroxide and electron beam melted sterilization techniques of 3D printing and injecting to mold highly dense polyethylene and polyamide materials to analyze their thermal, mechanical, and chemical performance [37].

12.3 3D-PRINTED CONDUCTING POLYMERS

Conducting polymers, including PEDOT, PPy, and polyaniline (PANi), are promising materials in a variety of applications, including fabrication for actuators, surface protection, energy storage flexible devices, data storage, wearable electronic devices, and biocompatible electronics (Figure 12.2). Rigorous experiments are performed for the up gradation of 3D printing strategies of conducting polymers, but only simple structures like isolated fibers have been operated in fuel cells [38]. PEDOT has been a very important conducting polymer for electronics and energy storage applications owing to its inherited characteristics of optical transparency and conductivity, along with its electrochemical and thermal stability in thin films. PEDOT:PSS-based materials can also be utilized as reinforcement materials for Ag-NWs by inkjet printing [39]. Jabeen et al. fabricated novel core–shell $NiCo_2O_4$@PANI nanorod arrays to work as electrode material for supercapacitors [40]. Pan et al., fabricated a PANI hydrogel with superb electronic conductance, which demonstrated the ability to work as an electrode for supercapacitor capacitance of ~480 $F{\cdot}g^{-1}$ [41]. PANI has also acted as a stabilizing agent of Ag-NWs to attain unique inks with improved electrical performance, viscosity (~4.4 cP), and aspect ratios [42]. The softness of 3D-printed conducting polymer hydrogels can provide beneficial lasting biomechanical interconnection with biotic soft tissue, which can possibly demonstrate specific advantages in the field of bio-electronic devices/applications and implants. Digital manufacturing 3D printing techniques have virtually boundless prospects to improve objects/structures by adapting suitable material configurations, fabrication environments, and geometrical technicality at each point in an object. PEDOT:PSS/Ag-NWs as a thin film for the potential 15–40 V were synthesized by 3D printing, and stability in temperature for the range 49–99 °C was observed with an interval of 30–50 seconds, confirming the constant heating and swift thermal reaction, while the surface temperature of PEDOT:PSS thin films remains steady to room temperature [43]. Conducting polymers like PPy, nanocellulose, and poly(glycerol sebacate) were mixed to fabricate inks for the DIW printing technique for the fabrication of 3D structures to be employed for drug delivery patches in therapies for myocardial infarction. PANI is a famous conducting polymer. PANI nanocomposite inks have

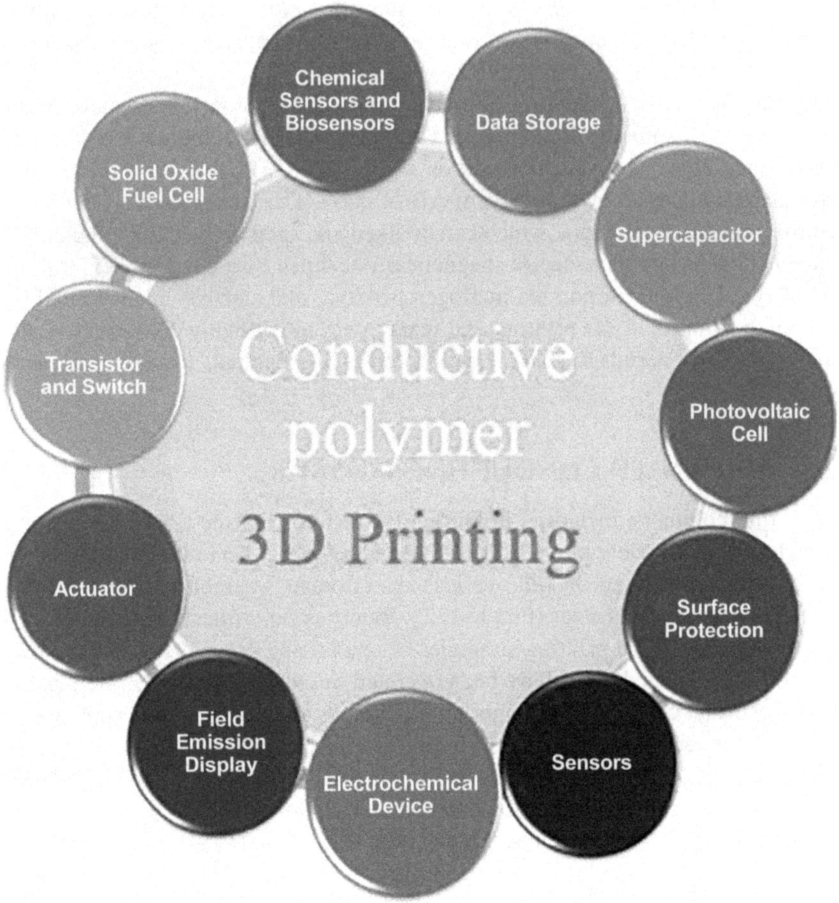

FIGURE 12.2 Utilization of 3D-printed conducting polymers for various advanced applications.

attained a reputation for 3D printing, and reduced graphene oxide has been blended with PANI to achieve printable PANI/GO inks that are shape dependent, with self stability and electrical conductance [44]. Conducting polymers have shown a significant ability to be utilized in the future generation of fuel cells.

12.4 FUEL CELLS

About 160 years ago, fuel cells were discovered, and the phenomenon of electrochemical process was first time observed by Alessandro Volta. Fuel cells follow a similar working principle as that of batteries. Instead of recharge, their only requirement is the constant availability of oxygen and fuel. Among different kinds of fuel cells, solid oxide fuel cells and proton exchange membranes are highly explored. Proton exchange membrane fuel cells usually work at a lower temperature range of

80 °C to 120 °C and are highly liable to have microscopic contaminations that might exist in fuels and demonstrate a low competence. SOFCs operate similarly to molten-carbonate fuel cells, involving high temperatures, and demonstrate superb tolerance to impurities that exist in fuels. Molten-carbonate fuel cells and SOFCs demonstrate higher efficiencies, but in molten-carbonate fuel cells, creep and corrosion are the main disadvantages due to the presence of molten alkaline salts in them [45].

Pristine graphene, reduced graphene oxide, and modified-graphene materials, due to their supporting material activity, have shown tremendous electro-catalytic potential for the methanol electro-oxidation route. Additionally, the working of graphene as a supporting material can be improved/enhanced by altering its surface morphologies and henceforth show abilities with their contribution to prospective applications/devices, like fuel cells, supercapacitors, solar cells, and batteries. Many studies have been conducted about the grain size, particle dispersion, morphological presentation, and catalytic nature of materials within graphene to act as supporting material for such materials that have demonstrated considerable enhancement in fuel cell performance [46]. In recent times, doping in graphene derivatives has become highly involved in oxygen reduction reactions for electrode/catalyst characteristics, which function at high temperatures, owing to the novel characteristics of graphene-based electrodes like mechanical elasticity, charge mobility, electrical conductivity, electrical current capacity, specific surface area, and flexibility. Graphene-supported electrodes, particularly nitrogen-doped graphene, have been reported to operate as a cathode to be operated at lower temperature for fuel cells functioning, like for polymer-electrolyte fuel cells, which has presented the significant catalytic activities. The high conductive nature and higher available surface area of graphene derivatives have demonstrated auspicious character to be utilized as electro-catalysts for fuel oxidation reactions. Polymer-based membranes with a combination of graphene have presented higher tensile strength, low fuel permeation, and improved ionic conductivity. Graphene possesses the ability to improve the corrosion resistance and conductivity of bipolar plates [47]. Chang et al. reported the fabrication of an 880-nm-thick nickel adhesion layer and 3.8-um-thick gold layer by a direct current sputter method. They used silver nanowires to enhance the conductive nature of fuel cells [48]. Yang et al. synthesized a 3D-printed water electrolyzer in one multi-functional plate, resulting in the reduction of interfacial interaction resistances. This analysis resulted in an efficiency of 86.48% at current density 2 A/cm^2 (80 °C), as well as a 61% variation in H_2 evolution as compared to conventional fuel cells [49].

12.5 3D-PRINTED CONDUCTING POLYMERS FOR SOLID OXIDE FUEL CELLS

SOFCs are electrochemical conversion devices for which electrolytes incorporate a ceramic metal oxide that fabricates electricity directly by oxidizing the fuel as an electrolyte. Walther Nernst, a German scientist, presented the concept of modern SOFCs. Walther established the "Nernst mass", which contained 85% ZrO_2 and 15% Y_2O_3. The prescribed composition was altered to develop the highly oxide-ion conductive electrolyte. Anode and cathode materials, along with the type of electrolytes, are considered the basic components of SOFCs. 3D printing techniques are widely employed for the fabrication of suitable electrolyte, anode, and cathode materials for SOFCs [50]. The function of an SOFC starts by delivering fuel and air to the anode and cathode, respectively. Bipolar plates on both sides of the fuel cell dispense gases and act as current conductors. The anode (negative electrode) donates electrons toward the external circuit and oxidizes the fuel throughout the electrochemical process. The cathode (positive electrode) collects the electrons from the external circuit and diminishes during the electrochemical process. The electrolyte serves as an intermediate source for ion transportation between the anode and cathode of the fuel cell [51]. The process is presented in Figure 12.3.

At Cathode:

$$\left(\tfrac{1}{2}\right)O_2 + (2)e^{-1} \rightarrow O^{2-}$$

At Anode:

$$H_2 + O^{2-} \rightarrow H_2O + (2)e^{-}$$

Inclusive:

$$H_2 + \left(\tfrac{1}{2}\right)O_2 \rightarrow H_2O + \Delta E$$

A carbon-aerogel lattice structure is printed by employing a direct ink writing technique; pressures were determined to analyze the ability to observe liquid pathways

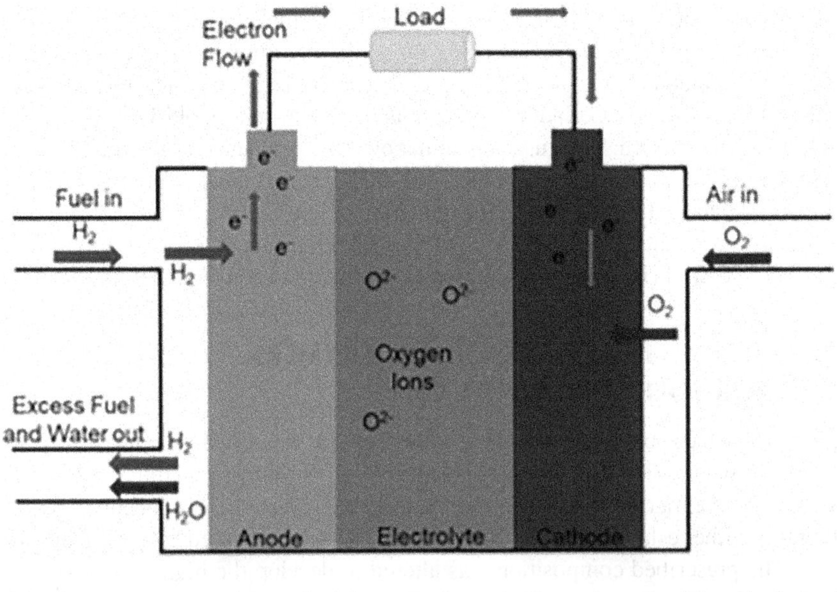

FIGURE 12.3 Basic working principles and schematic illustration of solid oxide fuel cell. Adapted with permission [14]. Copyright 2021, Elsevier.

to attain specific pressure thresholds. Mineshige et al. fabricated the Sm-doped ceria composite electrolyte; 3D printing has efficaciously presented the dense and thin Sm-doped ceria composite electrolyte layers to construct lower-temperature functional SOFCs [52]. Feng et al. reported the insertion of glass fiber into stearic acid, paraffin, and polyethylene wax binders to improve the proficiency of as-assembled SOFC; the assembly significantly enhanced the electrochemical activity of printed cells. Maximum power density at 550 °C has been reported to be ~448 mW.cm^{-2}, which is 20% more than a sample without glass fiber. Additionally, at 550 °C, the voltage for an open circuit was ~1.0 V, presented the excellent current-voltage character, improved power density, and electrochemical performance for the cell containing glass fiber to the printed electrolyte (Figure 12.4) [53].

Farandos et al. established steady dispersed inks of metal oxide particles with sub-micrometer sizes (i.e., NiO, $La_{1-x}Sr_x(MnO_{3-\delta})$, $(ZrO_2)_{0.92}(Y_2O_3)_{0.08}$) into liquid phases with appropriate solid fractions, which have revealed exceptional physical properties [54]. Kim et al. reported the influence of a polyethylene oxide (polymer) addition on the combustion properties of aluminum (Al; fuel)/copper oxide (CuO; oxidizer)-based nEMs. It was observed that by varying the PEO concentration, explosion-induced pressure was reduced considerably, resulting in the failure of propagation of combustion flame when PEO content surpassed 15 wt.% [55].

Melodia et al. employed a digital light processing 3D printing technique to fabricate solid polymer electrolytes, to analyze the tunable conductivity and mechanical

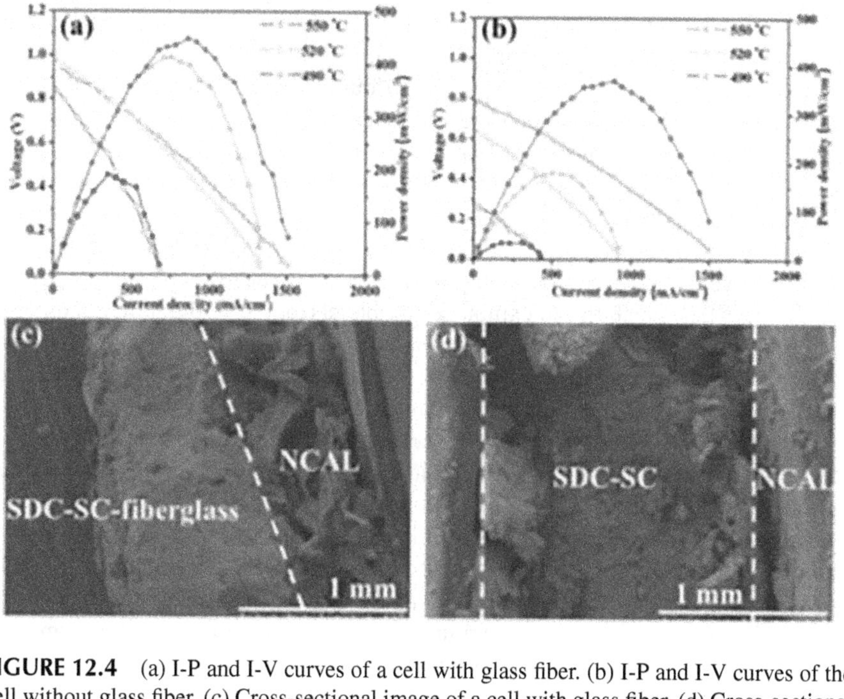

FIGURE 12.4 (a) I-P and I-V curves of a cell with glass fiber. (b) I-P and I-V curves of the cell without glass fiber. (c) Cross-sectional image of a cell with glass fiber. (d) Cross-sectional image of a cell without glass fiber. Adapted with permission [53], Copyright 2019, Elsevier.

properties to exploit the polymerization induced microphase separation progression to synthesize nano-structured conducting polymeric materials for energy storage devices/applications [56]. Jee et al. reported nitrogen-doped graphene as an air electrode to operate at lower-temperature at ~350 °C or less for SOFCs. After ~10 hours of functioning at 350 °C, the electrode's stability in performance started to reduce considerably owing to the oxidation of carbon [57].

12.6 APPLICATIONS, FUTURE PERSPECTIVES, AND CHALLENGES

The chapter presents an overview of various features of 3D printing technology and guides to choosing suitable printing techniques with appropriate printing materials for the targeted SOFC applications. Applications of SOFCs incorporate high power and high heat efficiency, solid electrolytes, low emissions, long-term stability, fuel flexibility, hybrid/gas turbine cycles, and relatively low-cost effectiveness. Less fuel consumption and better cell efficiency are also issues, as gas and heat must be carefully used to achieve excellent efficiency with no loss of heat or fuel. Integrated SOFC versions with alternative power generators, like hybrid power generators based on micro-turbine-SOFCs or combined heat and power devices, are suitable approaches to vary inclusive economic viability and energy efficiency [58]. Detailed research work and analysis, such as the opportunity to reduce the ohmic resistance of SOFCs and to bring out the improvement within the reforming process by carbon fuel, are going on currently to achieve the best performance. During the functioning of SOFCs, the existing major drawback is the higher operating temperature, which results extended starting times, a limited number of shutdowns, chemical and mechanical compatibility problems, corrosion, and breakdown of cell components. To increase the possible applications in various fields of technology and reduce SOFC cost, the fabrication of dense electrolytes and thin electrodes is one feasible method. SOFCs with cogeneration and hybrid modes contain the 80% plus efficiencies [59]. Nonetheless, 3D printing fabrication techniques of doping the derivatives of conducting polymers are anticipated to enhance the feasibility of employing such materials as air electrodes, even at high temperatures. Conducting polymer-related hybrid and composite material incorporation in SOFC energy conversion applications/devices has the ability to exhibit admirable performance durability and promising future aspects for viable devices. The utilization of conducting polymers may be as proton exchange membranes, porous electrodes, bipolar plates, and electrocatalysts to solve durability and performance-related challenges for practical applications. Such findings might inspire further research and development strategies for conducting polymer-based materials in the application of high-temperature SOFCs. Further work to decrease the ohmic resistance of SOFCs and enhance carbon fuel during the improvement process is presently ongoing to attain excellent performance [60]. A SOFC offers considerable benefits like higher efficiency, better activity of electrodes, and faster kinetics, along with additional expedient fuel flexibility owing to their higher functioning temperatures. Such outstanding properties can lead researchers to explore the hidden abilities of conducting polymers and utilize them for energy storage applications.

12.7 CONCLUSION

3D printing technology implementation on conducting polymer-based functional materials has revolutionized the renewable energy sources sector by producing novel functionalities and complex hybrid nanostructured shapes. Among several energy sources and devices like batteries and supercapacitors, conducting polymer-based electronic-based SOFCs are auspicious contenders to attain benefits from 3D printing techniques to develop novel merits that overcome the morphological limitations of the presently prevailing fabrication processes. In this chapter, a comprehensive analysis of different 3D printing techniques for the synthesis of conducting polymer-based materials for SOFCs was presented. It was found that the 3D printing of conducting polymers with other conducting fillers like graphene, carbon nanotubes, or silver nanostructure also presents promising features. It was concluded that solid oxide fuel cells constructed following 3D printing techniques offer 57% enhanced performance as compared to devices fabricated via conventional routes. This chapter also provides promising prospects for the future generation wearable electronic, bio-compatible, and energy storage applications.

REFERENCES

[1] Verma, Akash, Ruben Goos, Jurre De Weerdt, Patrick Pelgrims, and Eleonora Ferraris. "Design, fabrication, and testing of a fully 3D-printed pressure sensor using a hybrid printing approach." *Sensors* 22, no. 19 (2022): 7531.

[2] Tofail, Syed AM, Elias P. Koumoulos, Amit Bandyopadhyay, Susmita Bose, Lisa O'Donoghue, and Costas Charitidis. "Additive manufacturing: Scientific and technological challenges, market uptake and opportunities." *Materials Today* 21, no. 1 (2018): 22–37.

[3] Stansbury, Jeffrey W., and Mike J. Idacavage. "3D printing with polymers: Challenges among expanding options and opportunities." *Dental Materials* 32, no. 1 (2016): 54–64.

[4] Kumar, Sanjay, Pulak Bhushan, Mohit Pandey, and Shantanu Bhattacharya. "Additive manufacturing as an emerging technology for fabrication of microelectromechanical systems (MEMS)." *Journal of Micromanufacturing* 2, no. 2 (2019): 175–197.

[5] Macdonald, Eric, Rudy Salas, David Espalin, Mireya Perez, Efrain Aguilera, Dan Muse, and Ryan B. Wicker. "3D printing for the rapid prototyping of structural electronics." *IEEE Access* 2 (2014): 234–242.

[6] Stambouli, A. Boudghene. "Fuel cells: The expectations for an environmental-friendly and sustainable source of energy." *Renewable and Sustainable Energy Reviews* 15, no. 9 (2011): 4507–4520.

[7] Vaghari, Hamideh, Hoda Jafarizadeh-Malmiri, Aydin Berenjian, and Navideh Anarjan. "Recent advances in application of chitosan in fuel cells." *Sustainable Chemical Processes* 1, no. 1 (2013): 1–12.

[8] Sandstede, G., E. J. Cairns, V. S. Bagotsky, and K. Wiesener. "History of low temperature fuel cells." *Handbook of Fuel Cells* 1 (2010): 145–218.

[9] Dicks, Andrew L., and David A. J. Rand. *Fuel cell systems explained.* John Wiley & Sons, 2018.

[10] Sharma, Surbhi, and Bruno G. Pollet. "Support materials for PEMFC and DMFC electrocatalysts—A review." *Journal of Power Sources* 208 (2012): 96–119.

[11] Bose, Saswata, Tapas Kuila, Thi Xuan Hien Nguyen, Nam Hoon Kim, Kin-tak Lau, and Joong Hee Lee. "Polymer membranes for high temperature proton exchange membrane fuel cell: Recent advances and challenges." *Progress in Polymer Science* 36, no. 6 (2011): 813–843.

[12] Lanzini, Andrea, Hossein Madi, Vitaliano Chiodo, Davide Papurello, Susanna Maisano, and Massimo Santarelli. "Dealing with fuel contaminants in biogas-fed solid oxide fuel cell (SOFC) and molten carbonate fuel cell (MCFC) plants: Degradation of catalytic and electro-catalytic active surfaces and related gas purification methods." *Progress in Energy and Combustion Science* 61 (2017): 150–188.

[13] Andújar, José Manuel, and Francisca Segura. "Fuel cells: History and updating. A walk along two centuries." *Renewable and Sustainable Energy Reviews* 13, no. 9 (2009): 2309–2322.

[14] Singh, Mandeep, Dario Zappa, and Elisabetta Comini. "Solid oxide fuel cell: Decade of progress, future perspectives and challenges." *International Journal of Hydrogen Energy* 46, no. 54 (2021): 27643–27674.

[15] Criado-Gonzalez, Miryam, Antonio Dominguez-Alfaro, Naroa Lopez-Larrea, Nuria Alegret, and David Mecerreyes. "Additive manufacturing of conducting polymers: Recent advances, challenges, and opportunities." *ACS Applied Polymer Materials* 3, no. 6 (2021): 2865–2883.

[16] Charkhesht, Vahid, Begüm Yarar Kaplan, Selmiye Alkan Gürsel, and Alp Yurum. "Electrospun Nanotubular Titania and Polymeric Interfaces for High Energy Density Li-Ion Electrodes." *Energy & Fuels* 37, no. 8 (2023): 6197-6207.

[17] Vijayavenkataraman, Sanjairaj, Sathya Kannan, Tong Cao, Jerry YH Fuh, Gopu Sriram, and Wen Feng Lu. "3D-printed PCL/PPy conductive scaffolds as three-dimensional porous nerve guide conduits (NGCs) for peripheral nerve injury repair." *Frontiers in Bioengineering and Biotechnology* 7 (2019): 266.

[18] Ou, Canlin, Abhijeet L. Sangle, Thomas Chalklen, Qingshen Jing, Vijay Narayan, and Sohini Kar-Narayan. "Enhanced thermoelectric properties of flexible aerosol-jet printed carbon nanotube-based nanocomposites." *Apl Materials* 6, no. 9 (2018): 096101.

[19] Yoo, Ji Hoon, Yunkyung Kim, Mi Kyoung Han, Seonghwa Choi, Ki Yong Song, Kwang Choon Chung, Ji Man Kim, and Jeonghun Kwak. "Silver nanowire–conducting polymer–ITO hybrids for flexible and transparent conductive electrodes with excellent durability." *ACS Applied Materials & Interfaces* 7, no. 29 (2015): 15928–15934.

[20] Gu, Yifan, Yan Zhang, Yunhui Shi, Lifang Zhang, and Xinhua Xu. "3D all printing of polypyrrole nanotubes for high mass loading flexible supercapacitor." *ChemistrySelect* 4, no. 36 (2019): 10902–10906.

[21] Gnanasekaran, Karthikeyan, T. Heijmans, S. Van Bennekom, H. Woldhuis, S. Wijnia, G. De With, and H. Friedrich. "3D printing of CNT-and graphene-based conductive polymer nanocomposites by fused deposition modeling." *Applied Materials Today* 9 (2017): 21–28.

[22] Benedetti, M., Anton Du Plessis, R. O. Ritchie, M. Dallago, Seyed Mohammad Javad Razavi, and Filippo Berto. "Architected cellular materials: A review on their mechanical properties towards fatigue-tolerant design and fabrication." *Materials Science and Engineering: R: Reports* 144 (2021): 100606.

[23] Li, Le, Jian Meng, Xuran Bao, Yunpeng Huang, Xiu-Ping Yan, Hai-Long Qian, Chao Zhang, and Tianxi Liu. "Direct-ink-write 3D printing of programmable micro-supercapacitors from MXene-regulating conducting polymer inks." *Advanced Energy Materials* (2023): 2203683.

[24] Dasgupta, Archisman, and Prasenjit Dutta. "A comprehensive review on 3D printing technology: Current applications and challenges." *Jordan Journal of Mechanical & Industrial Engineering* 16, no. 4 (2022).

[25] Alshammari, Abdullah S., M. Shkunov, and S. Ravi P. Silva. "Inkjet printed PEDOT: PSS/MWCNT nano-composites with aligned carbon nanotubes and enhanced conductivity." *physica status solidi (RRL)–Rapid Research Letters* 8, no. 2 (2014): 150–153.

[26] Pan, Lijia, Guihua Yu, Dongyuan Zhai, Hye Ryoung Lee, Wenting Zhao, Nian Liu, Huiliang Wang et al. "Hierarchical nanostructured conducting polymer hydrogel with high electrochemical activity." *Proceedings of the National Academy of Sciences* 109, no. 24 (2012): 9287–9292.

[27] Distler, T., and A. R. Boccaccini. "3D printing of electrically conductive hydrogels for tissue engineering and biosensors-A review." *Acta Biomaterialia* 101 (2020): 1–13.
[28] Milojević, Marko, Uroš Maver, and Boštjan Vihar. "Recent advances in 3D printing in the design and application of biopolymer-based scaffolds." *Functional Biomaterials: Design and Development for Biotechnology, Pharmacology, and Biomedicine* 2 (2023): 489–559.
[29] Scordo, G., V. Bertana, A. Ballesio, R. Carcione, S. L. Marasso, M. Cocuzza, C. F. Pirri, M. Manachino, M. Gomez Gomez, and A. Vitale. "Effect of volatile organic compounds adsorption on 3D-printed PEGDA: PEDOT for long-term monitoring devices." *Nanomaterials* 11 (2021): 94.
[30] Agunbiade, Adedoyin O., Lijun Song, Olufemi J. Agunbiade, Chigozie E. Ofoedu, James S. Chacha, Haile T. Duguma, Sayed Mahdi Hossaini et al. "Potentials of 3D extrusion-based printing in resolving food processing challenges: A perspective review." *Journal of Food Process Engineering* 45, no. 4 (2022): e13996.
[31] Grossin, David, Alejandro Montón, Pedro Navarrete-Segado, Eren Özmen, Giovanni Urruth, Francis Maury, Delphine Maury et al. "A review of additive manufacturing of ceramics by powder bed selective laser processing (sintering/melting): Calcium phosphate, silicon carbide, zirconia, alumina, and their composites." *Open Ceramics* 5 (2021): 100073.
[32] Lupone, Federico, Elisa Padovano, Oxana Ostrovskaya, Alessandro Russo, and Claudio Badini. "Innovative approach to the development of conductive hybrid composites for Selective Laser Sintering." *Composites Part A: Applied Science and Manufacturing* 147 (2021): 106429.
[33] Wasley, Thomas J. "Digitally driven microfabrication of 3D multilayer embedded electronic systems." PhD diss., Loughborough University, 2016.
[34] Arias-Ferreiro, Goretti, Ana Ares-Pernas, M. Sonia Dopico-García, Aurora Lasagabaster-Latorre, and María-José Abad. "Photocured conductive PANI/acrylate composites for digital light processing. Influence of HDODA crosslinker in rheological and physicochemical properties." *European Polymer Journal* 136 (2020): 109887.
[35] Rasaki, S. A., C. Liu, C. Lao, H. Zhang, and Z. Chen. "The innovative contribution of additive manufacturing towards revolutionizing fuel cell fabrication for clean energy generation: A comprehensive review." *Renewable and Sustainable Energy Reviews* 148 (2021): 111369.
[36] Galati, Manuela. "Electron beam melting process: A general overview." *Additive Manufacturing* (2021): 277–301.
[37] Chen, Yuanyuan, Martin Neff, Brian McEvoy, Zhi Cao, Romina Pezzoli, Alan Murphy, Noel Gately, Michael Hopkins Jnr, Neil J. Rowan, and Declan M. Devine. "3D printed polymers are less stable than injection moulded counterparts when exposed to terminal sterilization processes using novel vaporized hydrogen peroxide and electron beam processes." *Polymer* 183 (2019): 121870.
[38] Ryan, Kirstie R., Michael P. Down, Nicholas J. Hurst, Edmund M. Keefe, and Craig E. Banks. "Additive manufacturing (3D printing) of electrically conductive polymers and polymer nanocomposites and their applications." *eScience* 2 (2022): 365-381.
[39] Sharma, Neha, Nitheesh M. Nair, Garikapati Nagasarvari, Debdutta Ray, and Parasuraman Swaminathan. "A review of silver nanowire-based composites for flexible electronic applications." *Flexible and Printed Electronics*7 (2022): 014009.
[40] Jabeen, Nawishta, Qiuying Xia, Mei Yang, and Hui Xia. "Unique core–shell nanorod arrays with polyaniline deposited into mesoporous NiCo2O4 support for high-performance supercapacitor electrodes." *ACS Applied Materials & Interfaces* 8, no. 9 (2016): 6093–6100.
[41] Pan, Lijia, Guihua Yu, Dongyuan Zhai, Hye Ryoung Lee, Wenting Zhao, Nian Liu, Huiliang Wang et al. "Hierarchical nanostructured conducting polymer hydrogel with high electrochemical activity." *Proceedings of the National Academy of Sciences* 109, no. 24 (2012): 9287–9292.

[42] Patil, Prathamesh, Suneha Patil, Prachi Kate, and Amol A. Kulkarni. "Inkjet printing of silver nanowires on flexible surfaces and methodologies to improve the conductivity and stability of the printed patterns." *Nanoscale Advances* 3, no. 1 (2021): 240–248.

[43] He, Xin, Ruihui He, Qiuming Lan, Weijie Wu, Feng Duan, Jundong Xiao, Mei Zhang, Qingguang Zeng, Jianhao Wu, and Junyan Liu. "Screen-printed fabrication of PEDOT: PSS/silver nanowire composite films for transparent heaters." *Materials* 10, no. 3 (2017): 220.

[44] Wang, Zishen, Qin' E. Zhang, Shichuan Long, Yangxi Luo, Peikai Yu, Zhibing Tan, Jie Bai et al. "Three-dimensional printing of polyaniline/reduced graphene oxide composite for high-performance planar supercapacitor." *ACS Applied Materials & Interfaces* 10, no. 12 (2018): 10437–10444.

[45] Yan, X. H., Ping Gao, Gang Zhao, Le Shi, J. B. Xu, and T. S. Zhao. "Transport of highly concentrated fuel in direct methanol fuel cells." *Applied Thermal Engineering* 126 (2017): 290–295.

[46 Hussain, Ahmad, Adeela Naz, Nawishta Jabeen, and Jazib Ali. "3D Graphene for Flexible Sensors." In *3D Graphene: Fundamentals, Synthesis, and Emerging Applications*, pp. 131-149. *Cham: Springer Nature Switzerland*, 2023..

[47] Jee, Y., A. Karimaghaloo, A. Macedo Andrade, H. Moon, Y. Li, J-W. Han, S. Ji et al. "Graphene-based oxygen reduction electrodes for low temperature solid oxide fuel cells." *Fuel Cells* 17, no. 3 (2017): 344–352.

[48] Chang, Ikwhang, Taehyun Park, Jinhwan Lee, Ha Beom Lee, Sanghoon Ji, Min Hwan Lee, Seung Hwan Ko, and Suk Won Cha. "Performance enhancement in bendable fuel cell using highly conductive Ag nanowires." *International Journal of Hydrogen Energy* 39, no. 14 (2014): 7422–7427.

[49] Yang, Gaoqiang, Jingke Mo, Zhenye Kang, Yeshi Dohrmann, Frederick A. List III, Johney B. Green Jr, Sudarsanam S. Babu, and Feng-Yuan Zhang. "Fully printed and integrated electrolyzer cells with additive manufacturing for high-efficiency water splitting." *Applied Energy* 215 (2018): 202–210.

[50] Masaud, Zubair, Muhammad Zubair Khan, Amjad Hussain, Hafiz Ahmad Ishfaq, Rak-hyun Song, Seung-Bok Lee, Dong Woo Joh, and Tak-Hyoung Lim. "Recent activities of solid oxide fuel cell research in the 3D printing processes." *Transactions of the Korean Hydrogen and New Energy Society* 32, no. 1 (2021): 11–40.

[51] Aaron, Doug, Costas Tsouris, Choo Y. Hamilton, and Abhijeet P. Borole. "Assessment of the effects of flow rate and ionic strength on the performance of an air-cathode microbial fuel cell using electrochemical impedance spectroscopy." *Energies* 3, no. 4 (2010): 592–606.

[52] Mineshige, Atsushi, Minoru Inaba, Shinji Nakanishi, Masafumi Kobune, Tetsuo Yazawa, Kenji Kikuchi, and Zempachi Ogumi. "Vapor-phase deposition for dense CeO_2 film growth on porous substrates." *Journal of the Electrochemical Society* 153, no. 6 (2006): A975.

[53] Feng, Zuying, Liang Liu, Lingyao Li, Jiahe Chen, Yuhong Liu, Yan Li, Liang Hao, and Yan Wu. "3D printed Sm-doped ceria composite electrolyte membrane for low temperature solid oxide fuel cells." *International Journal of Hydrogen Energy* 44, no. 26 (2019): 13843–13851.

[54] Farandos, Nick, Lisa Kleiminger, Anna Hankin, Chin Kin Ong, and Geoff H. Kelsall. "3D printing of functional layers for solid oxide fuel cells and electrolysers." In *Electrochemical Society Meeting Abstracts sofc2015*, no. 1, pp. 343–343. The Electrochemical Society, Inc., 2015.

[55] Kim, Ho Sung, and Soo Hyung Kim. "Additive manufacturing and combustion characteristics of polyethylene oxide/aluminum/copper oxide-based energetic nanocomposites for enhancing the propulsion of small projectiles." *Nanomaterials* 13, no. 6 (2023): 1052.

[56] Melodia, Daniele, Abhirup Bhadra, Kenny Lee, Rhiannon Kuchel, Dipan Kundu, Nathaniel Corrigan, and Cyrille Boyer. "3D printed solid polymer electrolytes with bicontinuous nanoscopic domains for ionic liquid conduction and energy storage." *Small* (2023): 2206639.

[57] Jee, Y., A. Karimaghaloo, A. Macedo Andrade, H. Moon, Y. Li, J-W. Han, S. Ji et al. "Graphene-based oxygen reduction electrodes for low temperature solid oxide fuel cells." *Fuel Cells* 17, no. 3 (2017): 344–352.

[58] Al-Khori, Khalid, Yusuf Bicer, and Muammer Koc. "Integration of solid oxide fuel cells into oil and gas operations: Needs, opportunities, and challenges." *Journal of Cleaner Production* 245 (2020): 118924.

[59] Zakaria, Zulfirdaus, Saiful Hasmady Abu Hassan, Norazuwana Shaari, Ahmad Zubair Yahaya, and Yap Boon Kar. "A review on recent status and challenges of yttria stabilized zirconia modification to lowering the temperature of solid oxide fuel cells operation." *International Journal of Energy Research* 44, no. 2 (2020): 631–650.

[60] Liu, Renzhu, Chunhua Zhao, Junliang Li, Fanrong Zeng, Shaorong Wang, Tinglian Wen, and Zhaoyin Wen. "A novel direct carbon fuel cell by approach of tubular solid oxide fuel cells." *Journal of Power Sources* 195, no. 2 (2010): 480–482.

13 Electrochemical Properties of 3D-Printed Conducting Polymers in Relation to Supercapacitors

Tingting Jiang and George Z. Chen

13.1 SUPERCAPACITORS AND CONDUCTING POLYMER ELECTRODES

Electrochemical capacitors (ECs) or supercapacitors are promising electrochemical energy storage (EES) devices for portable electronic devices, microsensors, electrical vehicles, and renewable energy stations. Compared with lithium-ion or -metal batteries, ECs offer favorable power density, energy efficiency, safety and cycle life, and options for device configurations which translate to manufacturing simplicity and flexibility. These advantages, in combination with materials of lower resource and environmental restrictions such as porous carbon and metal oxides, offer greater sustainability.[1–4]

There are three types of ECs: electric double-layer (EDL) capacitors, pseudocapacitors (or redox EC), and their asymmetrical or hybrid devices. EDL capacitors store charges at the electrode/electrolyte interface and provide higher power density with fast ion adsorption/desorption but limited energy density. Since the 1950s,[5] a large family of EDL capacitors based on porous or nano-structured carbon of high specific areas has been commercially applied in many fields. Pseudocapacitors realize energy storage through redox reactions resulting from delocalized valence electrons, differing from those due to localized valence electrons.[4] Pseudocapacitance occurs not only at the surface but also in the electrode materials and can reach beyond the limit of EDL capacitance. However, due to ion transport restrictions, redox reactions are usually slower than ion absorption/desorption, resulting in inferior power performance. Transition metal oxides and electronically conducting polymers (ECPs), such as RuO_2 and polypyrrole, are among the most used electrode materials in pseudocapacitors. Hybrid or asymmetrical supercapacitors have been gaining attention recently, which combines the benefits of both pseudo- and EDL capacitors for better performance.

For high-power/energy applications, electrode materials with larger surface areas or redox capacities should be developed. For wearable devices, flexible and

DOI: 10.1201/9781003415985-13

stretchable electrodes are needed to maintain structural integrity after a large number of distortion/restoration and charging/discharging cycles.[6–8] On account of energy storage for miniature devices, unique planar or interdigital electrodes at micro- or nanoscales are needed. On the other hand, solid-state supercapacitors could meet the demand for miniaturization and high safety. Based on these requirements, appropriate electrode materials and the relevant manufacturing methods should be developed through, for instance, 3D printing of ECPs as discussed in this chapter.

Polyaniline (PANi), polypyrrole (PPy), polythiphene (PTh), and their derivatives have all been utilized for charge storage through pseudocapacitance and de-/doping of ions. Different from traditional polymers, ECPs are conductive due to electron delocalization or conjugation over the alternating single and double bonds formed by the overlap of carbon p_z orbitals along the polymer chain. ECPs can be p-doped where an electron is removed from the valence band to form a positive hole as the polymer is oxidized or n-doped where an electron is injected into the unfilled band as the polymer is reduced. Because conjugation results more often from oxidation or dehydrogenation, p-doping is more likely to be achieved among all reported ECPs.

Due to the conjugated bonds, ECPs in the oxidized or conductive states are rigid at molecular scales. However, their combination with a dopant, surfactant, or binder of flexible chain-like molecules or with a soft or gel-like material to form a composite can lead to satisfactory flexibility at the macroscale, as exemplified by polystyrene sulfonate-doped poly(3,4-ethylenedioxythiophene) (PSS:PEDOT). In the following text, flexibility and related properties refer to that of the respective composite.

In supercapacitors, ECPs deliver multiple advantages, including low densities, good processibility, low cost, and large charge capacity via reversible redox reactions. In addition, the conductivity can be tailored by controlling the doping level. An ECP supercapacitor may consist of two electrodes of the same or the opposite doping types, resulting in differences in cell voltage and charge capacity.

PANi and derivatives are among the most investigated in supercapacitors due to their low costs, ease of preparation, high and tailored doping level, large specific capacitance, and good stability. PANis can deliver higher capacitance than the others in both aqueous and nonaqueous electrolytes and form a hybrid cell with another type of electrode made from carbon materials such as carbon nanotubes and graphenes. PPy is one of the most important p-doped electrode materials and has the advantages of being nontoxic and highly conducting. Besides, it presents good electrochemical activity in various electrolytes including aprotic, aqueous, and nonaqueous ones. However, PPy can only be p-doped, limiting the application in type III and IV of ECP supercapacitor.[9] PThs, especially PEDOT, also promises long durability, low cost, and high conductivity. It is easy to realize both p-doping and n-doping, resulting in a wider application range in different types of supercapacitors. Besides, all these three types of ECPs can be designed to form composites with each other or other polymers or inorganics to achieve electrodes with better flexibility and performance.[10, 11]

Different from traditional manufacturing methods, additive manufacturing or three-dimensional (3D) printing provides a unique solution for polymer material fabrication in micro- or nano-sized, functional, and flexible devices. 3D printing can be computer designed and accurately controlled for complex structures. It can use different materials in the bottom-up process to achieve an integrated device consisting of all parts with fewer and smaller amounts of raw materials to benefit the environment. The main methods for 3D printing include extrusion, such as direct ink writing (DIW) and fused deposition modeling, and polymerization, such as powder-bed fusion selective laser sintering (SLS).[7, 12, 13]

Combining the unique advantages of 3D printing and the high flexibility and tailorability of ECPs, flexible and wearable supercapacitors and micro- or nano-electronic devices can be designed and manufactured without compromising electrochemical and electrical performances (Figure 13.1).

In this chapter, 3D-printed ECP electrodes for supercapacitors and their electrochemical properties will be introduced and analyzed. The involved materials, structures, and their performances in EES devices and techniques will be summarized, followed by a discussion on possible challenges.

13.2 3D-PRINTED CONDUCTING POLYMERS IN SUPERCAPACITORS

To meet the demand for wearable or miniaturized supercapacitors with high energy and power densities, the key is to design and fabricate flexible and stretchable electrode materials. 3D printing of ECP electrodes is an attractive strategy. Due to their specific conducive and chemical properties, 3D printing of ECPs differs from that of normal polymeric, metallic, or inorganic nonmetallic materials. Besides, manufacturing processes should be designed and optimized separately for different ECPs.

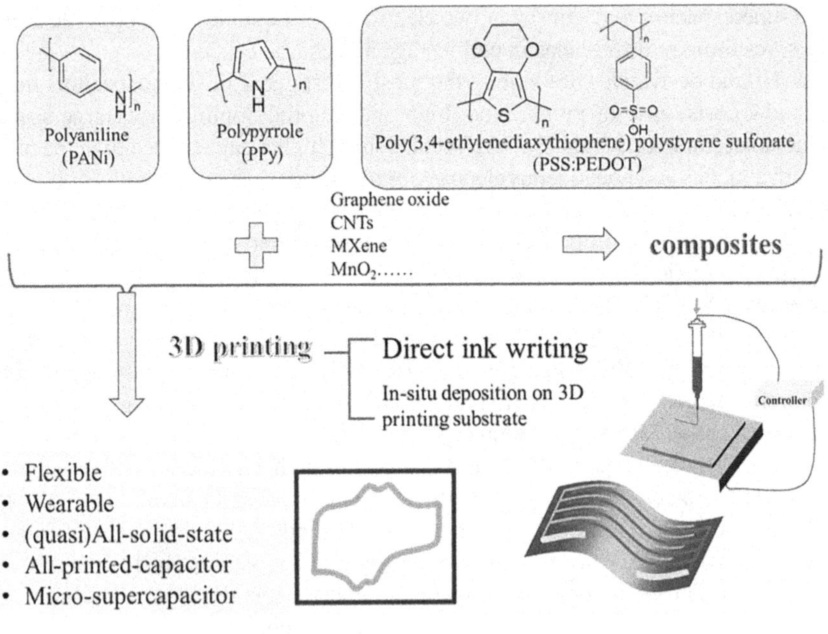

FIGURE 13.1 Schematic diagram of 3D printed ECP-based electrode materials in supercapacitors.

The structure of electrodes, including unique morphology, size, surface area, and shape may all influence the electrode performance and hence the whole device. Investigation of the influences of 3D printing techniques, and selection and processing of ECPs on electrode structures and performance could help improve the supercapacitors for wearable or micro-electronic applications.[14]

13.2.1 Techniques for 3D Printing of Conducting Polymer Electrodes

To achieve 3D structured ECP electrodes, in addition to the previously mentioned extrusion based DIW and fused deposition modeling, in-situ formation of the ECP is also effective on a 3D printed substrate such as carbon materials or metals.

DIW is the most investigated technique in 3D printing of ECP electrodes, the set-up is shown in Figure 13.2a.[15–18] For high energy density and wearable supercapacitors, DIW can fabricate electrodes with a high degree of flexibility and stretchability in both geometries and patterns with a rational design. It is easy to control the thickness of electrodes through the layer-by-layer process, enabling a shorter ion-transport distance and a higher electron transfer rate.[19] In addition, by changing the ink of electrode materials and electrolytes, DIW can realize the roll-to-roll preparation mode and build a whole supercapacitor device, both symmetric and asymmetric.[20] There are a series of reports on EDL capacitor material fabrication by DIW, such as carbon or metal carbides, which are further fabricated into all-solid-state supercapacitors (Figure 13.2b).

In the process of DIW, ink with appropriate rheological properties is essential to achieve printing success in terms of stability and performance in energy storage devices. First, the ink should be shear-thinning, which ensures the ink is homogenously dispersive, so that the extruded ink can be a pseudo-plastic fluid and flow through the micro-sized nozzle easily and continuously. On the other hand, polymeric inks are usually viscoelastic, which means once extruded from the nozzle, the ink should be dispersed on the stage in the desired pattern and size accurately and then quickly become self-standing and maintain the designed pattern, particularly for manufacturing micro-devices.

To meet the rheological requirements for ink in DIW of ECP electrodes, the solvents, usually organic, should be selected carefully to minimize toxicity or flammability. The demand for both eco-friendliness and rheological benefits should be balanced. In some recent works, aqueous inks have received more attention because of their lower costs and better bio-compatibility.[19, 21] Then, some polymer stabilizers or surfactants may be added to the solution or gel ink to maintain the stability of the ink. However, most stabilizers and surfactants are insulating, which may further deteriorate the carrier transfer efficiency in electrodes. A series of 2D inorganic materials have been selected to composite with ECPs to form more stable inks, including GO,[19] carbon nanotubes,[23] black phosphorous nanosheets,[24] and MXenes.[25, 26]

Eco-friendly polymers, such as poly (N-isopropylacrylamide) (PNIPAM) and polyethylene oxide (PEO),[21] have also been added to the ink to improve the viscosity. In addition, 3D-printed ECP electrodes are usually self-standing, without the need for a current collector. The weight ratio of active material in the ink determines

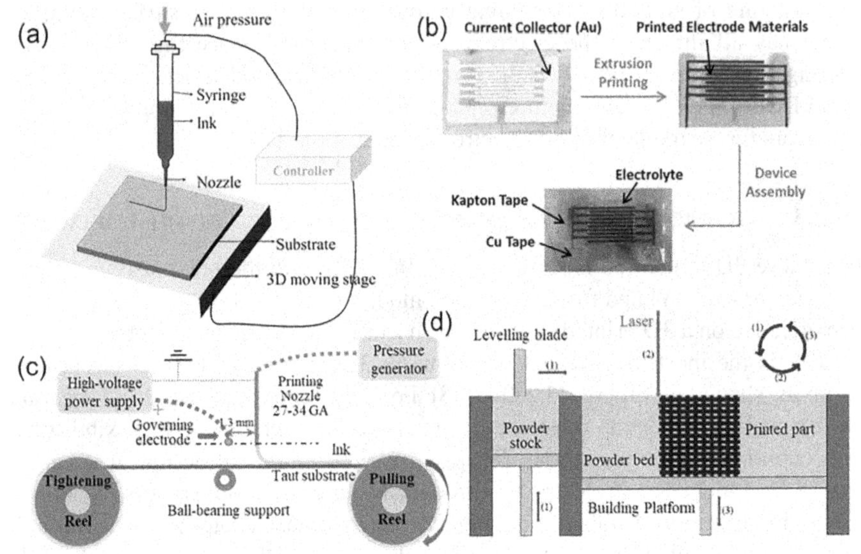

FIGURE 13.2 (a) Schematic diagram of DIW. (b) Schematic and photographic illustrations of the DIW fabrication of an all-solid-state supercapacitor on a flexible PET substrate. Adapted with permission.[19] Copyright (2018) Wiley. (c) The schematic of the eDIW prototype. Redrawn according to [21]. (d) The process of powder-bed technology. Adapted with permission.[22] Copyright (2018) Elsevier.

the capacitance of the final electrodes. The concentration of active material and the final accuracy in ink composition and printing outcome should also be balanced.

Modifications of the traditional DIW technique were proposed by Wang *et al.*, who used an electric field-assisted DIW (eDIW) to produce PSS:PEDOT-based electrodes in supercapacitors.[21] By integrating an electric field, higher speed, more stable printing and greater resolution could be achieved with the same ink (Figure 13.2c). The printing of PSS:PEDOT aqueous ink by eDIW reached a high speed of 1.72 m s^{-1} with micro-scale resolution, which is much higher than the traditional DIW methods.

There are other strategies to achieve a 3D structured electrode. For example, 3D metal printing is a mature additive manufacturing technology, such as powder-bed technology, and selected laser melting.[22] First, a 3D structured porous metal scaffold is manufactured by 3D printing with controlled laser power and scan speed (Figure 13.2d). Then, the ECP material is deposited on the outer surface of the metal scaffold to form the electrodes for supercapacitors. Lu *et al.*[22] proposed an approach to building a porous iron-nickel current collector with low density and high specific surface area via a layer-by-layer power-bed technique. Then, the PANi material was deposited on the porous alloy by in-situ anodic polymerization. In another case, a porous stainless steel (316L) scaffold was 3D printed, exhibiting high strength and surface area. Then, MnO_x and PSS:PEDOT were co-electrodeposited into the pores of the stainless steel scaffold.[27] Similarly, carbon material can also act as both EDL capacitive electrode material and current collectors for a faradic electrode material like ECPs. Graphene was first direct-ink-written into an aerogel template in the presence of a viscosifier and a gel former, which could then be removed by heating at elevated temperatures. Then, PPy was chemically deposited on the aerogel template to achieve a hybrid electrode material for supercapacitors.[28, 29] As an excellent conductor, graphene can also be 3D printed into unique structures by fused deposition modeling (FDM) in the form of a conductive filament.[30] After this rapid and cost-effective process, PPy can be electrodeposited on the surface of graphene. In these two examples, PPy-GA composite electrodes could provide a more porous, flexible electrode than those with metal scaffolds and hence offer better EES performance.

13.2.2 Materials of 3D-Printed Conducting Polymer Electrodes

As mentioned in section 13.1, the three main ECPs fabricated by additive manufacturing for supercapacitor electrodes are PANi, PPy, and PSS:PEDOT.

As one of the most widely used pseudocapacitive materials, PPy could provide high electrochemical activity, low density, good mechanical properties, and good flexibility. However, the disordered structure may result in limited electrolyte penetration and hence the impeded ion diffusion path, leading to poor rate capability and cycling stability. At the same time, its poor dispersion in the gel or paste of PPy hinders the ink-based 3D printing process. It is difficult to prepare the 3D-printing ink with pure PPy, which is usually powdery in nature if prepared via chemical polymerization. 2D materials such as black phosphorus nanosheets can be used to composite with PPy to increase the contact area between the active material and electrolyte and create an extra buffer space to improve the electrochemical performance of PPy. A non-ionic surfactant was additionally used to modify the viscoelasticity of the solution and achieve a more printable ink, which was then directly written to build 3D PPy-black phosphorus composite electrodes. PPy has also been reported as an additional deposited layer on the 3D-printed graphene substrate that showed enhanced specific capacitance.[11, 24, 28]

PANi attracts much attention as an electrode material for pseudocapacitors due to its unique doping and dedoping chemistry and high electric conductivity. Although PANi presents moderate solubility in some polar organic solvents and is easy to prepare, there are rare reports on the 3D printing of pure PANi to make electrodes in supercapacitors. DIW can be applied to construct a 3D structured electrode after blending the polymer with GO in solution due to the higher concentration of the shear-thinning solution, which is attributed to the polymer/GO supramolecular interactions. The composite electrode provides higher electric conductivity and better mechanical strength. PANi can also be deposited on pre-printed 3D scaffolds as an electrode, such as the alloy scaffold prepared by the powder-bed technique.[11, 31]

PSS:PEDOT has tunable electrical and mechanical properties, structural stability, good conductivity, biocompatibility, and ease of processing. Different from PPy and PANi, PSS:PEDOT exhibits good water dispersibility, which provides more probability of its utilization in ink-based 3D printing techniques. It is perhaps the most reported ECP for 3D-printed electrodes in energy storage devices. Additionally, PSS:PEDOT can be made into 3D electrodes by DIW without compositing with other materials. However, the low-viscosity solution of PSS:PEDOT needs extra modifications with additives to improve the shear viscosity and other processing properties. Yang *et al.* performed DIW with an aqueous or blend solution of PSS:PEDOT nanofibrils as printable ink and obtained a series of 3D negative Poisson's ratio (NPR) structures with different patterns. A free-standing stretchable electrode was further assembled in the quasi-solid-state supercapacitor. Ovhal *et al.*[32] selected a dimethyl sulfoxide (DMSO)–doped PSS:PEDOT soluble ink to produce 3D electrodes. Based on the roll-to-roll 3D DIW process, a flexible micro-supercapacitor can be fabricated. A composite based on PSS:PEDOT printed by DIW and post-deposited PSS:PEDOT on pre-3D-printed scaffolds have also been reported. MXene has also been proven as a good partner of PSS:PEDOT in gel ink, resulting in a much higher electric conductivity of printable ink than other inks.[25] CNTs have been added to the ink formation process to further improve energy storage performance with high conductivity and stretchability.[11, 21]

13.2.3 Structure and Properties of 3D-Printed Conducting Polymer Electrodes

Apart from the selection of materials, the diversity in structure, pattern, and micromorphology of 3D-printed polymeric electrodes also affect the properties. Generally, mechanical durability and electric conductivity are of great concern in printed polymer-based electrodes.

The NPR structure provides unique mechanical properties of expansion in all directions under unidirectional stress.[33] Thus, once the loads are subjected to the NPR structure, the bulk modulus will be reduced by the rotation of the elastic cell, which may prevent the fracture of the material and lead to improved stretchability and flexibility. Yang *et al.*[23] introduced an arc-shaped microstructure into DIW-printed PSS:PEDOT-based 3D electrodes. Various microstructures, including S-hinged, re-entrant, chiral, and wavy mesh, demonstrate the broadened strain range (Figure 13.3a). By finite-element analysis and the uniaxial tensile test, the electrodes with a wavy mesh structure and S-hinged geometry exhibited the highest and lowest maximum stress, respectively. According to the experimental details and simulation analysis, the ultimate tensile rates of the S-hinged electrode reached as high as 150%, and the rates were 30%, 18%, and 11.5% in re-entrant, chiral, and wavy mesh electrodes, respectively.

Jain *et al.*[34] performed DIW 3D printing with a conductive ink based on modified cellulose fibers (dialcohol cellulose, DALC) and PSS:PEDOT and obtained four different patterns: 20-layer hollow cube, hollow cylinder, 5-layer serpentine pattern, and 5-layer mesh structure (Figure 13.3b). By modifying the structure and the ratio between DALC and PSS:PEDOT, the printed structure exhibited a high conductivity of 30 S cm^{-1} and high tensile strains (>40%). Li *et al.*[25] fabricated a composite of PSS:PEDOT/MXene by DIW with a gel ink, which was further utilized in an all-printed flexible micro-supercapacitor. Benefitting from the flexibility of both the PSS:PEDOT chains and MXene nanosheets, several 3D electrodes with different patterns have been printed with abundant porosity (Figure 13.3c).

Pattern design and the corresponding mechanical properties of 3D-printed conducting polymeric electrodes for supercapacitors, which may be affected by the printing process, rheology of ink, and deposition substrate, could demonstrate the

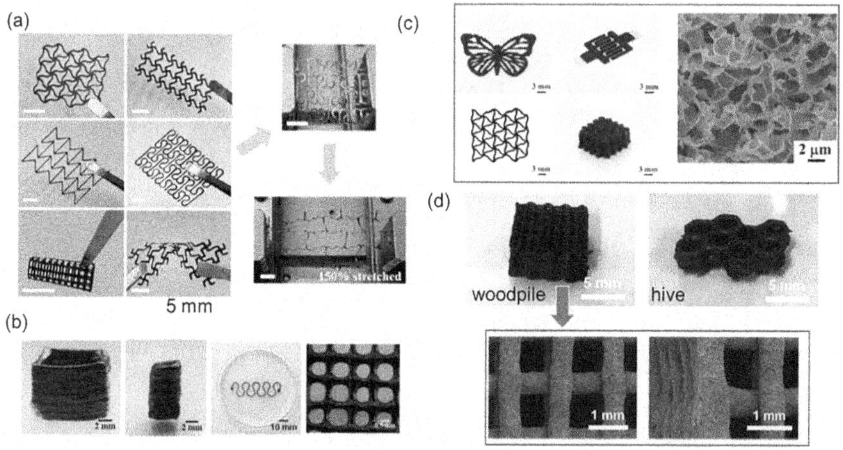

FIGURE 13.3 (a) Images of 3D-printed PSS:PEDOT electrodes with different NPR structures. Adapted with permission from.[23] Copyright (2021) The Royal Society of Chemistry. (b) Images of printed 3D and 2D patterns of (from left to right) a 20-layer hollow cube, a hollow cylinder, a 5-layer serpentine pattern, and a 5-layer mesh structure with modified cellulose fiber/PSS:PEDOT ink. Adapted with permission.[34] Copyright (2023) Wiley. (c) Images of the printed butterfly pattern, interdigital electrode, NPR structure, and thick microlattice based on PSS:PEDOT/MXene/ethylene glycol (EG) gel composite inks. Adapted with permission.[25] Copyright (2023) Wiley. (d) Optical and SEM images of the woodpile and hive structures printed by DIW with black phosphorus/PPy ink. Adapted with permission.[24] Copyright (2020) The Royal Society of Chemistry.

possibility of manufacturing printable electrodes and all-printable supercapacitors. Xing *et al.*[24] printed a black phosphorus nanosheet (BPNS)/PPy lattice by DIW with two designed patterns, woodpile and hive (Figure 13.3d). In particular, it can be seen that the woodpile pattern of the BPNS/PPy composite exhibited a circular shape of filament, indicating the exact maintenance of BPNS/PPy ink after the directed writing process.

The interdigital pattern is also a common design in 3D printing of ECP electrodes. 2D patterns including interdigit, antennas, and electronic circuits and 3D patterns like cylinders and honeycombs have been fabricated from GO/PANi composite gel ink.

13.2.4 Electrochemical Performance of 3D-Printed Conducting Polymer Electrodes in Supercapacitors

Supercapacitor electrodes based on 3D-printed ECPs, including PSS:PEDOT, PPy, and PANi, have been reported since 2016.[27] The electrochemical performance of both the single electrode and the whole energy storage devices such as flexible devices, all-solid-state devices, and supercapacitors with the same or different electrodes have been extensively explored. The electrochemical characteristics and the charge/energy storage ability are influenced by the nature, structures, and morphology of the ECPs. The electrochemical performance of 3D-printed ECP-based electrodes in supercapacitors are summarized in Table 13.1.

TABLE 13.1
Electrochemical Performances of 3D-printed ECP-Based Electrodes for Supercapacitors

Material	Technique	Electrochemical Performance (Electrode)	Electrochemical Performance (Device)	Retention	Energy, Power
PANi/rGO[31]	Direct ink writing	423 F g^{-1} @0.8 A g^{-1}	1329 mF cm^{-2} @4.2 mA cm^{-2}	75% after 1000 cycles	
PSS:PEDOT/CNTs[23]	Direct ink writing	990 mF cm^{-2} @1 mA cm^{-2}	730 mF cm^{-2} @1 mA cm^{-2}	74.7% after 14,000 cycles	
PSS:PEDOT/MXene[25]	Direct ink writing		889 mF cm^{-2} @1 mA cm^{-2} (ten layers)	83% after 6000 cycles	
PSS:PEDOT/DALC[34]	Direct ink writing	197 F g^{-1} @ 10 A g^{-1}			
PSS:PEDOT[32]	Direct ink writing		31.6 mF cm^{-2}, 15.6 F g^{-1} @5 mA cm^{-2}		0.36 mWh cm^{-2}, 80 mW cm^{-2}
Conductive polyethylene[35]	3D printing of supercapacitor fiber		220 mF cm^{-2} @4 mA cm^{-2}	≈100% over 13,000 cycles	306 μWh cm^{-2} @3 V
MnOx/PSS:PEDOT[27]	3D printing of stainless steel as scaffold+ co-electrodeposition	1.52 F cm^{-2} @1 mA cm^{-2}			
Cellulose/PEDOT:TOS[36]	3D printing of aerogel+ in-situ polymerization	78 F g^{-1} @0.1 A g^{-1}			
Black phosphorous/PPy[24]	Direct ink writing	417 F g^{-1} @0.2 A g^{-1}			6.5 Wh kg^{-1}, 0.0374 W kg^{-1}
GO/PANi[19]	Direct ink writing		153.6 mF cm^{-2}, 19.2 F cm^{-3} @5 mV s^{-1} (symmetric device)	100% over 5000 cycles	4.83 mWh cm^{-3}, 25.3 W cm^{-3}
PPy/graphene[28]	3D printing of graphene aerogel+ in-situ deposition	2 F cm^{-2} @6 A g^{-1}	1.1 Fcm^{-2} @6 A g^{-1}	75% after 5000 cycles	0.78 mWh cm^{-2}
PANi[22]	3D printing of Fe-Ni alloy+ in-situ deposition	540.68 F g^{-1} @5 mVs^{-1}			
Ppy/rGO[29]	3D printing of graphene aerogel+ in-situ deposition		98.37 F g^{-1} @0.5 Ag^{-1}		

The electrochemical performance of direct ink–written PSS:PEDOT-based electrodes with a unique NPR structure was investigated.[23] Figure 13.4a–b shows the cyclic voltammograms (CVs) at a scan rate of 20 mV s^{-1} and galvanostatic charge-discharge (GCD) curves at 1 mA cm^{-2} of a PSS:PEDOT electrode and PSS:PEDOT/CNT electrode in a three-electrode cell in a 1.0 M H_2SO_4 electrolyte with a Ag/AgCl reference electrode and a Pt foil counter electrode. With the addition of CNTs, the electrode exhibited a more rectangular shape, due to the EDL capacitive storage characteristics of CNTs. The larger integral area of CV reflected an improved capacitance of 990 mF cm^{-2} by adding CNTs due to more efficient carrier transportation and many more active sites for charge storage, which is attributed to the interconnected CNTs in the ECP framework. The PSS:PEDOT/CNTs composite was further utilized in a quasi-solid-state symmetric supercapacitor with the PVA/H_2SO_4 gel electrolyte (Figure 13.4c–d). Similar to the electrochemical performance of PSS:PEDOT/

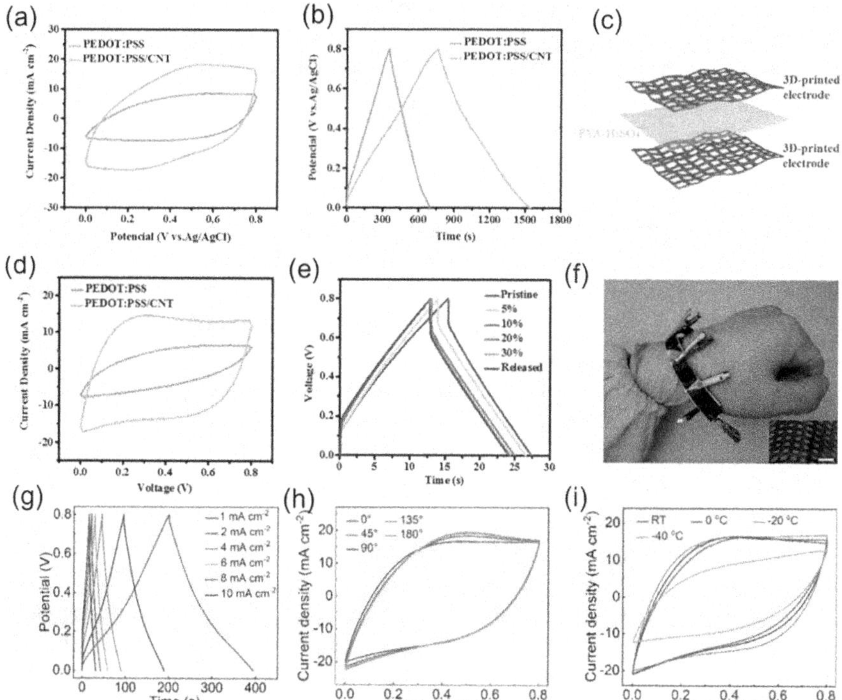

FIGURE 13.4 (a) CVs at 20 mV s^{-1} and (b) GCDs at 1 mA cm^{-2} of 3D-printed PSS:PEDOT and PSS:PEDOT/CNT electrodes. (c) Schematic of symmetric supercapacitor consisting of two identical PSS:PEDOT/CNT electrodes. (d) CVs at 20 mV s^{-1} of PSS:PEDOT supercapacitor and PSS:PEDOT/CNT supercapacitor. (e) GCDs of the PSS:PEDOT/CNT stretchable supercapacitor under different stretching strains. (f) Photographs of a wearable PSS:PEDOT/CNT supercapacitor. The inset shows a magnified optical image of the electrode (scale bar: 1 mm). Adapted with permission.[23] Copyright (2021) The Royal Society of Chemistry. (g) GCDs of PSS:PEDOT/MXene/EG micro-supercapacitor at current densities from 1 to 10 mA cm^{-2}. CVs of PSS:PEDOT/MXene/EG micro-supercapacitor at different (h) bending degrees and (i) low temperatures. Adapted with permission.[22] Copyright (2023) Wiley.

CNTs electrodes, symmetric supercapacitors consisting of PSS:PEDOT/CNTs also achieved an areal capacitance of 730 mF cm^{-2} at 1 mA cm^{-2}, which is much larger than that of the PSS:PEDOT electrode. It could still maintain over 90% of its initial electrochemical performance under different stretching strains (Figure 13.4e). Therefore, this flexible symmetric supercapacitor demonstrated good properties for wearable devices (Figure 13.4f).

Apart from the symmetric structure, interdigital electrodes have proven beneficial for charge storage and transfer by shortening the diffusion length of ions. The gel ink of PSS:PEDOT/MXene/EG (PME) was printed into 3D electrodes and further assembled in micro-supercapacitors.[22] MXene nanosheets can act as a bridge to connect PEDOT and PSS chains in gel ink to form a 3D interconnect framework, resulting in remarkable improvements in both the viscosity of the ink and the electrochemical properties of the printed electrode. By means of DIW 3D printing, it is possible to fabricate thick electrodes with higher mass loading of the active material. Therefore, the micro-supercapacitor of PME exhibited a high areal capacitance of 242.4 mF cm^{-2} at a current density of 1 mA cm^{-2} (Figure 13.4g). The PME supercapacitor also demonstrated good stretchability, flexibility, and low-temperature properties. When subjected to a large bending degree, the supercapacitor exhibited CVs nearly the same as the initial one (Figure 13.4h). Also, EG in the PME electrode could improve its anti-freezing property, which is necessary for outdoor wearable applications. Figure 13.4i shows the good stability in the electrochemical performance of PME supercapacitors from room temperature to −20 °C.

DIW of the PANi/GO ink was performed with various stabilizers or additives, and the 3D electrode obtained represented the potential in wearable energy storage applications. Wang *et al.*[31] prepared interdigital electrodes of PANi/GO by DIW with different mass ratios of GO and PANi. CVs of the PANi0.4/rGO electrode were recorded at different scan rates in a three-electrode cell with 1.0 M H_2SO_4 as the electrolyte, the rGO aerogel on a Pt foil as the counter electrode, and an SCE reference electrode (Figure 13.5a). These CVs are broadly capacitive in nature, where the two couples of redox peaks at 0.3 and 0.5 V were associated with the redox couples of leucoemeraldine/emeraldine and hydreoxyl/quinonyl (and/or amino/imine) terminal groups of oligoanilines, respectively. The capacitance derived from the GCDs of the printed electrode was 423 F g^{-1} at 0.8 Ag^{-1} (Figure 13.5b). Based on this PANi0.4/rGO composite ink, an interdigital electrode and the all-printed planar supercapacitor with the interdigital electrode were fabricated (Figure 13.5c). The H_2SO_4-PVA gel electrolyte in this solid supercapacitor was also 3D printed into the gap between two interdigital electrodes. CVs at various scan rates of the all-printed PANi/rGO supercapacitor indicate capacitive dominance with minor Nernstian (battery-like) contributions in accordance with the combined charge storage mechanisms in PANi and GO (Figure 13.5d). According to the GCD in Figure 13.5e, the areal capacitance could reach as high as 1329 mF cm^{-2} at a current density of 4.2 mA cm^{-2}, which is much higher than that of the planar supercapacitor with graphene electrodes.

Liu *et al.*[19] also prepared a GO/PANi composite electrode by DIW with aqueous ink and further fabricated the all-solid-state symmetric flexible micro-supercapacitors. During the DIW process, by repetitive layer-by-layer printing of the GO/PANi ink, interdigitated electrodes with different layers were obtained. The

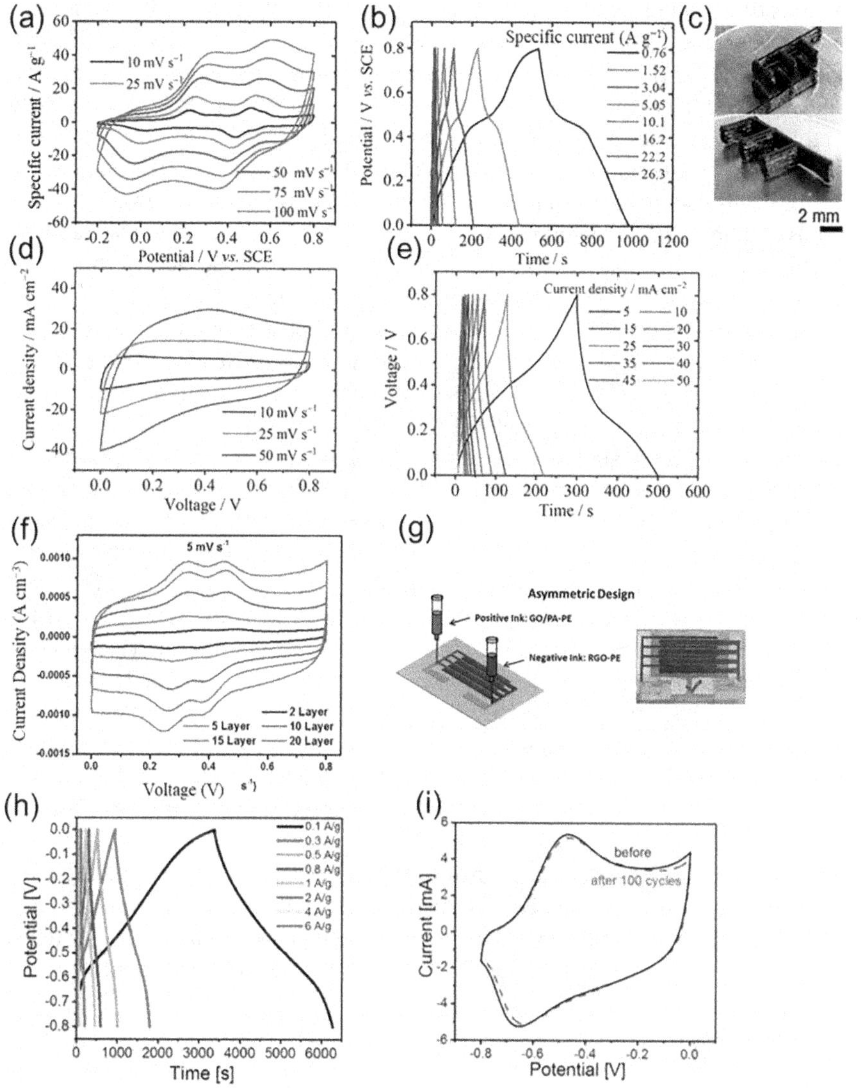

FIGURE 13.5 (a) CVs at different scan rates in the potential range from –0.2 to 0.8 V and (b) GCDs at different specific currents of a 3D-printed PANi0.4/rGO electrode. (c) Photographs of printed interdigital PANi0.4/rGO electrodes. (d) CVs at different scan rates and (e) GCDs at different current densities of a planar supercapacitor with interdigital PANi0.4/rGO electrodes. Adapted with permission.[31] Copyright (2018) American Chemical Society. (f) CVs at 5 mV s^{-1} of symmetric flexible supercapacitors with different numbers of printing layers and hence different thicknesses and mass loading of GO/PANi electrode materials. (g) Schematic illustration of DIW printing and the photograph of a printed asymmetric flexible supercapacitors. Adapted with permission.[22] Copyright (2018) Wiley. (h) The GCDs of PPy–GA-0.1 electrode at different specific currents. (i) CVs before and after 100 compression cycles of a PPy–GA micro-lattice electrode. Adapted with permission.[28] Copyright (2018) Wiley.

CVs of DIW-printed GO/PANi electrodes with different layers at the scan rate of 5 mV s^{-1} exhibited a rectangular shape with two pairs of small but notable redox peaks of the PANi (Figure 13.5f). As the number of layers increased, the areal capacitance increased. The highest areal capacitance was 153.6 mF cm^{-2} in the 20-layer electrode. By depositing the ink during the printing process on the metallic current collector and casting the PVA-H_3PO_4 polymer gel electrolyte onto the interdigitated electrodes, all-solid-state flexible micro-supercapacitors could be successfully assembled. Based on a similar technique, an asymmetric supercapacitor was also fabricated by adding an extra separate syringe for the negative electrode during the printing process (Figure 13.5g). Due to the appropriate charge balance design, the GO/PANi-PSS:PEDOT/rGO-PSS:PEDOT asymmetric supercapacitor with interdigital electrodes demonstrated high capacitance retention at nearly 100% after 5000 cycles.

Manufactured by DIW of GO and self-assembly of conducting polymers, super-elastic PPy-coated graphene aerogel (GA) electrodes could deliver good charge storage properties as well as flexibility and elasticity. Qi *et al.*[28] performed 3D printing with a GO ink and obtained a GA template after freeze-drying and annealing. After in-situ polymerization of PPy on GA, a composite electrode was made for a supercapacitor with flexibility and elasticity. The GCDs of PPy-GA at various specific currents in Figure 13.5h show a triangular shape that deviates from the ideal one, indicating the presence of Nernstian performance of PPy. By comparing composites with different PPy ratios, the sample treated in the 0.1 M PPy solution achieved the best electrochemical performance of a high specific capacitance of 395 F g^{-1} at 0.1 Ag^{-1} and a capacitance retention above 90% after 5000 cycles. The compressive test was performed on the best electrode. After 100 compression cycles (to 90% strain), CVs showed good repeatability with good anti-compression properties and less compression fatigue (Figure 13.5i).

Generally, 3D printing of ECP provides new prospects for the fabrication of both electrodes and whole cells of micro-supercapacitors, especially for applications in wearable and flexible devices. The redox active nature of ECPs could provide high specific capacitance in energy storage devices. Benefitting from various designs and accurate manufacturing, more structures of supercapacitors can be realized with efficient carrier transfer and high flexibility, providing the possibility for flexible and wearable applications.

13.3 CHALLENGES AND OUTLOOK

Based on the previously discussed ECP and 3D printing methods, supercapacitors with wearable, micro, and solid-state characteristics have been widely investigated. There are four directions for the future application and development of 3D-printed ECPs in supercapacitors.

(1) The modification of electrodes fabricated by 3D printing in supercapacitors. Ease of fabrication to achieve tailorable structure and function is one of the most visible characteristics of 3D-printed ECP electrodes. However, the limited performance in cycling stability, specific capacitance, ion diffusion property, solubility in ink solvents, and temperature dependence of ECPs should be further improved. Compositing with one or more other electrode materials with higher specific capacitance, structure regulation, and

micro-structure design may achieve better capacitive energy storage performance. Furthermore, the rheological properties of ECP inks should be controlled more precisely during 3D printing. The amounts of currently used additives, which are used to balance the printing efficiency, accuracy, and electrochemical performance but are inactive in charge storage, should also be minimized.

(2) The construction of supercapacitor devices with 3D-printed ECP electrodes. ECP electrodes can be constructed in different geometries with high resolution by 3D printing. The design of electrodes can be further combined with the structure modification of energy storage devices. For example, the traditional sandwich-structured devices require single-directional mass and charge transfer, while planar supercapacitors are assembled with interdigital electrodes with a more compact structure without compromising the efficiency of charge transfer. Also, solid or quasi-solid electrolytes could also be 3D printed along the printed electrodes to form all-solid-state supercapacitors. The ionic conductivity, 3D printing properties, and compatibility of electrolytes with the two electrodes should also be investigated. Last, ECPs are usually directly ink-written on a substrate, which may be later directly used as the electrodes of supercapacitors. The adhesive property of the ECPs on a substrate during 3D printing and energy storage process, and the mechanical strength, electric conductivity, and stability of electrodes should be considered.

(3) Demands from the supercapacitors. 3D-printed ECPs are important candidates for micro-supercapacitors. Except for the high specific capacitance, the printing accuracy of 3D printed electrodes should meet the requirements of micro devices, which is dependent on the development of both the DIW technology and ECP ink composition. On the other hand, all-printed supercapacitors with 3D-printed electrodes and electrolytes are also the trend of the future supercapacitors, which can be directly obtained through one fabrication technique.

(4) Requirements of application scenarios. As discussed, 3D-printed ECP-based supercapacitors have shown great application potentials in wearable micro-supercapacitors and multi-functional sensors. Therefore, the conducting polymeric electrodes produced by 3D printing should be mechanically flexible and stable, water or moisture resistant, and low/high temperature adaptable to meet the potentially complex working environment.

13.4 SUMMARY

Recent advances in the developments of 3D-printed ECPs and their electrochemical performance in supercapacitors have been discussed. DIW is the most widely used technique for producing ECP electrodes based on PPy, PANi, PSS:PEDOT and, more importantly, their composites with other capacity and performance enhancing materials. By regulating the ink composition, electrodes with various structures and tunable mechanical properties can be printed and further assembled in supercapacitors. The influences of electrode design and supercapacitor structure on the electrochemical performance of ECPs were summarized. Then, four aspects of the future development of 3D-printed conducting polymeric electrodes for supercapacitors were discussed.

In summary, 3D printing is expected to offer more opportunities for novel structures of ECP electrode materials in supercapacitors, which could promote the development of wearable, micro, and highly integrated devices. It is believed that by combining advanced and precise fabrication methods with tailorable ECPs as electrodes, solid-state supercapacitors compatible with complex situations will become available in the near future.

ACKNOWLEDGMENTS

The authors acknowledge financial support from the Natural Science Foundation of Hubei Province (2021CFB434), the EPSRC (GR/R68078), and Innovate UK (10017140).

REFERENCES

[1] Simon P and Gogotsi Y, Materials for electrochemical capacitors. *Nat. Mater*. 2008. **7**, 845–854

[2] Conway B E, Transition from "supercapacitor" to "battery" behavior in electrochemical energy storage. *J. Electrochem. Soc*. 1991. **138**, 1539

[3] Winter M and Brodd R J, What are batteries, fuel cells, and supercapacitors? *Chem. Rev*. 2004. **104**, 4245–4270

[4] Chen G Z, Supercapacitor and supercapattery as emerging electrochemical energy stores. *Int. Mater. Rev*. 2017. **62**, 173–202

[5] Becker H I, *Low voltage electrolytic capacitor*. 1957, US patent (US2800616).

[6] Zhang S, Liu Y, Hao J, Wallace G G, Beirne S, and Chen J, 3D-printed wearable electrochemical energy devices. *Adv. Funct. Mater*. 2021. **32**, 2103092

[7] Liang J, Jiang C, and Wu W, Printed flexible supercapacitor: Ink formulation, printable electrode materials and applications. *Appl. Phys. Rev*. 2021. **8**, 021319

[8] Patel K K, Singhal T, Pandey V, Sumangala T P, and Sreekanth M S, Evolution and recent developments of high performance electrode material for supercapacitors: A review. *J. Energy Storage*. 2021. **44**, 103366

[9] Lu M, *Supercapacitors: Materials, systems, and applications*. 2013, John Wiley & Sons, Singapore.

[10] Loganathan N N, Perumal V, Pandian B R, Atchudan R, Edison T N J I, and Ovinis M, Recent studies on polymeric materials for supercapacitor development. *J. Energy Storage*. 2022. **49**, 104149

[11] Banerjee S and Kar K K, Conducting polymers as electrode materials for supercapacitors, in *Handbook of nanocomposite supercapacitor materials II: Performance*, Springer Series in Materials Science, vol. 302. Kar, K (ed.). 2020, Springer, Cham. pp. 333–352.

[12] Martinelli A, Nitti A, Giannotta G, Po R, and Pasini D, 3D printing of conductive organic polymers: Challenges and opportunities towards dynamic and electrically responsive materials. *Mater. Today Chem*. 2022. **26**, 101135

[13] Shen K, Ding J, and Yang S, 3D printing quasi-solid-state asymmetric micro-supercapacitors with ultrahigh areal energy density. *Adv. Energy Mater*. 2018. **8**, 1800408

[14] Yuk H, Lu B, Lin S, Qu K, Xu J, Luo J, and Zhao X, 3D printing of conducting polymers. *Nat. Commun*. 2020. **11**, 1604

[15] Jordan R S and Wang Y, 3D printing of conjugated polymers. *J. Poly. Sci. Part B: Poly. Phys*. 2019. **57**, 1592–1605

[16] Ryan K R, Down M P, Hurst N J, Keefe E M, and Banks C E, Additive manufacturing (3D printing) of electrically conductive polymers and polymer nanocomposites and their applications. *eScience*. 2022. **2**, 365–381

[17] Ligon S C, Liska R, Stampfl J, Gurr M, and Mulhaupt R, Polymers for 3D printing and customized additive manufacturing. *Chem. Rev*. 2017. **117**, 10212–10290

[18] Dou P, Liu Z, Cao Z, Zheng J, Wang C, and Xu X, Rapid synthesis of hierarchical nanostructured Polyaniline hydrogel for high power density energy storage application and three-dimensional multilayers printing. *J. Mater. Sci.* 2016. **51**, 4274–4282
[19] Liu Y, Zhang B, Xu Q, Hou Y, Seyedin S, Qin S, Wallace G G, Beirne S, Razal J M, and Chen J, Development of graphene oxide/polyaniline inks for high performance flexible microsupercapacitors via extrusion printing. *Adv. Funct. Mater.* 2018. **28**, 1706592
[20] Melodia D, Bhadra A, Lee K, Kuchel R, Kundu D, Corrigan N, and Boyer C, 3D Printed solid polymer electrolytes with bicontinuous nanoscopic domains for ionic liquid conduction and energy storage. *Small.* 2023. e2206639
[21] Wang X, Plog J, Lichade K M, Yarin A L, and Pan Y, Three-dimensional printing of highly conducting PEDOT: PSS-based polymers. *J. Manuf. Sci. Eng.* 2023. **145**, 011008
[22] Lu X, Zhao T, Ji X, Hu J, Li T, Lin X, and Huang W, 3D printing well organized porous iron-nickel/polyaniline nanocages multiscale supercapacitor. *J. Alloys Compd.* 2018. **760**, 78–83
[23] Yang J, Cao Q, Tang X, Du J, Yu T, Xu X, Cai D, Guan C, and Huang W, 3D-Printed highly stretchable conducting polymer electrodes for flexible supercapacitors. *J. Mater. Chem. A.* 2021. **9**, 19649–19658
[24] Xing R, Xia Y, Huang R, Qi W, Su R, and He Z, Three-dimensional printing of black phosphorous/polypyrrole electrode for energy storage using thermoresponsive ink. *Chem. Commun.* 2020. **56**, 3115–3118
[25] Li L, Meng J, Bao X, Huang Y, Yan X P, Qian H L, Zhang C, and Liu T, Direct-ink-write 3D printing of programmable micro-supercapacitors from MXene-regulating conducting polymer inks. *Adv. Energy Mater.* 2023. **13**, 2203683
[26] Luo W, Ma Y, Li T, Thabet H K, Hou C, Ibrahim M M, El-Bahy S M, Xu B B, and Guo Z, Overview of MXene/conducting polymer composites for supercapacitors. *J. Energy Storage.* 2022. **52**, 105008
[27] Liu X, Jervis R, Maher R C, Villar-Garcia I J, Naylor-Marlow M, Shearing P R, Ouyang M, Cohen L, Brandon N P, and Wu B, 3D-printed structural pseudocapacitors. *Adv. Mater. Technol.* 2016. **1**, 1600167
[28] Qi Z, Ye J, Chen W, Biener J, Duoss E B, Spadaccini C M, Worsley M A, and Zhu C, 3D-printed, superelastic polypyrrole–graphene electrodes with ultrahigh areal capacitance for electrochemical energy storage. *Adv. Mater. Technol.* 2018. **3**, 1800053
[29] Foo C Y, Lim H N, Mahdi M A, Wahid M H, and Huang N M, Three-dimensional printed electrode and its novel applications in electronic devices. *Sci. Rep.* 2018. **8**, 7399
[30] Srinivasan K V S, Santo J, and Penumakala P K, Effect of surface modification of printed electrodes on the performance of supercapacitors. *J. Energy Storage.* 2022. **56**, 106043
[31] Wang Z, Zhang Q E, Long S, Luo Y, Yu P, Tan Z, Bai J, Qu B, Yang Y, Shi J, Zhou H, Xiao Z Y, Hong W, and Bai H, Three-dimensional printing of polyaniline/reduced graphene oxide composite for high-performance planar supercapacitor. *ACS Appl. Mater. Interfaces.* 2018. **10**, 10437–10444
[32] Ovhal M M, Kumar N, and Kang J-W, 3D direct ink writing fabrication of high-performance all-solid-state micro-supercapacitors. *Mol. Crystals Liq. Cryst.* 2020. **705**, 105–111
[33] Huang C and Chen L, Negative Poisson's ratio in modern functional materials. *Adv. Mater.* 2016. **28**, 8079–8096
[34] Jain K, Wang Z, Garma L D, Engel E, Ciftci G C, Fager C, Larsson P A, and Wågberg L, 3D printable composites of modified cellulose fibers and conductive polymers and their use in wearable electronics. *Appl. Mater. Today.* 2023. **30**, 101703
[35] Khudiyev T, Lee J T, Cox J R, Argentieri E, Loke G, Yuan R, Noel G H, Tatara R, Yu Y, Logan F, Joannopoulos J, Shao-Horn Y, and Fink Y, 100 m long thermally drawn supercapacitor fibers with applications to 3D printing and textiles. *Adv. Mater.* 2020. **32**, e2004971
[36] Françon H, Wang Z, Marais A, Mystek K, Piper A, Granberg H, Malti A, Gatenholm P, Larsson P A, and Wågberg L, Ambient-dried, 3D-printable and electrically conducting cellulose nanofiber aerogels by inclusion of functional polymers. *Adv. Funct. Mater.* 2020. **30**, 1909383

14 3D-Printed Conducting Polymers for Supercapacitors

Nidhi, Ramesh C. Thakur, Elyor Berdimurodov, Alok Kumar, Ashish Kumar, and Praveen K. Sharma

14.1 INTRODUCTION

Conducting polymers, a type of polymer with inherent electrical conductivity, are appealing materials for a variety of applications, including bioelectronics, flexible electronics, and energy storage. However, traditional conducting polymer production procedures such as electron-beam lithography, screen printing, and ink-jet printing have stifled rapid innovation and widespread usage of conducting polymers. For conducting polymer 3D printing, we report a high-performance 3D-printable conducting polymer ink based on poly(3,4-ethylenedioxythiophene):polystyrene sulfonate (PEDOT:PSS) [1]. Conducting polymers, for instance, are more easily moldable into high-resolution and high-aspect-ratio microstructures due to their better printability. These molecules can then be joined with other materials, like dielectric elastomers, through multi-material 3D printing. Hydrogel microscopic structures that are both soft and incredibly conductive can be created by 3D printing conducting polymers. 3D printing is described as "the process of joining materials to make objects from 3D model data, usually layer upon layer, as opposed to subtractive manufacturing technologies," in accordance with the American Society for Testing and Materials. Rapid manufacturing (RM), additive manufacturing, additive fabrication (AF), additive processes, additive layer fabrication (ALF), layered manufacturing (LM), rapid prototyping, additive techniques (AT), or solid freeform fabrication are all terms for 3D printing [2]. Engineers prefer the phrase "additive manufacturing," whereas the public is more familiar with "3D printing." The terms "additive manufacturing" and "3D printing" are used interchangeably in this work. The term "additive manufacturing" suggests a distinction from subtractive manufacturing. Power-driven machine instruments, such as lasers, drill presses, milling machines, toothed broaches, and lathes combined with an acute cutting tool are used in subtractive manufacturing to eliminate material [3]. These techniques can reach a high degree of intricacy by employing cutting-edge methods, which include electron beam machining, electrochemical machining, electrical discharge machining, ultrasonic machining, and photochemical machining. Other subtractive manufacturing techniques include shaping (such as ball bearing, rotary swaging, sheet drawing, forging, thread rolling, and extrusion) and casting. However, as one might expect, this production procedure is

DOI: 10.1201/9781003415985-14

wasteful because a significant amount of material is eliminated from the finished product. Additionally, genetic modification tooling is costly and time consuming [4].

Using additive manufacturing processes has numerous benefits beyond cost and material savings. The main benefit is the abundance of machining challenges may be easily addressed. For example, material variations can be precisely adjusted to create minor details (such as voids, interior geometries, and complex) [5–7]. Dentist caps and bridges are two examples of items that can be made by additive printing that need to be highly customized. Subtractive manufacturing, on the other hand, can only machine one biomedical part with a strict mapping for one patient during a defined time. Due to these benefits, manufacturing using additives has significantly surpassed subtractive production in popularity [8, 9].

The use of 3D printing in multilayer micromanufacturing and fast tooling has great potential. It will need a significant amount of fundamental and applied research in the field of 3D printing to create novel manufacturing processes that combine several materials for multiscale and diverse behaviors [10]. Particles with thermal, mechanical, optical, electrical, and other functional qualities can be used in a wide range of applications, including energy storage, thermal packaging, purification, filtration, medical implants, optoelectronics, and electrical devices [11, 12]. Then, along with an overview of the processing variables, merits, demerits, and upcoming difficulties of each printing technique, we will introduce a range of printing mechanisms, such as powder bed vat polymerization, and a few other less popular 3D printing techniques [13].

In addition, we will demonstrate the use of common polymers and components in this session on 3D printing, including viscous inks, liquid monomers, loosely packed pellets, stiff filaments, and compliant gels, containing micro as well as nanoscale particles. The general printing techniques used in particle and polymer processing will be the focus of this review. This chapter concludes by identifying and discussing potential future directions for a few new application situations. The creation of multifunctional systems for a variety of applications will be facilitated by identifying obstacles in materials science and manufacturing techniques, specifically when utilizing multi-materials like polymer and particles at multiple scales (such as macroscale structures and nanoscale morphologies) [14, 15].

This approach constructs products by combining components in order to minimize waste while keeping acceptable geometric precision [16, 17]. The process starts with a meshing three-dimensional computer model, which can be created using CAD software or image data that has been gathered. It is common to create a surface Tessellation Language file. The mesh data is then sliced into a build file with 2D layers and sent to the 3D printer. It is possible to deal with thermosetting polymer materials like epoxy resins as well as thermoplastic polymer materials like polyamide, acrylonitrile butadiene styrene, polylactic acid, and polycarbonate utilizing 3D printing technology [18, 19]. Epoxy resins are reactive polymers that must be cured by heat or UV light. As a result, epoxy resins are perfect for printing procedures that use heat or UV light. The aerospace and engineering sectors can use polymer 3D printing to create intricate portable structures, the art world can use it to replicate artifacts or teach students, and the medical field can use it to manufacture tissues and organs depending on the material chosen [20]. However, as pure polymer objects produced by 3D printing lack the strength and functionality of fully functional and load-bearing parts, the majority of 3D-printed polymer things are still used as hypothetical prototypes rather than actual components. Such limitations prevent 3D-printed polymers from being widely used in industry [21].

In order to overcome these difficulties, 3D printing of polymer composites combines the matrix to build a system with desirable functional or structural features that none of the pieces could provide on their own. High mechanical strength and functional composites made from polymer matrix can be produced by adding particle, fiber, or nanomaterial reinforcements to polymers [22, 23]. Traditional nanocomposite fabrication methods including casting, molding, and machining yield complexly shaped objects through material removal processes. Although these systems have a well-established and controlled manufacturing process and composite performance, they have limited ability to control the intricate internal structure. 3D printing can produce complex composite structures with minimum waste. Composites' size and shape can be precisely modified using computer-aided design [24]. Consequently, composites produced through 3D printing achieve an unmatched combination of process flexibility and high-performance results. Although there has been a lot of discussion about 3D printing over the past 30 years, most review articles that have been published have concentrated on the development of method of processing and the printing of pure polymer materials [25, 26]. Nonetheless, significant advancements in the production of printable polymer composites with enhanced performance have been made recently. We start by briefly outlining the properties of polymer composites and the 3D printing technology that is employed to create them. Following that, we investigate the deployment of comprehensive printing technology and the enhancement of polymer composite property enhancements [27, 28].

Dental crowns and bridges, for example, are excellent candidates for additive printing because they require a high degree of customization. Subtractive manufacturing, on the other hand, can only build one biomedical part with a rigorous design for one patient within a set time frame. These benefits have resulted in a significant shift away from subtractive manufacturing and towards additive manufacturing [29]. Material structure and composition changes are also feasible with 3D printing. In functionally graded structures, regular lattices, patterned dots, depth-changing motifs, thin-diameter lines, porous membranes, and low-thickness films can be built with different levels of variability or continuity. Tunable topology, foam density, form optimizations, and surface roughness are all available. Across the 3D printing spectrum, some platforms (e.g., binder and material jetting) can provide for discrete control inside a layer or between layers (vat polymerization and powder bed fusion) and can also allow for point-by-point material modifications [30]. Some of these techniques are shown in Figure 14.1.

14.1.1 Emerging Applications and Potential for 3D Printing

3D printing offers various benefits in composite manufacturing, including high precision, low cost, and customized shape [31]. 3D printing has found extensive use in several fields, like heat dissipation, electrical conductivity, optical manipulation, structural material conservation, and energy storage in cells and supercapacitors.

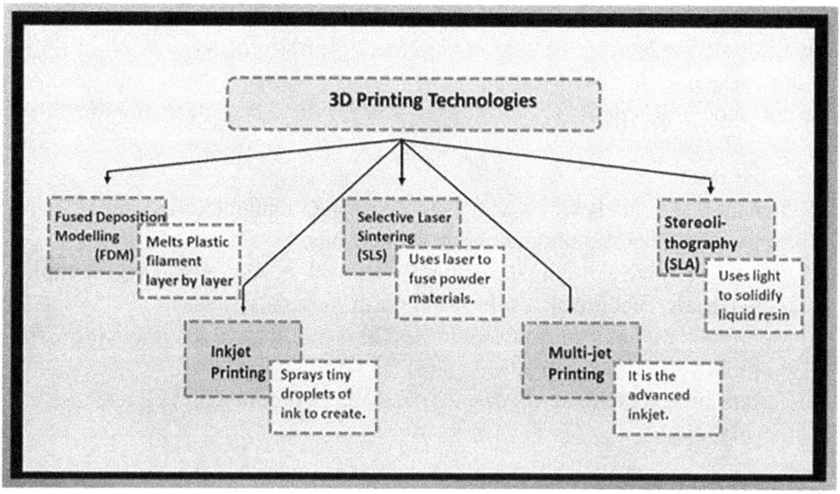

FIGURE 14.1 Some of the main 3D printing technologies used.

Using computer numerical control printers, 3D printing technology enables the direct fabrication of 3D objects from CAD models. Before being separated into control codes that may be utilized to operate 3D printers to print layer by layer, CAD models are turned into computer-readable forms. One layer of 3D objects is deposited or designed at a time using laser optics or printheads, depending on the materials and printing processes used. During printing, the patterned areas—which are composed of powders, resins, inks, or filaments—are crosslinked or cemented to create objects with some strength or functionality [32]. Many printing processes have been created since the introduction of 3D printing technology to accommodate different types of materials. Based on their respective printing principles, the seven printing technologies are categorized as follows: powder bed fusion, energy deposition, material jetting, material extrusion, vat photopolymerization, sheet lamination, and binder jetting. Binder jetting, for example, prints metals and ceramics by selectively fusing powder components with liquid bonding agents. Sheet lamination is a technique that was originally designed for aesthetic and visual structures made of paper or metals that unites sheets of material to make an object. Energy deposition, on the other hand, uses focused heat energy to fuse materials as they are deposited and is currently only utilized for metals. In the field of medicine, 3D printing has also been very useful in the areas of bionics, prosthetics as well as digital dentistry. All aspects of medicine are inevitably being positively impacted and changed by this. Although most of the work is still in the exploratory stage, experts believe the use of 3D printing as a tool will completely change medicine in the future.

The apparel sector has not been exempted either. Clothes built with 3D printing are being made. 3D-printed gowns, shoes, and bikinis are being experimented with by fashion designers. Parts for cars and airplanes are created with 3D printing technology. The rapid and effective printing of parts is making a significant contribution to the value chain. To keep up with the development of new materials and evolving needs for products, 3D printing should never be viewed as an independent process. Rather, it is becoming an essential component of multi-process systems or an integrated process of numerous systems.

To conclude, the benefits of 3D printing over classical manufacturing include:

1) Rapid prototyping.
2) Increased design elasticity and customization in manufacturing.
3) Support for complex geometry and dimensions.
4) Ability to achieve optimal regional chemical breakdowns (for example, biomaterials, functional grading, and multi-materials).
5) Modification of physical morphologies (basic building block orientation, for example).
6) Zero or little scrap material (e.g., no machining required and high feedstock recyclability).

14.1.2 3D Printing of Composite Materials

3D printing can be utilized to process particles, polymers, and their composite or blended forms. Optical devices, thermal exchange, electrical micromachines, surface alterations, structural supports, and biological applications are all made possible by the integration of soft matter and rigid body systems. But polymers are pliable materials, and most particles show irregularities, ineffective interactions with polymers, or brittleness in large amounts. The matrix and additives are combined in composite 3D printing to form a system with better functional or structural qualities than any of the constituents alone. Because of their superior mechanical, chemical, structural, and physical properties, composite or reinforced materials—especially those belonging to the polymer class—are getting more and more attention as materials for a wide range of technical and scientific applications. Such composites are designed ahead of time and printed in the same way as pure materials. When 3D printing soft polymer goods, inorganic particles or fibers are utilized as functional additives in addition to blending two or more distinct soft polymers [33]. The purpose of 3D printing with composites is to offer a versatile and efficient manufacturing method that enhances design possibilities and addresses specific material and performance requirements across various industries, and Figure 14.2 shows some of these purposes.

Attempts to print composites are motivated by four variables, as follows:

1) Enhancing the matrix material's printability.
2) Mechanically reinforcing the matrix material.
3) Adding additional features (including thermal, electrical, and magnetic capabilities) to the material system.
4) Constructing a permeable structure with sacrificial elements.

14.1.3 3D Printing with Multiple Materials

The printing process is used in 3D printing of numerous materials to combine multiple materials into a single functional entity. Various functional inorganic or organic

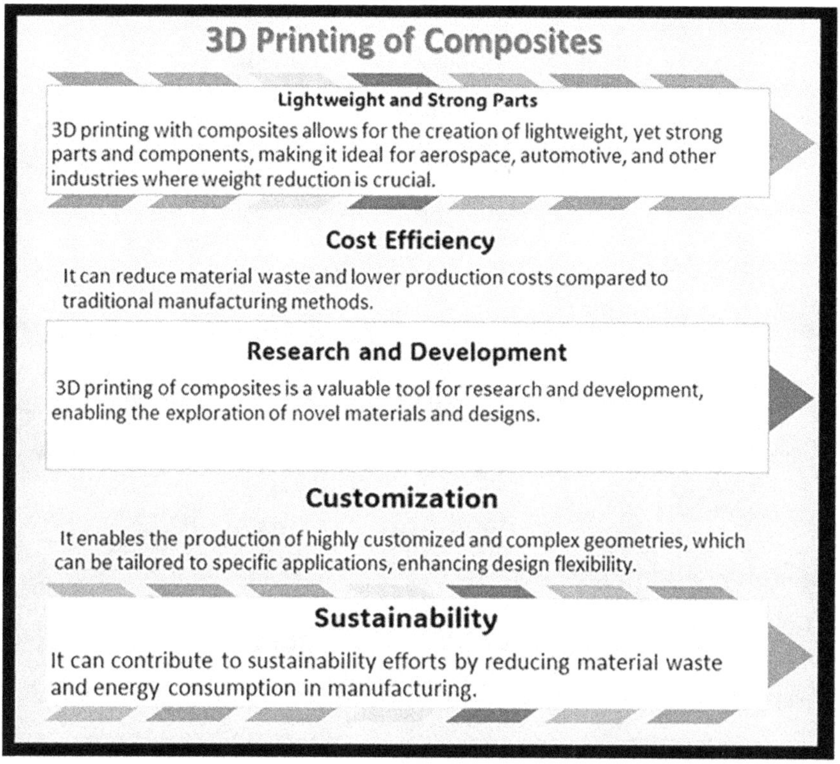
3D Printing of Composites
Lightweight and Strong Parts
3D printing with composites allows for the creation of lightweight, yet strong parts and components, making it ideal for aerospace, automotive, and other industries where weight reduction is crucial.
Cost Efficiency
It can reduce material waste and lower production costs compared to traditional manufacturing methods.
Research and Development
3D printing of composites is a valuable tool for research and development, enabling the exploration of novel materials and designs.
Customization
It enables the production of highly customized and complex geometries, which can be tailored to specific applications, enhancing design flexibility.
Sustainability
It can contribute to sustainability efforts by reducing material waste and energy consumption in manufacturing.

FIGURE 14.2 Purposes of 3D printing of composites.

components are employed with soft polymer polymers to enhance the efficiency of printed systems. The print materials are mainly segregated prior to printing, and several methods for dispersing the multiple materials have been recorded, including multi-nozzle multivat printing embedded printing and coaxial printing [34]. The purpose of 3D printing with multiple materials is to expand the range of possibilities in design, functionality, and aesthetics, making it a versatile tool across various industries and applications, and few of the purposes are shown in Figure 14.3.

Multi-vat printing is the most basic class of multi-material printing technologies used in numerous applications. Zhang et al. [35] achieved multi-material printing of soft actuators by combining DIW printing and inkjet printing. Inkjet printing was used for low-viscosity UV curable silicone rubber, while DIW printing was used for high-viscosity inks such shape memory polymers and conductive silver nanoparticle ink [36].

Switchable single-nozzle printing, which allows for the printing of various materials with unambiguous material boundaries, was used to directly build stiffness-tunable soft actuators. Switchable vats are an excellent choice for printing numerous materials using vat photopolymerization. Kowsari et al. [37] created an in vitro hepatic model, and Ma et al. manufactured a multi-material DLP printer along with switchable material supply sources (vats). with different cell-containing hydrogels.

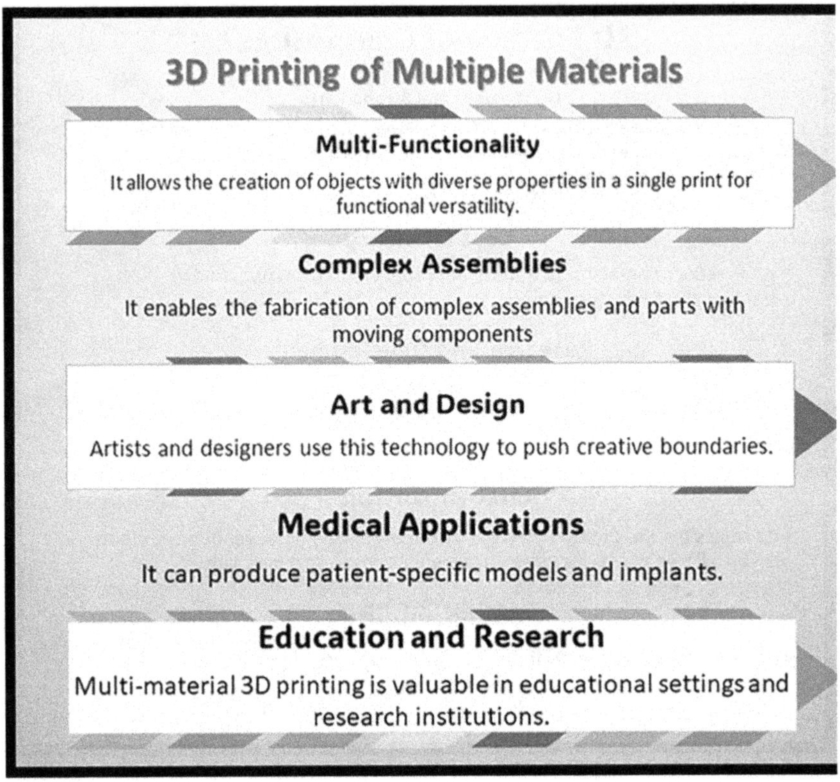

FIGURE 14.3 Purpose of 3D printing of multiple materials.

This quick multi-material 3D bioprinting approach holds a lot of promise in personalized medicine. Core-shell/coaxial printing distributes several materials using core-shell/coaxial nozzles, allowing the fabrication of functional structures. Printing Ga-based liquid metal, low-viscosity liquid metal-based electronics, and complex vessels with several components have all been accomplished using this technology. Embedded printing is a low-viscosity ink printing process that has been used to print flexible sensors, soft protein and polysaccharide hydrogels, and low-viscosity silicone rubber. Creating multifunctional systems, providing mechanical reinforcement, and providing sacrificial support are the main goals of integrating different materials in printing.

14.1.4 Properties of 3D Printing

3D-printed conducting polymers have electrical conductivity of 155 $S.cm^{1}$ in the dry state and 28 $S.cm^{1}$ in the hydrogel state, which is analogous to early demonstrated high-performance conducting polymers [38]. Higher electrical conductivity is obtained with a smaller nozzle diameter, possibly as a result of shear-induced changes in the PEDOT:PSS nanofibril alignment [39]. With maximal strains of 13% in the dry state and 20% in the hydrogel state, mechanical twisting with 3D-printed conduction polymers is feasible, most likely due to shear-induced changes in the PEDOT:PSS nanofibril alignment. 3D-printed conducting polymers exhibit a narrow variation in conductivity to electricity over an extensive range of compressive and longitudinal bending scenarios. It's possible that 3D-printed conducting polymers will still have good electrical conductivity after 10,000 cycles of repeated bending [40].

A viable approach to the quick and easy creation of multi-material, high-resolution conducting polymer structures and devices is by 3D printing to conduct polymer components. [41]. The rapid production of more than 100 circuit layouts on a flexible polyethylene terephthalate (PETE) substrate in less than 30 minutes with feature sizes of 100 μm or less using a single continuous printing process is enabled by highly repeatable 3D printing of conducting polymers at high resolution [42]. In terms of design selection determined by applicational criteria, this programmable, high-resolution, and high-throughput fabrication of conducting polymer motifs is poised to be a more adaptable alternative to screen and ink-jet printing [43].

14.1.5 3D Polymers in Supercapacitors

Wearable devices that sense and respond to external stimuli, such as human activity monitoring, wearable sensors, smart electronic skin, artificial organs, prostheses, and continuous health monitoring have led to a paradigm change in consumer electronics [44]. Interest in developing intelligent wearable systems has increased because of the development of next-generation stretchable electronic devices with excellent performance and an elastic mechanical reaction. Energy storage devices have to be able to continuously and reliably generate electrical power, even when curved into intricate designs to ensure long-term operation on non-planar and dynamic surfaces such as the human body [45].

Supercapacitors are a class of energy storage device that has historically been well known for having high power densities, extended cycle lives, and high charge-discharge rates. Supercapacitors' low energy density constraint, as opposed to batteries', is the primary cause of their limited use in practical applications. Currently, there are two primary methods for increasing supercapacitors' electrode-based energy density. One is the discovery of materials for electrodes, such as carbon-based substances and metal oxides of Ru, Mn, Ni, and Co as well as their composites, with high specific surface areas and high energy storage activities. Creating a suitable electrode structure is the second step in increasing the electrolyte and electrode's contact area. The development of an independent flexible and stretchable supercapacitor is critical for preserving structural integrity through repeated severe deformations that exceed the capabilities of rigid conventional devices.

Currently, numerous standard ways for creating stretchable electrodes have been presented, including active compounds to substrates that are flexible and stretchable, like polydimethylsiloxane (PDMS), polyurethane (PU), and thermoplastic copolyester (Ecoflex). But conventional stretchy electrodes are limited by the substrate's deformability during the stretching or deformation process, as well as by the drawbacks of high resistance and low capacitance, which can lead to subpar electrochemical performance [46, 47].

Supercapacitors with intricate predesigned designs and promising electrochemical performance can be produced using 3D printing technology. This is mostly explained by the ink's delicate structure and formulation, which are based on nanoparticles and can help with mass loading and ion transport. Supercapacitors that are printed using 3D technology are still in their infancy, and there are several issues that need to be resolved in later research. The ability to precisely manage the target objects' thicknesses and sizes through 3D printing makes it ideal for modification. More specifically, printing precision is critical for supercapacitors. When compared to traditional microfabrication techniques, most current printing procedures often have lower printing resolutions. Few technologies can reach the nanoscale scale, while most printing techniques can easily achieve resolutions in the micrometer range. Therefore, present micrometer-scale printing resolutions restrict the compact and full utilization of space; in other words, more surface area is taken up to attain desirable performance. Typically, printing variables such as printing operations, ink materials (material type), interface engineering of substrates (structure, reactivity), and afterwards (temperature, period) with minimal throughput and process yield loss are required to precisely optimize the resultant resolution and morphology (e.g., the thickness of electrodes) of printed patterns.

Using structural engineering techniques to reduce material stresses—such as reshaping non-stretchable materials into helical, serpentine, sponge, wavy, and net shapes—is an additional technique for producing bendable electrodes. Fiber-shaped electrodes produce distinctive patterns in fabrics that enable them to adapt to strain throughout the stretching deformation process. The use of structural engineering technology is generally dependent on the electrodes' structural design, which can convert linear strain that the electrode receives into buckling or bending strain. This enables the development of stretchable electrodes made of hard materials.

Finally, consumer electronics have undergone a paradigm shift because of the introduction of smart wearable systems, with the goal of developing next-generation electronics that are stretchable and have a high energy density, quick charge and discharge speeds, and extended cycling lives. Stretchable electrodes with conventional substrate deformability have limitations such as high resistance and low capacitance, which results in poor electrochemical performance [48, 49]. To generate stretchable electrodes, structural engineering approaches such as altering the geometric structure of non-stretchable materials into sponge, helical, wavy, net shapes, and serpentine can be used to lessen stresses exerted on the material. Fabrics with unique patterns created by fiber-shaped electrodes can adjust to strain throughout the stretching deformation process as well.

Non-polarized (NPR) electrodes have shown considerable improvements in stretchability and flexibility. When applied to vertical stress, conventional materials expand vertically and compress laterally; however, auxetic materials with an NPR structure inflate in all directions when tugged in a single direction. Electrodes with complicated NPR structures, on the other hand, can be challenging to fabricate using traditional methods [50].

Ink-based 3D printing is a productive and efficient method for creating a free-standing stretchable electrode with multidimensional property tunability and an NPR structure. A high-performance conductive polymer called PEDOT:PSS ink is produced by combining structural mapping design with direct ink writing (DIW) technology. PSS pens are viscoelastic, high concentration, additive-free, and have high performance. The demerits of non-tunable mechanical performance are addressed, and the narrow strain ranges of conventional supercapacitor electrodes are expanded by incorporating an arc-shaped microstructure into the standard NPR structure. The electrode's maximal strain is reduced by the effective arc-shaped microstructure, which also provides a homogenous stress zone, which leads to remarkable adaptability and extraordinary stretchability per finite-element analysis results. The hybrid polymer/CNT electrode that has been optimized has an area capacitance of 990 mF/cm^2. Unusual long-term cycle stability and potential areal capacitance are found in a 3D-printed quasi-solid-state symmetric supercapacitor. Figure 14.4 shows a few of the main applications of 3D printing in supercapacitors.

Finally, a flexible conducting polymer electrode was created by combining extrusion 3D printing technique with logical structural patterning. When stretched to 150% or twisted to 180°, the optimized arc-shaped NPR structure provides good strain and stress alleviation for electrodes while retaining structural integrity. The fabrication process is simple and highly scalable, and the superior electrochemical and mechanical performance of stretchy conducting polymer electrodes should pave the way for the use of diverse flexible electronic devices [14, 31]. Extrusion 3D printing technology and logical structural patterning were used to create a flexible conducting polymer electrode. When stretched or bent, the optimized arc-shaped NPR structure provides effective strain and stress reduction while retaining structural integrity. The electrodes also perform well electrochemically, having a high areal capacitance of 990 mF.cm^1. The areal capacitance of the quasi-solid-state symmetric supercapacitor built from 3D-printed electrodes is promising, as is the long-term cycling stability. The production procedure is simple and highly scalable, and the stretchable conducting polymer electrodes' superior mechanical and electrochemical performance

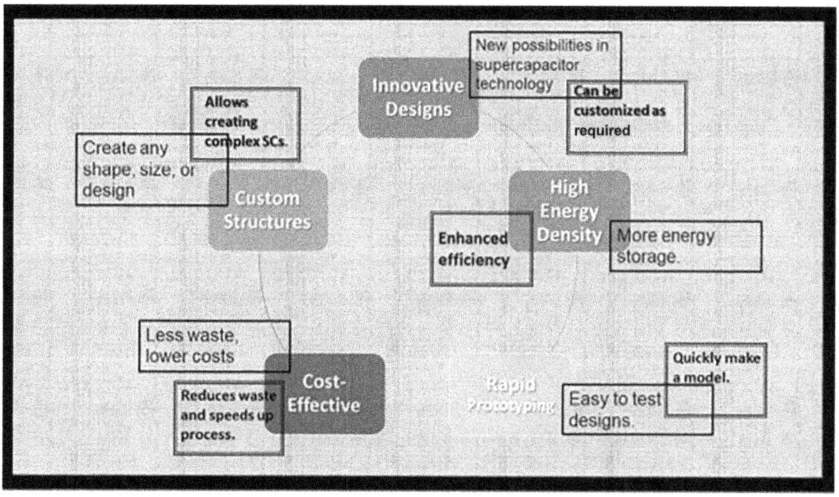

FIGURE 14.4 Applications of 3D printing in supercapacitors.

should pave the way for the implementation of flexible electronic devices. 3D printing advancements, particularly multi-material 3D printing of soft polymer materials, have made considerable strides in biological applications. However, numerous difficulties remain before practical applications, including devices, materials, and design. This chapter includes a thorough overview of typical 3D printing techniques, including vat polymerization, jetting, and rheology-based approaches.

14.2 CONCLUSION

This chapter addresses the properties and functionality of 3D-printed composite parts, along with their possible uses in aircraft, electronics, and biomedical engineering. It also covers polymer composite 3D printing procedures. Stereolithography, inkjet 3D printing, selective laser sintering fused deposition modeling, and 3D plotting are a few of the methods covered. 3D printing technologies have effectively handled a wide range of industrial applications, such as autos, airplanes, and smart devices. However, adequate feedstock systems for generating PMCs via 3DP technologies are frequently lacking. Extrusion 3D printing technology and logical structural patterning were used to create a flexible conducting polymer electrode. When stretched or bent, the optimized arc-shaped NPR structure provides optimum strain and stress reduction while retaining structural integrity. The electrodes also perform well electrochemically, having a high areal capacitance of 990 mF/cm^2. A quasi-solid-state symmetric supercapacitor made of 3D-printed electrodes also demonstrates promising areal capacitance and remarkable long-term cycling stability. Stretchable conducting polymer electrodes' simple and highly scalable production process, as well as their superior mechanical and electrochemical performance, offer a positive step in the direction of the use of a variety of adaptable electronic devices. Researchers are researching novel materials and uses for 3D printing of polymer composites. This chapter lays the groundwork for future research into materials, process control, scalability, and product performance in 3D printing of polymer composites.

REFERENCES

[1] S. Singh, S. Ramakrishna, and F. Berto, "3D Printing of polymer composites: A short review," *Mater. Des. Process. Commun.*, vol. 2, no. 2, pp. 1–13, 2020, doi: 10.1002/mdp2.97.

[2] J. Yang *et al.*, "3D-Printed highly stretchable conducting polymer electrodes for flexible supercapacitors," *J. Mater. Chem. A*, vol. 9, no. 35, pp. 19649–19658, 2021, doi: 10.1039/d1ta02617h.

[3] B. Caulfield, P. E. McHugh, and S. Lohfeld, "Dependence of mechanical properties of polyamide components on build parameters in the SLS process," *J. Mater. Process. Technol.*, vol. 182, no. 1–3, pp. 477–488, 2007, doi: 10.1016/j.jmatprotec.2006.09.007.

[4] K. Chatterjee and T. K. Ghosh, "3D printing of textiles: Potential roadmap to printing with fibers," *Adv. Mater.*, vol. 32, no. 4, pp. 1–24, 2020, doi: 10.1002/adma.201902086.

[5] R. C. Thakur and R. Sharma, "Viscometric studies of divalent transition metal sulphates in mixtures of water–diethylene glycol at 298.15–318.15 K," *Russ. J. Phys. Chem. A*, vol. 91, no. 9, pp. 1703–1709, 2017, doi: 10.1134/S0036024417090254.

[6] S. Bashir, A. Thakur, H. Lgaz, I. M. Chung, and A. Kumar, "Corrosion inhibition efficiency of bronopol on aluminium in 0.5 M HCl solution: Insights from experimental and quantum chemical studies," *Surf. Interfaces*, vol. 20, no. May, p. 100542, 2020, doi: 10.1016/j.surfin.2020.100542.

[7] H. Kaur, R. C. Thakur, and H. Kumar, "Effect of proteinogenic amino acids L-serine/L-threonine on volumetric and acoustic behavior of aqueous 1-butyl-3-propyl imidazolium bromide at T = (288.15, 298.15, 308.15, 318.15) K," *J. Chem. Thermodyn.*, vol. 150, p. 106211, 2020, doi: 10.1016/j.jct.2020.106211.

[8] S. C. Ligon, R. Liska, J. Stampfl, M. Gurr, and R. Mülhaupt, "Polymers for 3D Printing and Customized Additive Manufacturing," *Chem. Rev.*, vol. 117, no. 15, pp. 10212–10290, 2017, doi: 10.1021/acs.chemrev.7b00074.

[9] L. Y. Zhou *et al.*, "Multimaterial 3D printing of highly stretchable silicone elastomers," *ACS Appl. Mater. Interfaces*, vol. 11, no. 26, pp. 23573–23583, 2019, doi: 10.1021/acsami.9b04873.

[10] H. Jena, J. K. Katiyar, and A. Patnaik, Editors, *Composites Science and Technology Tribology of Polymer and Polymer Composites for Industry 4.0.* [Online]. Available: http://www.springer.com/series/16333

[11] R. Sharma and R. C. Thakur, "Study of thermodynamic and acoustic behaviour of nicotinic acid in binary aqueous mixtures of D-lactose," *AIP Conf. Proc.*, vol. 1860, 2017, doi: 10.1063/1.4990353.

[12] G. Parveen, S. Bashir, A. Thakur, S. K. Saha, P. Banerjee, and A. Kumar, "Experimental and computational studies of imidazolium based ionic liquid 1-methyl-3-propylimidazolium iodide on mild steel corrosion in acidic solution," *Mater. Res. Express*, vol. 7, no. 1, 2019, doi: 10.1088/2053-1591/ab5c6a.

[13] R. De Tayrac *et al.*, "In vitro degradation and in vivo biocompatibility of poly(lactic acid) mesh for soft tissue reinforcement in vaginal surgery," *J. Biomed. Mater. Res.—Part B Appl. Biomater.*, vol. 85, no. 2, pp. 529–536, 2008, doi: 10.1002/jbm.b.30976.

[14] K. Chi *et al.*, "Freestanding graphene paper supported three-dimensional porous graphene-polyaniline nanocomposite synthesized by inkjet printing and in flexible all-solid-state supercapacitor," *ACS Appl. Mater. Interfaces*, vol. 6, no. 18, pp. 16312–16319, 2014, doi: 10.1021/am504539k.

[15] H. Im, K. Lee, H. Lee, C. M. Castro, and R. Weissleder, "Lab on a Chip detection and profiling," *Lab Chip*, vol. 17, pp. 2892–2898, 2017. [Online]. Available: http://dx.doi.org/10.1039/C7LC00247E

[16] N. George *et al.*, "Microwave accelerated green approach for tailored 1,2,3–triazoles via CuAAC," *Sustain. Chem. Pharm.*, vol. 30, no. July, p. 100824, 2022, doi: 10.1016/j.scp.2022.100824.

[17] N. George *et al.*, "Click modified bis-appended Schiff base 1,2,3-triazole chemosensor for detection of Pb(II)ion and computational studies," *J. Mol. Struct.*, vol. 1288, no. January, p. 135666, 2023, doi: 10.1016/j.molstruc.2023.135666.

[18] Y. Zhu, W. Xu, D. Ravichandran, S. Jambhulkar, and K. Song, "A gill-mimicking thermoelectric generator (TEG) for waste heat recovery and self-powering wearable devices," *J. Mater. Chem. A*, vol. 9, no. 13, pp. 8514–8526, 2021, doi: 10.1039/d1ta00332a.

[19] B. J. Adzima, C. J. Kloxin, C. A. Deforest, K. S. Anseth, and C. N. Bowman, "3D photofixation lithography in Diels-Alder networks," *Macromol. Rapid Commun.* vol. 33, no. 24, pp. 2092[0][0]–2096, 2012, doi: 10.1002/marc.201200599.

[20] S. Dul, L. Fambri, and A. Pegoretti, "Fused deposition modelling with ABS-graphene nanocomposites," *Compos. Part A Appl. Sci. Manuf.*, vol. 85, no. March, pp. 181–191, 2016, doi: 10.1016/j.compositesa.2016.03.013.

[21] T. Gao *et al.*, "3D printing of tunable energy storage devices with both high areal and volumetric energy densities," *Adv. Energy Mater.*, vol. 9, no. 8, pp. 1–10, 2019, doi: 10.1002/aenm.201802578.

[22] H. Yuk *et al.*, "3D printing of conducting polymers," *Nat. Commun.*, vol. 11, no. 1, pp. 4–11, 2020, doi: 10.1038/s41467-020-15316-7.

[23] H. Li, L. Song, J. Sun, J. Ma, and Z. Shen, "Dental ceramic prostheses by stereolithography-based additive manufacturing: Potentials and challenges," *Adv. Appl. Ceram.*, vol. 118, no. 1–2, pp. 30–36, 2019, doi: 10.1080/17436753.2018.1447834.

[24] N. Shahrubudin, T. C. Lee, and R. Ramlan, "An overview on 3D printing technology: Technological, materials, and applications," *Procedia Manuf.*, vol. 35, pp. 1286–1296, 2019, doi: 10.1016/j.promfg.2019.06.089.

[25] Q. Shi *et al.*, "Recyclable 3D printing of vitrimer epoxy," *Mater. Horizons*, vol. 4, no. 4, pp. 598–607, 2017, doi: 10.1039/c7mh00043j.

[26] W. Xu *et al.*, "3D printing for polymer/particle-based processing: A review," *Compos. Part B Eng.*, vol. 223, no. June, p. 109102, 2021, doi: 10.1016/j.compositesb.2021.109102.

[27] L. Y. Zhou, J. H. Ye, J. Z. Fu, Q. Gao, and Y. He, "4D printing of high-performance thermal-responsive liquid metal elastomers driven by embedded microliquid chambers," *ACS Appl. Mater. Interfaces*, vol. 12, no. 10, pp. 12068–12074, 2020, doi: 10.1021/acsami.9b22433.

[28] L. Long *et al.*, "3D printing of recombinant Escherichia coli/Au nanocomposites as agitating paddles towards robust catalytic reduction of 4-nitrophenol," *J. Hazard. Mater.*, vol. 423, no. PA, p. 126983, 2022, doi: 10.1016/j.jhazmat.2021.126983.

[29] R. Pugliese, B. Beltrami, S. Regondi, and C. Lunetta, "Polymeric biomaterials for 3D printing in medicine: An overview," *Ann. 3D Print. Med.*, vol. 2, p. 100011, 2021, doi: 10.1016/j.stlm.2021.100011.

[30] A. M. Pekkanen, R. J. Mondschein, C. B. Williams, and T. E. Long, "3D printing polymers with supramolecular functionality for biological applications," *Biomacromolecules*, vol. 18, no. 9, pp. 2669–2687, 2017, doi: 10.1021/acs.biomac.7b00671.

[31] B. Zhang, B. Seong, V. D. Nguyen, and D. Byun, "3D printing of high-resolution PLA-based structures by hybrid electrohydrodynamic and fused deposition modeling techniques," *J. Micromech. Microeng.*, vol. 26, no. 2, p. 25015, 2016, doi: 10.1088/0960-1317/26/2/025015.

[32] X. Wang, M. Jiang, Z. Zhou, J. Gou, and D. Hui, "3D printing of polymer matrix composites: A review and prospective," *Compos. Part B Eng.*, vol. 110, pp. 442–458, 2017, doi: 10.1016/j.compositesb.2016.11.034.

[33] F. Gong, X. Cheng, Q. Wang, Y. Chen, Z. You, and Y. Liu, "A review on the application of 3D printing technology in pavement maintenance," *Sustain.*, vol. 15, no. 7, pp. 1–14, 2023, doi: 10.3390/su15076237.

[34] T. D. Ngo, A. Kashani, G. Imbalzano, K. T. Q. Nguyen, and D. Hui, "Additive manufacturing (3D printing): A review of materials, methods, applications and challenges," *Compos. Part B Eng.*, vol. 143, no. February, pp. 172–196, 2018, doi: 10.1016/j.compositesb.2018.02.012.

[35] Y. Zhang, N. Zhang, H. Hingorani, N. Ding, and D. Wang, "Fast-response, stiffness-tunable soft actuator by hybrid multimaterial 3D printing," *Adv. Funct. Mater.*, vol. 1806698, pp. 1–9, 2019, doi: 10.1002/adfm.201806698.

[36] N. Kumar, P. K. Jain, P. Tandon, and P. M. Pandey, "Additive manufacturing of flexible electrically conductive polymer composites via CNC-assisted fused layer modeling process," *J. Brazilian Soc. Mech. Sci. Eng.*, vol. 40, no. 4, 2018, doi: 10.1007/s40430-018-1116-6.

[37] H. Korhonen *et al.*, "Fabrication of graphene-based 3D structures by stereolithography," *Phys. Status Solidi Appl. Mater. Sci.*, vol. 213, no. 4, pp. 982–985, 2016, doi: 10.1002/pssa.201532761.

[38] M. Korger, J. Bergschneider, M. Lutz, B. Mahltig, K. Finsterbusch, and M. Rabe, "Possible applications of 3D printing technology on textile substrates," *IOP Conf. Ser. Mater. Sci. Eng.*, vol. 141, no. 1, 2016, doi: 10.1088/1757-899X/141/1/012011.

[39] J. Saroia *et al.*, "A review on 3D printed matrix polymer composites: Its potential and future challenges," *Int. J. Adv. Manuf. Technol.*, vol. 106, no. 5–6, pp. 1695–1721, 2020, doi: 10.1007/s00170-019-04534-z.

[40] F. Paquin, J. Rivnay, A. Salleo, N. Stingelin, and C. Silva, "Multi-phase semicrystalline microstructures drive exciton dissociation in neat plastic semiconductors," *J. Mater. Chem. C*, vol. 3, pp. 10715–10722, 2015, doi: 10.1039/b000000x.

[41] M. Revilla-León, M. Sadeghpour, and M. Özcan, "An update on applications of 3D printing technologies used for processing polymers used in implant dentistry," *Odontology*, vol. 108, no. 3, pp. 331–338, 2020, doi: 10.1007/s10266-019-00441-7.

[42] M. Guvendiren, J. Molde, R. M. D. Soares, and J. Kohn, "Designing biomaterials for 3D printing," *ACS Biomater. Sci. Eng.*, vol. 2, no. 10, pp. 1679–1693, 2016, doi: 10.1021/acsbiomaterials.6b00121.

[43] S. Chakraborty and M. C. Biswas, "3D printing technology of polymer-fiber composites in textile and fashion industry: A potential roadmap of concept to consumer," *Compos. Struct.*, vol. 248, no. May, p. 112562, 2020, doi: 10.1016/j.compstruct.2020.112562.

[44] C. Zhu *et al.*, "Supercapacitors based on three-dimensional hierarchical graphene aerogels with periodic macropores," *Nano Lett.*, vol. 16, no. 6, pp. 3448–3456, 2016, doi: 10.1021/acs.nanolett.5b04965.

[45] S. Zhang *et al.*, "Hydrogel-enabled transfer-printing of conducting polymer films for soft organic bioelectronics," *Adv. Funct. Mater.*, vol. 30, no. 6, pp. 1–8, 2020, doi: 10.1002/adfm.201906016.

[46] V. Wood *et al.*, "Inkjet-printed quantum dot-polymer composites for full-color AC-driven displays," *Adv. Mater.*, vol. 21, no. 21, pp. 2151–2155, 2009, doi: 10.1002/adma.200803256.

[47] T. J. Wallin *et al.*, "Click chemistry stereolithography for soft robots that self-heal," *J. Mater. Chem. B*, vol. 5, no. 31, pp. 6249–6255, 2017, doi: 10.1039/c7tb01605k.

[48] S. Zhang *et al.*, "Hydrogel-enabled transfer printing: Hydrogel-enabled transfer-printing of conducting polymer films for soft organic bioelectronics (Adv. Funct. Mater. 6/2020)," *Adv. Funct. Mater.*, vol. 30, no. 6, p. 2070038, 2020, doi: 10.1002/adfm.202070038.

[49] M. Fera, F. Fruggiero, A. Lambiase, and R. Macchiaroli, "State of the art of additive manufacturing: Review for tolerances, mechanical resistance and production costs," *Cogent Eng.*, vol. 3, no. 1, p. 1261503, 2016, doi: 10.1080/23311916.2016.1261503.

[50] C. R. Chen, H. Qin, H. P. Cong, and S. H. Yu, "A highly stretchable and real-time healable supercapacitor," *Adv. Mater.*, vol. 31, no. 19, pp. 1–10, 2019, doi: 10.1002/adma.201900573.

15 3D Printing Conducting Polymers for Flexible Supercapacitors

Haitao Zhang, Xiang Chu, and Xinglin Jiang

15.1 INTRODUCTION

The recent rise in flexible electronics such as flexible displays, portable electronic devices, and wearable devices has attracted great attention and holds the potential to promote the coming era of the Internet of Things (IoT) [1]. Flexible electronics are inseparable from a continuous energy supply. However, the development of flexible energy storage devices lags far behind the energy demand from billions of flexible electronics. To meet the various design and power needs of modern equipment, researchers have invested more and more attention in ultra-thin, flexible, and safe energy storage equipment. Flexible supercapacitors (FSCs), as one kind of energy storage device, are an efficient flexible energy storage device due to their excellent performance, including mechanical flexibility, long cycle life, fast charging/discharging rates, light weight, safety, and reliability [2].

In recent years, we have witnessed rapid development of FSCs, including electrode materials, electrolytes, and micro/nano-processing technology. Also, with the increasing number of micro/nano-processing technologies and their increased processing accuracy, a series of low-cost, high-efficiency, and multifunctional manufacturing technologies have been provided for FSCs, especially various printing technologies [3]. These technologies are particularly important in manufacturing FSCs, such as laser engraving, laser direct writing, embossing, spray coating, and 3D printing [4]. Among them, 3D printing technology is a competitive method because of high efficiency, high accuracy, structural customization, and low cost [5, 6]. Hence, we will focus on discussing 3D printing conducting polymers for FSCs in this chapter.

15.2 CONDUCTING POLYMERS FOR FSCs

15.2.1 Fundamentals of FSCs

FSCs combine the advantages of mechanical flexibility and electrochemical energy storage capability. High energy density (E) and power density (P) are the critical performance factors in FSCs to support the power and energy for flexible electronics. E and P for FSCs are calculated by the following equations:

$$E = 1/2C\Delta V^2 \tag{1}$$

$$P = E/\Delta t \tag{2}$$

 DOI: 10.1201/9781003415985-15

where C is the capacitance, ΔV is the voltage window, and Δt is the discharge time [7]. For FSCs, the current collectors, electrode materials, electrolytes, and architecture are the main factors that bring outstanding electrochemical performance for FSCs.

Among these elements, flexible electrodes and electrolytes are critical to high-performance FSCs. For flexible electrodes, the selection and treatment of substrate materials have a significant impact, such as the bonding strength between electrode materials and substrates. Also, the electrode materials determine the performance of FSCs, including capacitance characteristics, cycle life, and rate performance. To obtain FSCs with high mechanical strength and good flexibility, on one hand, a flexible current collector and electrode materials with good mechanical properties are highly desired. On the other hand, gel or solid-state electrolytes with high strength are designed. Dependence on the mechanical strength of the electrode is reduced by stacking them layer by layer.

15.2.2 Structures of FSCs

In terms of structure, researchers have developed 3D stacked (sandwich-like), 2D planar (interdigital-like), and 1D fibrous (fiber-like) FSCs, as shown in Figure 15.1 [8]. Sandwich-like stacked FSCs are usually packaged with a layer of gel electrolyte sandwiched between two flexible electrodes (Figure 15.1a). Due to simple manufacturing process, wide applicability, and excellent mechanical and electrochemical properties, a sandwich-like structure is currently the most frequently used structure. A 2D planar supercapacitor is composed of an electrolyte layer and two electrodes. With the development of printing electronics technology, micro-supercapacitors (MSCs) have also been widely studied. Generally speaking, MSCs include flexible substrates, interdigital electrodes, and electrolyte layers (Figure 15.1b). Planar interdigital MSCs have a small volume and are very suitable for supplying power to microcircuits. Due to the tight arrangement between electrodes, they have high-volume energy density and power density. Fiber-like FSCs consist of an inner and outer electrode with a coaxial structure and two electrodes placed parallel and coated with an electrolyte layer (Figure 15.1c). They usually have excellent mechanical flexibility and stretchability that give them wide application in flexible wearable electronics. However, fiber substrates typically exhibit electrochemical inactivity and a limited electrode/electrolyte interface and pose a risk of detachment from each other during deformation, on one hand; on the other hand, flexible fiber-like supercapacitors commonly share relatively low capacitance and energy density. In this regard, it is significant to develop novel technology like 3D printing to realize strong interface linkage.

15.2.3 Composition of FSCs

Similar to traditional supercapacitors, electrode materials in FSCs are the most critical in storing energy. Nanocarbons, transition metal oxides, sulfides, nitrides or carbides, and conducting polymers are all exploited to use as electrochemical active materials [9]. When mechanical flexibility is taken into consideration, conducting polymers hold an additional advantage. Therefore, conducting polymers are widely investigated in FSCs. In FSCs, conducting polymers can be used as both

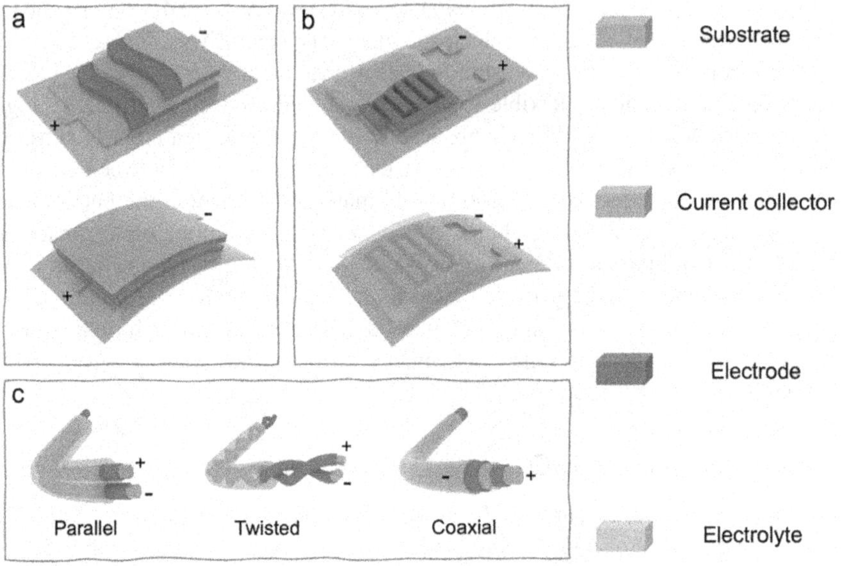

FIGURE 15.1 The structure of FSCs. (a) Sandwich-like FSCs. (b) Interdigital-like FSCs. (c) Fiber-like FSCs. Adapted with permission [8]. Copyright (2023), The Royal Society of Chemistry.

electrochemical active materials like electronic conducting polymers and flexible electrolytes like ionic conducting polymers.

15.2.3.1 Electronic Conducting Polymers for Electrode Materials

Conducting polymers are organic polymers that conduct charge through a conjugated bond system along the polymer chains. Typical conducting polymers include polyaniline (PANI), polypyrrole (PPy), polythiophene (PTh), and their derivatives. Their good conductivity is due to the delocalization of electrons in the conjugated polymer skeleton, which can transport electrons in a doped state. Their energy storage mechanism is to achieve high energy density through rapid redox reaction [10]. The oxidation-reduction process includes reversible n-type or p-type doping/dedoping processes that allow conducting polymers to store a large amount of charge. The redox reaction not only occurs on the surface but also in the whole-body phase, resulting in a large Faradaic capacitance (up to 1000 F g^{-1}). Also, this redox process is relatively reversible, with the structure remaining unchanged, leading to long-cycling stability [11].

Among these common types of conducting polymers, PANI, shown in Figure 15.2a1, is a hot material at present. It has many advantages, including high theoretical specific capacitance (2000 F g^{-1}), simple acid doping/dedoping mechanism, relatively stable chemical properties in the environment, simple polymerization method, and low price. Through tailoring the morphology, size, dimensionality, and crystallinity like construing PANI nanofibers, nanorods, nanosheets, and 2D PANI materials with hierarchical structure, PANI materials possess both good mechanical flexibility and excellent electrochemical stability. PPy, shown in Figure 15.2a2, is a star material in the conductive polymer family. Compared with PANI, it can not only work in acidic media but also in neutral electrolyte. Moreover, PPy has good mechanical processability, environmental friendliness, and excellent biocompatibility. PTh, shown in Figure 15.2a3, is a common conducting polymer, but it is not widely used in the field of SCs because of its relatively low specific capacitance and accordingly low energy density. One kind of PTh derivatives, poly (3,4-ethylenedioxythiophene)

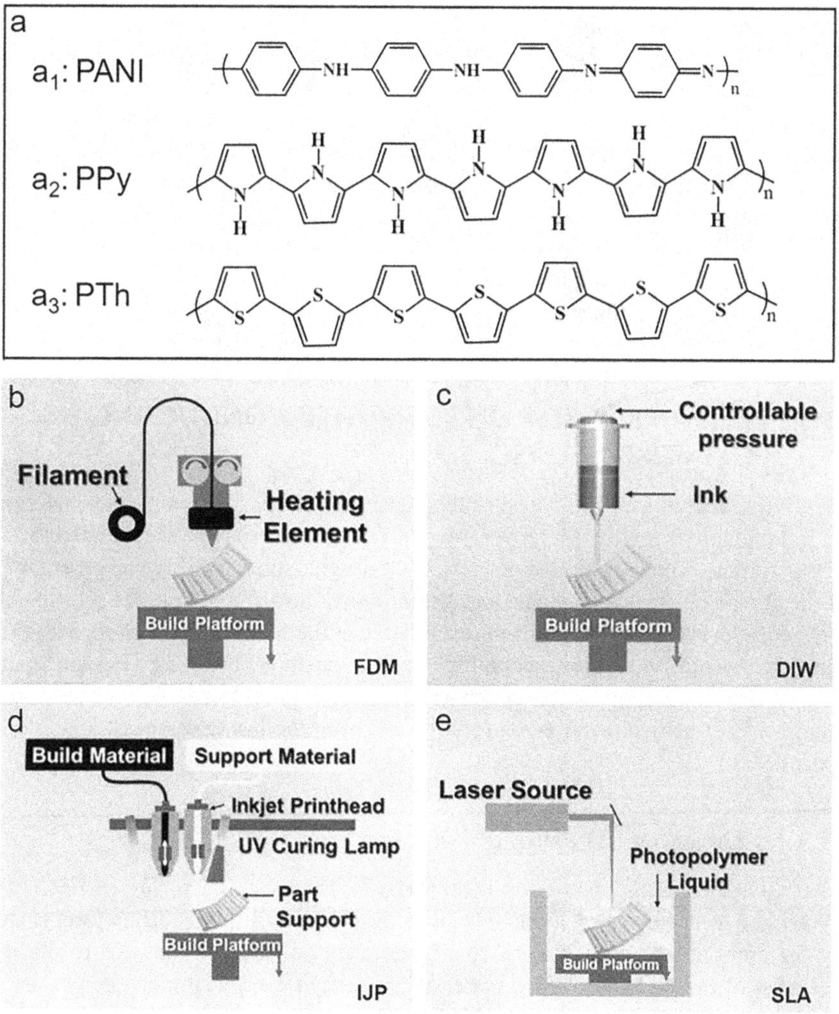

FIGURE 15.2 Some typical conducting polymers and 3D printing methods used in FSCs. (a) Molecular structures of polyaniline (PANI), polypyrrole (PPy), and polythiophene (PTh). (b) Fused deposition modeling (FDW) method. (c) Direct ink writing (DIW) method. (d) Inkjet printing (IJP) method. (e) Stereolithography (SLA) method. Adapted with permission [12]. Copyright (2022), Wiley.

doped polystyrene sulfonic acid (PEDOT: PSS), has the potential to construct high-power FSCs because of its extraordinary electrical conductivity.

15.2.3.2 Ionic Conducting Polymers for Flexible Electrolytes

To realize the whole flexibility of SCs, it is necessary to develop gel and solid-state electrolytes to replace conventional liquid electrolytes and polymer separators. In comparison, gel electrolytes with good flexibility and ionic conductivity have a great impact on the safety, temperature resistance, and fast charging/discharging capability of wearable supercapacitors.

At present, gel electrolytes can be divided into three types: hydrogel electrolytes, organic gel electrolytes, and ionic liquid gel electrolytes [13]. Hydrogel electrolytes usually include polyvinyl alcohol (PVA)—an acid/base/salt system, such as PVA/H_3PO_4, PVA/H_2SO_4, PVA/KOH, or PVA/LiCl. Hydrogel electrolytes have high security, but their operating voltage window is slightly low (0.8~1.6 V). The organic solvent gel electrolyte has a large operating voltage window (up to 3.0 V), but the organic solvent is volatile, and the device is prone to explosion at high temperatures. Ionic liquid gel electrolyte has low vapor pressure and a wide electrochemical window (up to 5.0 V); however, it is expensive, and its ionic conductivity is relatively low. Improving the compatibility between polymer electrode materials and electrolytes is crucial to high-performance FSCs that meet the demands of flexible electronics.

15.3 RECENT PROGRESS ON 3D-PRINTED CONDUCTING POLYMER-BASED FSCs

3D printing techniques could ensure the fabrication process is simple, scalable, and environmentally friendly. Moreover, several features such as precise manufacture at the micrometer or even nanometer scale, light weight, and flexibility can be provided by the choice of various printable and flexible substrates. Therefore, 3D printing has received widespread attention from researchers in the field of FSCs. In recent years, many 3D printing methods, including fused deposition modeling (Figure 15.2b), direct ink writing (Figure 15.2c), inkjet printing (Figure 15.2d), stereolithography (Figure 15.2e), screen printing, and roll-to-roll printing, have been exploited to construct FSCs [12].

15.3.1 Design of 3D Printing Inks

When 3D printing technology is incorporated in the manufacturing of FSCs, the ink printed is not traditional metal, plastic, or organic materials. Therefore, selecting the appropriate concentration of ink becomes rather important due to the fact that most of the electrode active materials in current energy storage devices are in powder form [12]. Generally, printable flexible electrode inks must meet the following requirements: (1) good electrochemical performance, high power density, high energy density, and long cycle stability. (2) High conductivity for fast charge transfer. (3) Appropriate physical properties (such as viscosity, rheological properties, surface tension, and drying speed) and high printing resolution. (4) Excellent dispersibility and no precipitation or agglomeration. In fact, good mechanical strength and plasticity are mandatory in order to obtain high-performance FSCs with good adhesion to the substrate.

Rheology, determining the deformation and flow of materials, is the key to ink flow behavior and printing characteristics. As pressure is applied to the printing ink, the ink will flow to reduce the strain caused by the external force. Different ink systems have different resistance to flow behavior, which is related to the viscosity of the ink. Viscosity, defined as the ratio of shear stress to shear rate, is the most commonly used rheological parameter to describe the friction force in the ink system. According to flow behavior, ink can be divided into Newtonian fluid and non-Newtonian fluid. Newtonian fluid refers to the liquid whose viscosity remains unchanged at different shear rates. In contrast, the viscosity of non-Newtonian fluid changes with the shear rate. When a material is sheared at a constant rate, its apparent viscosity decreases with the increase of shear time, which is called thixotropy. Thixotropy is mainly reflected in the reversible change of ink viscosity from high to low. During the printing process, the spatial structure of the ink is destroyed under external pressure, manifested as a decrease in viscosity with an increase in ink fluidity, flowing from the needle tip to the substrate. When printing stops, the applied stress disappears, and the internal spatial structure accompanying with the viscosity of the ink is restored, thus forming a desired pattern. Inks with good thixotropy plastic behavior are essential for 3D printing to build high-performance FSCs [5].

To meet the requirements of proper rheological and viscosity characteristics in 3D printing, conducting polymers can be designed to be hydrogel, which is a kind of extremely hydrophilic cross-linked polymer with 3D network structure. Conductive polymer hydrogel (CPH), as a new type of hydrogel, combines a 3D network structure, high specific surface area, biocompatible interface of the hydrogel, and conducting polymer's high electrolyte permeability and good electrical and optical properties. Hence, CPH has been commonly used in many fields such as sensors, artificial skin, tensile electronics, medical and health, and FSCs [14].

15.3.2 3D Printing Technologies

In a typical 3D printing process, FSCs are formed by the following steps: drawing the graphic structure of the device to be prepared, modulating the electrode material into fine particles of uniform viscous ink, then directly printing it on the flexible substrate, and finally dropping gel on its surface to form FSCs. Compared with traditional industrial manufacturing, 3D printing has the following main advantages: (1) 3D printing can directly shape complex structures without the need for additional processes and hence has great advantages in multi-layer embedded structures. (2) 3D printing does not require additional molds. (3) 3D printers are much smaller than traditional injection molding machines, making them suitable for use in special narrow space situations. (4) 3D printing products have the characteristics of rapid prototyping, which reduces the industrial design process and indirectly speeds up the production cycle of the entire industry. (5) Cost advantage. (6) 3D printing can produce a large number of identical products. (7) Expanding the range of material forming [15].

3D-printed FSCs share higher specific capacity/capacitance, energy density, and power density than those of traditional sandwich-like supercapacitors. 3D printing technology also makes 1D fibrous FSCs possible. Compared with 2D and 3D FSCs, 1D fibrous FSCs have higher feasibility and can be woven into flexible textiles for wearable applications [16]. Moreover, 3D printing technology has universality, flexibility, and a wide range of material choices, which can complete the manufacturing of the entire FSCs in one step, as well as the integration of all accompanying electronic products, which is more cost effective than manufacturing components and assembling them separately.

15.3.3 3D-Printed Conducting Polymer-Based FSCs

In contrast to traditional formative techniques, such as spin casting, injection molding, and machining, 3D printing technology can accomplish almost any desired stereoscopic geometry without the need for so-called templates, molds, or photolithographic masks. This allows accurate tuning of the geometry and structure of FSCs, resulting in significant improvements in energy and power density. Considering the preparation requirements of supercapacitors themselves, DIW, FDM, IJP, and SLA are the most commonly used techniques in this field [13]. In this section, we will summarize 3D-printed conductive polymer FSCs and provide printing principles in terms of ink composition, formulation, and electrode material selection.

15.3.3.1 IJP-Printed Conductive Polymer-Based FSCs

IJP is a high-precision, non-contact, fully dosing process that does not require a mask for patterning. More importantly, when printing capacitive components, IJP allows many parameters to be systematically varied to achieve the necessary control over thickness and microstructure. Therefore, these advantages make IJP a widely used material injection technology for FSCs. For liquid-phase treated PANI, they can be deposited on a variety of substrates suitable for IJP printing. For example, Diao et al. fabricated GO (graphene oxide) @PANI inks via covalent binding and prepared FSCs on flexible polymer substrates via IJP. The printed FSCs provided high specific capacitance and excellent cycling stability [17]. Delekta et al. reported fully printed FSCs using an ink based on PANI and MnO_2 passivated graphene nanocomposites. Thanks to this double passivation process, the ink has good stability in mild glycol solvents. The fully inkjet-printed FSCs have both high energy density and power density [18].

In addition to PANI, PEDOT:PSS with tunable conductivity, high flexibility, and good biocompatibility also attracts attention for IJP-printed FSCs. Fan et al. prepared FSCs by using PEDOT:PSS as both current collector and active material. The obtained FSCs have a high specific capacitance and good long-term stability [19]. In addition, an aqueous inkjet printable MXene/PEDOT:PSS (MP) composite ink was designed by Ma et al., as shown in Figure 15.3a. The hybrid ink showed excellent stability and printability to the extent that the printed electrodes had good uniformity (Figure 15.3b). With the help of IJP, various thicknesses of MP electrodes were printed from the micro-nozzle, and therefore the mass loading and electrode thickness were efficiently regulated (Figure 15.3c). The high conductivity of PEDOT:PSS effectively relieved the repackaging of MXene sheets and facilitated electron/ion transfer on the electrodes. The IJP-printed symmetrical FSCs achieved a volumetric

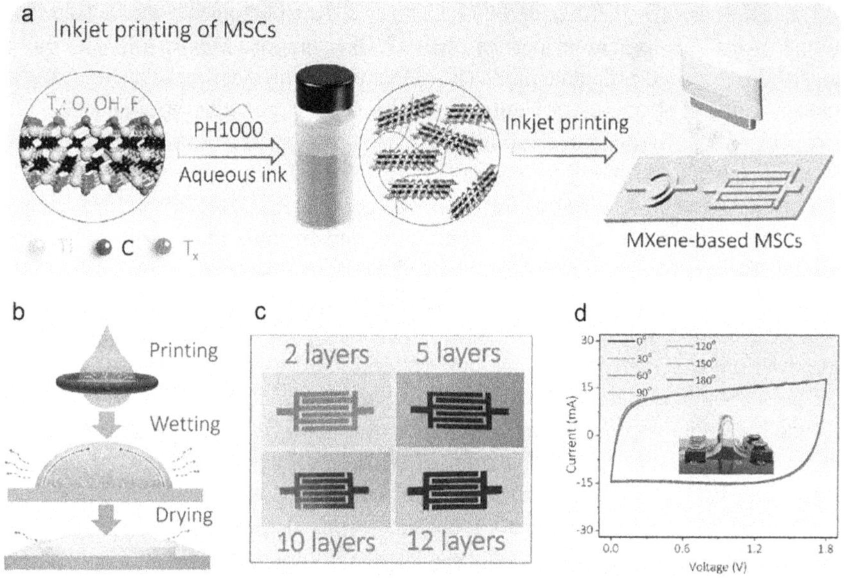

FIGURE 15.3 IJP 3D printing for FSCs. (a) Schematic of MP hybrid ink for inkjet printing MP-MSCs. (b) Schematic of inkjet-printing process of MP inks. (c) Digital photographs of MP-MSCs with different printed layers. (d) CV curves tested in bending states of 30°, 60°, 90°, 120°, and 180°. Inset is a photograph of MP-MSCs bent at 180°. Adapted with permission [20]. Copyright (2021), Wiley.

capacitance of up to 754 F cm^{-3}. In addition, the series-connected FSCs exhibited good structural stability and mechanical flexibility, with fully overlapping cyclic voltammetry (CV) curves at different bending angles (Figure 15.3d) [20]. IJP-printed FSCs can be prepared on paper substrates with high printing resolution. For example, Li et al. applied the IJP technique to prepare high-performance FSCs without any post-processing using a water/glycol solution mixture based on PEDOT:PSS, graphene quantum dots, and graphene as the ink [21]. The printed paper-based FSCs obtained a large area capacitance of >2 mF cm^{-2} at a high scan rate of 1000 mV s^{-1}. Liu et al. prepared a PEDOT:PSS@ carbon nanotubes (CNTs)/Ag ink and fabricated IJP-printed flexible MSCs with high rate capability [22].

15.3.3.2 DIW-Printed Conducting Polymer-Based FSCs

DIW is currently the most commonly used 3D printing technique for fabricating FSCs, thanks to the ease of operation, versatility, material versatility, and the ability to achieve high-quality loading of the active material. PANI is a widely used material for DIW-printed FSCs. However, conductive polymers are used in the form of liquid monomers or polymer solutions, resulting in slow recovery of viscosity to ensure shape retention when deposited onto the substrate. To overcome this obstacle, it is necessary to endow PANI with the ideal rheological properties required for DIW printing. In this regard, Wang et al. developed a PANI/GO gel ink that can be used for DIW printing. The added GO acts as a thixotropic agent, which effectively modulates the rheological properties of the PANI solution. Meanwhile, GO can be converted to reduced graphene oxide (RGO) that has high electrical conductivity and good mechanical strength, thus enhancing the electrical conductivity of PANI [23]. DIW-printed PANI/RGO interdigital electrodes can be used to construct a planar FSCs that has a specific capacitance of up to 1329 mF cm^{-2}. Liu et al. synthesized a highly concentrated, viscous and water-dispersible GO/PANI ink for extrusion printing by adjusting the synthesis conditions and formulation composition. After printing, the micro-electrodes did not need any post-treatment. The DIW-printed asymmetric flexible MSCs with GO/PANI and graphene as positive and negative electrodes, respectively, deliver a voltage window up to 1.2 V, high energy density and power density, and 100% capacitance retention after 5000 cycles [24].

PPy is another conducting polymer with high electrochemical activity, low density, and high volumetric specific capacitance (400–500 F cm^{-1}). Due to its flexibility, PPy has also been used for DIW printing. For example, a printable black phosphorus nanosheet (BPNA)/PPy composite ink was prepared by Xing et al. Relying on the tri-block polymer Pluronic F127 nonionic surfactant, the poor dispersion of BPNS in aqueous solution and the viscoelasticity of BPNS/PPy ink were solved. The mass specific capacitance of the DIW-printed electrode reached 417 F g^{-1}, and it had excellent cycling stability after 10,000 charge/discharge cycles [26]. In addition, Gu et al. fabricated high-performance FSCs on flexible carbon cloth by the DIW technique using PPy nanotubes as the electrode material. The printed devices have excellent mechanical stability, maintaining 93% capacitance at a 120° bending angle [27]. Qi et al. fabricated a PPy-coated graphene aerogel (GA) electrode by combining DIW printing and polymer self-assembly method. After coating with PPy, the specific capacitance of the PPy@GA electrode significantly increased from 14 to 395 F g^{-1}. The PPy coating also contributes to the compressive strength of the GA electrode, and DIW-printed FSCs have excellent stability under repeated compression cycles [28].

In addition to the previously mentioned materials, PEDOT:PSS has also attracted much attention for 3D printing supercapacitors. This is mainly attributed to the special properties of PSS chains that can act both as dopants and surfactants for PEDOT. As a result, PEDOT:PSS has superior water solubility, easy processing, and tunable mechanical properties compared to other conductive polymers (e.g., PANI and PPy). As shown in Figure 15.4a, Li et al. prepared a PEDOT:PSS/MXene/EG (PME) gel ink that can be DIW printed into thick interdigitated electrodes. The uniformly distributed MXene nanosheets improved the printability and also tuned the interconnected electronic structure of PEDOT:PSS. DIW-printed large-scale interfinger electrodes showed high resolution and ensured fast ion/electron transport (Figure 15.4b). The constructed all-gel-state devices have almost coincident CV curves at a large bending degree of 180° (Figure 15.4c) and maintain close to 100% initial capacitance after undergoing a series of bending deformations (Figure 15.4d) [25]. Cheng et al. reported an optimized PEDOT:PSS ink for DIW-printed FSCs. Through dual additive-induced physical cross-linking, PEDOT:PSS hydrogels simultaneously show high electrical conductivity and excellent mechanical stretchability. The constructed FSCs maintain similar CV curves during severe bending and even twisting [29]. In addition, Yang et al. combined structural design and 3D printing to

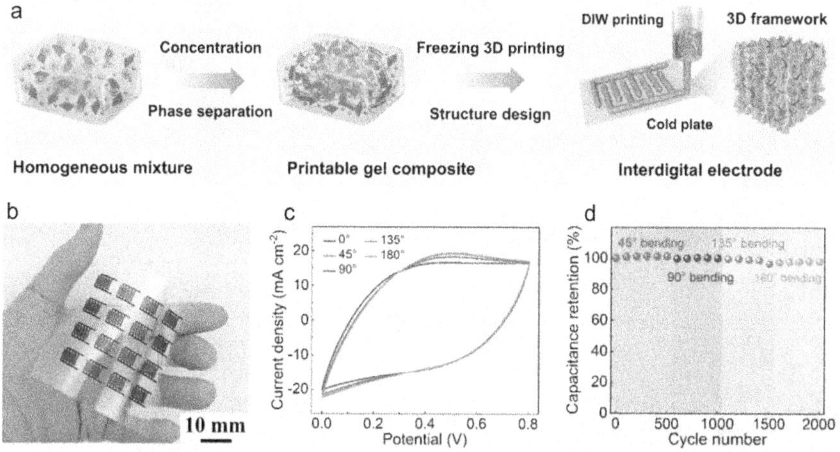

FIGURE 15.4 DIW 3D printing for FSCs. (a) Fabrication process of PEDOT:PSS/MXene/EG (PME) gel composite inks and interdigital electrodes. (b) Digital image of a 4 × 4 electrode array. (c) CV curves of PME MSCs under different bending angles. (d) Capacitance retention of PME MSCs under different bending angles. Adapted with permission [25]. Copyright (2023), Wiley.

develop additive-free and freestanding stretchable electrodes with different negative Poisson's ratio structures, which are based on a highly concentrated, viscoelastic, and additive-free PEDOT:PSS ink. By further integrating the carbon nanotubes (CNT). DIW-printed PEDOT:PSS/CNT electrodes can provide satisfactory electrochemical performance, as well as stable output power even under extreme deformation [30].

15.3.3.3 Other 3D-Printed Conducting Polymer-Based FSCs

In addition to the 3D printing technologies mentioned previously, some technologies, such as FDM, SLA, selective laser melting, and selective laser sintering, are also attracting attention in the field of FSCs. Among them, FDM is a representative material extrusion technology, where 3D micro-periodic polyelectrolyte structures have a solidification vessel to cure the extruded ink before application. SLA applies reductive photopolymerization to produce 3D products by selectively curing the liquid values using UV light. Unfortunately, FDM and SLA are not applicable to conductive polymers because conductive polymers are neither melt nor light curable. Therefore, for FSCs constructed by these 3D printing technologies, conductive polymers are often used as coating layer to provide additional capacitance. For example, Foo et al. fabricated FDM 3D-printed electrodes using conductive graphene filaments and a commercial Alpha 3D printer (Figure 15.5a), and then electrodeposited PPy on the electrode surface (Figure 15.5b) [31]. Graphene@PPy-based FSCs exhibited good capacitive performance (Figure 15.5c–e) with a specific capacitance of 98.4 F g^{-1}. In addition, Vaghasiya et al. used graphene/polylactic acid filaments to create FDM electrodes with any desired shapes, such as 3D cylindrical (3Dcy), disk (3Ddc), and 3D rectangular (3Drc) electrodes. Electrodes coated with Ti_3C_2@PPy complex have

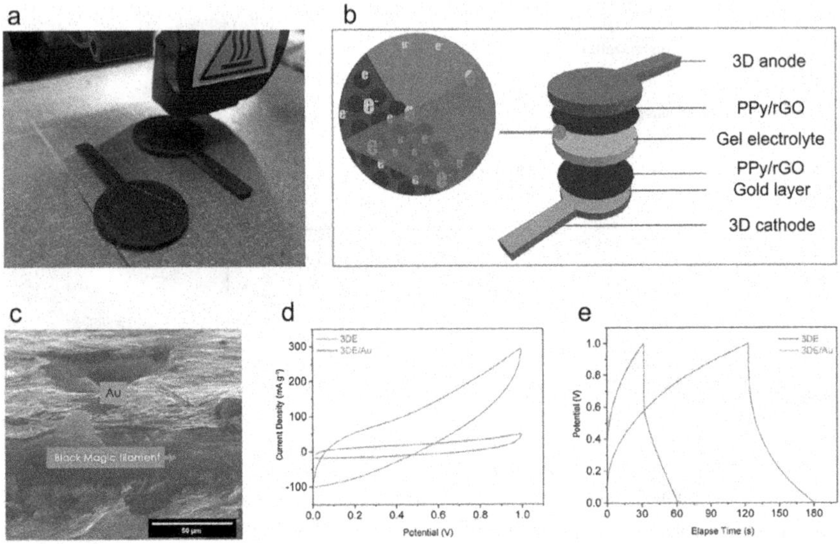

FIGURE 15.5 FDM 3D printing for FSCs. (a) Optical image of FDM 3D printing process. (b) Schematic illustration of solid-state flexible supercapacitor fabrication. (c) Cross-sectional SEM image of 3D-printed electrode. (d) CV analysis of FSCs over 1.0 V potential range at scan rate of 50 mV s^{-1}. (e) Galvanostatic charge/discharge profile of FSCs at current density of 0.5 A g^{-1}. Adapted with permission [31]. Copyright (2018), Springer Nature.

excellent conductivity, capacitive performance, cycle life, and power density. 3Ddc Ti_3C_2@PPy supercapacitor electrodes showed a specific capacitance of 118.2 F g^{-1} and an excellent cycling stability up to 6000 cycles [32].

As discussed, several common 3D printing methods, including DIW, IJP, FDM, and SLA, have been successfully utilized in constructing conducting polymer-based FSCs. However, conducting polymers can hardly be processed using other 3D printing methods because they are generally neither meltable nor light curable. In this regard, it is crucial to develop easily processable conducting polymer-based ink that is versatile for different printing methods through molecular design or modification. Additionally, researchers have a long way to go still in developing scalable and cost-effective conducting polymer inks.

15.4 POTENTIAL APPLICATIONS OF 3D-PRINTED CONDUCTING POLYMER-BASED FSCs

FSCs using conducting polymer-based electrodes are envisaged to bridge the gap between carbonaceous SCs and batteries, leading to high power density, along with improved energy density [8]. While 3D printing technologies are properly employed in the fabrication procedure of FSCs, 3D nanostructured electrodes with high mass loading can be easily constructed [6]. Therefore, the energy density of devices with

3D electrodes can be drastically improved in comparison with conventional FSCs using thin film electrodes [33, 34]. 3D-printed FSCs with conducting polymer electrodes hold the merits of high-power density, improved energy density, and considerable flexibility. They can thus be potentially utilized in many applications, including implantable devices [35, 36], IoT systems [37, 38], wearable electronics [39, 40], and self-powered systems [20, 41] (Figure 15.6).

15.4.1 3D-Printed Conducting Polymer-Based FSCs for Implantable Devices

Since the beginning of human history, the demand for effective healthcare systems for diagnosis and treatment of health problems has grown steadily. Implantable devices (Figure 15.6a), such as cardiovascular implantable electronic devices and neuroimplantable devices play a crucial role in maintaining human health by reducing morbidity and mortality of common diseases, which can consequently give rise to in-time and long-term healthcare of patients [42].

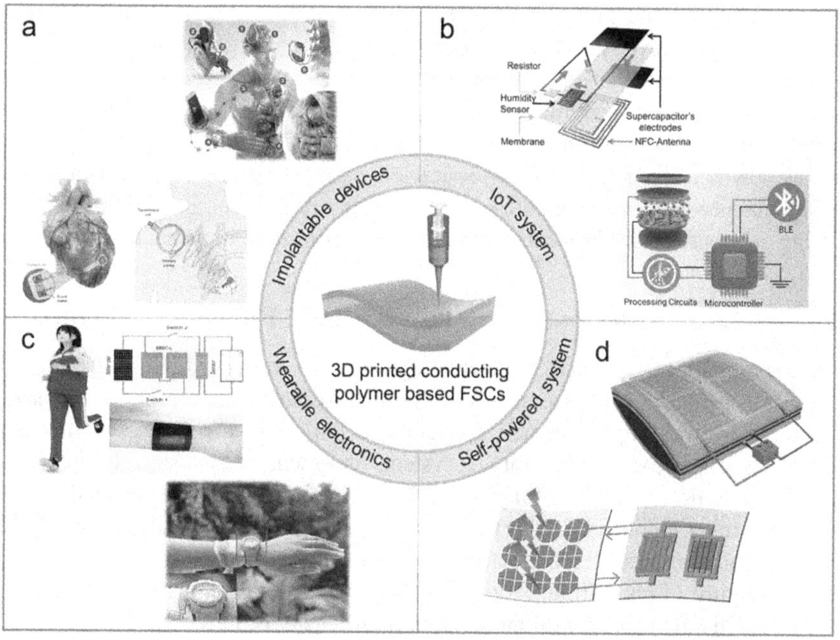

FIGURE 15.6 Potential applications of 3D-printed conducting polymer-based FSCs. (a) Implantable devices. Adapted with permission [35]. Copyright (2020), Elsevier. Adapted with permission [36]. Copyright (2021), Springer Nature. (b) IoT system. Adapted with permission [37]. Copyright (2022), Wiley. Adapted with permission [38]. Copyright (2022), Elsevier. (c) Wearable electronics. Adapted with permission [39]. Copyright (2016), Springer Nature. Adapted with permission [40]. Copyright (2023), Wiley. (d) Self-powered system. Adapted with permission [20]. Copyright (2021), Wiley. Adapted with permission [41]. Copyright (2015), Springer Nature & Tsinghua University Press.

A power supply system is one of the most significant components of implantable devices, which can drive implantable devices sustainably and stably. Generally, the longevity of implantable devices is to a large extent determined by the capacity of power sources. From the historical point of view, the evolution of implantable devices was always accompanied by the development of power sources [36]. The first fully implantable pacemaker was placed in Sweden, driven by a nickel-cadmium battery with an output voltage of 1.25 V and a capacity of 190 mAh to keep the patient's heart rhythm for 3 h. Due to the short longevity, this nickel-cadmium battery was soon replaced by series-connected mercury-zinc batteries in the 1960s. However, the gas generation during discharging and short-circuit problem hindered the mercury-zinc battery from further application. To address the limited longevity of previous implantable batteries, the nuclear battery was invented successfully with an unprecedented lifespan of over 30 years (the first isotope-based (^{238}Pu) pacemaker). With the rapid development of lithium-based batteries in the 1970s, lithium primary batteries became the standard power source for modern pacemakers due to the high energy density delivered in limited volume [43].

Generally speaking, the energy density of state-of-the-art FSCs is still much lower than that of lithium-based batteries. Numerous efforts have been dedicated in recent years to improving the energy density of FSCs. Conducting polymer-based FSCs store charges via highly reversible surface redox reactions, namely pseudocapacitance, which is much higher than that of conventional electrical double-layer SCs. Additionally, through employing 3D printing technology, the loading mass of electrodes and overall energy density of device can be drastically enhanced. Therefore, the 3D-printed conducting polymer-based flexible supercapacitor is envisaged to possess promising energy density that is comparable with lithium batteries. In this regard, future 3D-printed conducting polymer-based FSCs could be potentially utilized in implantable devices.

15.4.2 3D-Printed Conducting Polymer-Based FSCs for IoT Systems

With the rapid development of 5G technologies and booming proliferation of distributed electronics, the IoT system is developing at a soaring rate, which is envisaged to bring our lives unprecedented convenience [44]. The fabrication of miniaturized electrochemical energy storage systems is essential for the development of future electronic devices for IoT applications where billions of connected devices are increasingly employed in our daily life. Lithium-ion micro-batteries and MSCs are two typical miniaturized electrochemical energy storage systems used for IoT devices (Figure 15.6b). Lithium-ion micro-batteries generally offer high energy density (~1 mWh cm^{-2}) along with moderate power density (<5 mW cm^{-2}), while MSCs deliver ultrahigh power density (>10 mW cm^{-2}) but suffer from limited energy density (<0.1 mW h cm^{-2}). In this part, we summarize recent advances of FSCs utilized in IoT devices and give some perspectives on the development of FSCs for future IoT devices.

Recently, Mousavi and co-workers invented an integrated chip with MSCs, sensor, resistor, and near field communication (NFC) antenna circuit components for IoT application. In this integrated circuit, laser-scribed graphene (LSG) was employed as the basic component of electrodes for all the individual elements, including MSCs, sensor, resistor, and NFC antenna, that are densely integrated in one membrane. Nanostructured PANI with promising pseudocapacitance is then deposited onto LSG layers to increase the energy density of MSCs. As a result, the as-prepared flexible MSCs deliver high energy density of 0.407 mW h cm^{-3} along with a considerable power density of 196 mW cm^{-3}. Due to the remarkable electrochemical performance of this flexible MSC, the humidity sensor and NFC antenna integrated on this circuit can be successfully driven with sub-second response time and excellent impedance matching. This work unambiguously demonstrates that PANI-based MSCs can be used as paradigm power supply system that underpin the on-chip electronics in the IoT era [37]. Similarly, Park et al. proposed one kind of self-charging SCs by integrating SCs and a triboelectric nanogenerator (TENG) onto a single device. Consequently, the energy generated by the TENG can be efficiently stored in SCs without adding extra power management or rectifier circuits. The as-prepared FSCs deliver high areal capacitance of 25.60 mF cm^{-2} along with an energy density of 0.0278 mWh cm^{-2}. More importantly, FSCs can be successfully charged to 210 mV within 9 s via the TENG. It can thus be utilized in IoT system to control the augmented reality game machine [38].

Based on the mentioned research works, it can be easily concluded that FSCs can be rationally utilized as a promising power source candidate to underpin IoT devices, attributed to high power density and ultralong cycling life. However, conventional FSCs suffer from limited energy density and a poor technological readiness level. In this regard, 3D-printed conducting polymer FSCs can be employed to address the mentioned limitations due to the following merits. On the one hand, 3D printing technology can construct high mass-loading electrodes with 3D nanostructured frameworks, leading to greatly improved capacitance and energy density. On the other hand, 3D printing technology is highly efficient and versatile, which can give rise to low-cost and scalable fabrication.

15.4.3 3D-Printed Conducting Polymer-Based FSCs for Wearable Electronics

Wearable electronics is one of the most attractive research hotspots, and they can be utilized in pulse diagnosis, blood pressure assessment, cardiovascular system diseases monitoring, and sleep apnea identification [45]. Wearable electronics play a crucial role in real-time monitoring of hospital patients and family health care, which can substantially improve people's health and quality of life. Wearable electronics are expected to be bendable and deformable to fit different body position changes. Correspondingly, the energy supply system of wearable electronics should be flexible to accommodate various deformations. In this regard, FSCs can be regarded as an ideal class of energy supply for wearable electronics [46].

To this date, conducting polymer-based FSCs have been widely utilized in wearable electronics (Figure 15.6c). Recently, Chen et al. proposed polyaniline-based flexible MSCs using a facile spray-coating method by which PANI ink can be easily spray-coated onto a current collector to form interdigitated electrodes. These PANI-based MSCs exhibit a high areal capacitance of 96.6 mF cm^{-2} along with a remarkable volumetric capacitance of 26.0 F cm^{-3}. Additionally, the as-prepared MSCs exhibit considerable flexibility with no performance decay under different bending angles. Thanks to the promising flexibility of this PANI-based MSCs, they can be properly integrated and utilized in wearable electronics to power different IoT devices [47]. Wang and co-workers designed all-printed, sweat-based wearable biosupercapacitors by properly integrating a biofuel cell unit and SC unit. In this bifunctional biosupercapacitor, the biofuel cell unit converts chemical energy from human perspiration to electrical energy via enzymatic electrochemical reactions. As a result, the SC unit can be fully charged to drive other wearable devices. The self-charging hybrid wearable device obtained high power of 1.7 mW cm^{-2} in vitro and 343 μW cm^{-2} on the body during exercise, suggesting considerable potential as a power source for future wearable electronics [48].

15.4.4 3D-Printed Conducting Polymer-Based FSCs for Self-Powered Systems

Developing self-powered systems without the need for an external power supply is of great importance for future electronics [49]. Energy harvesting and energy storage units are two crucial components for self-powered systems, in which the energy harvested by the energy harvesting unit can be properly stored by the energy storage unit. Therefore, the self-powered system can be continuously and stably operated without an external power supply.

FSCs can be properly used as energy storage units for self-powered systems attributed to the high power density and ultralong cycling life (Figure 15.6d). For instance, Wang and co-workers developed FSCs using laser scribed graphene (LIG) as active electrodes by direct laser writing onto the surface of the Kapton film. Serial connected LIG-based FSCs are then integrated with TENG to form a self-powered system. Through controlling the gap distance of the TENG, the output voltage can be properly controlled at ~30 V. Therefore, the FSC array can be efficiently charged by the TENG. This flexible self-charging power unit consisting of FSCs and TENG can harvest mechanical energy and convert it into electrical energy to drive different IoT electronics [41]. In addition to TENG, solar cells can also be integrated with FSCs to construct self-powered systems. Wu's group developed flexible MSCs by directly printing MXene/PEDOT:PSS hybrid ink onto paper substrate using inkjet printing method. The as-prepared flexible MSCs exhibit an unprecedented volumetric capacitance of 754 F cm^{-3} along with 9.4 mWh cm^{-3} energy density[20].

FSCs are going to prompt the prosperity of the approaching IoT era, in which a smart and convenient life is enabled by billions of connected and distributed electronics. It can also be easily envisaged that 3D-printed conducting polymer-based FSCs with high energy density and long lifespan will become a popular class of power supply for future IoT devices.

15.5 PERSPECTIVES

SCs are designed to form a unique class of electrochemical energy storage devices with high power delivery and intermediate energy density to bridge the gap between batteries and capacitors. Conducting polymers such as PANI and PPy store charges via a pseudocapacitive mechanism, which delivers higher energy density than electrical double-layer capacitors. Conducting polymers can be 3D printed to form high mass-loading electrodes for FSCs with improved energy density, which is regarded as a competitive and promising electrochemical energy storage candidate for future electronics. However, to date, it still remains a great challenge to develop high performance and cost-effective 3D-printed conducting polymer-based FSCs. This goal can be achieved by comprehensively addressing the following crucial issues.

1) Conducting polymers such as PANI and PPy suffer from limited electronic conductivity. Therefore, an extra current collector with high electronic conductivity is often employed during the fabrication of 3D-printed conducting polymer-based FSCs, which consequently increases the fabrication cost of devices. In this regard, developing highly conductive polymer ink can be a feasible route towards current collector-free FSCs.
2) The poor cyclic stability arising from repeated volumetric expansion/shrinkage during galvanostatic charge/discharge procedure is recognized as the Achilles' heel of conducting polymers. Therefore, 3D-printed conducting polymer-based FSCs usually suffer from inferior cycling stability due to the structural collapse and polymeric chain degradation. The poor cyclic stability of conducting polymer-based FSCs can be addressed by introducing conductive frameworks or designing proper surface nanostructures.
3) FSCs will be subjected to various external stresses during different application scenarios. Poor interfaces of devices will lead to electrode delamination from the substrate, which will further lead to device failure. Therefore, it is necessary to optimize the interfaces of 3D-printed conducting polymer-based FSCs to enable good integrity and contact of different functional layers during repeated deformation.
4) The next generation of 3D-printed conducting polymer-based FSCs is supposed to be easily fabricated and cost effective. The top priority to achieve this goal is to devise scalable and low-cost conducting polymer ink that is suitable for 3D printing. Therefore, we argue here that more research attention should be devoted to developing novel conducting polymer ink that can be scalably synthesized yet remain cost effective.

REFERENCES

[1] P. Simon, Y. Gogotsi, Perspectives for electrochemical capacitors and related devices, Nat. Mater. 19 (2020) 1151–1163.
[2] H. Zhang, H. Su, L. Zhang, B. Zhang, F. Chun, X. Chu, W. He, W. Yang, Flexible supercapacitors with high areal capacitance based on hierarchical carbon tubular nanostructures, J. Power Sources 331 (2016) 332–339.
[3] Q. Xue, J. Sun, Y. Huang, M. Zhu, Z. Pei, H. Li, Y. Wang, N. Li, H. Zhang, C. Zhi, Recent progress on flexible and wearable supercapacitors, Small 13 (2017) 1701827.
[4] Y. Z. Zhang, Y. Wang, T. Cheng, L. Q. Yao, X. Li, W. Y. Lai, W. Huang, Printed supercapacitors: Materials, printing and applications, Chem. Soc. Rev. 48 (2019) 3229–3264.
[5] H. Li, J. Liang, Recent development of printed micro-supercapacitors: Printable materials, printing technologies, and perspectives, Adv. Mater. 32 (2020) 1805864.

[6] M. Sha, H. Zhao, Y. Lei, Updated insights into 3D architecture electrodes for micropower sources, Adv. Mater. 33 (2021) 2103304.
[7] P. Yang, W. Mai, Flexible solid-state electrochemical supercapacitors, Nano Energy 8 (2014) 274–290.
[8] X. Chu, W. Yang, H. Li, Recent advances of polyaniline-based micro-supercapacitors, Mater. Horiz. 10 (2023) 670–697.
[9] P. Simon, Y. Gogotsi, Materials for electrochemical capacitors, Nat. Mater. 7 (2008) 845–854.
[10] J. Wang, J. Wang, Z. Kong, K. Lv, C. Teng, Y. Zhu, Conducting-polymer-based materials for electrochemical energy conversion and storage, Adv. Mater. 29 (2017) 1703044.
[11] G. A. Snook, P. Kao, A. S. Best, Conducting-polymer-based supercapacitor devices and electrodes, J. Power Sources 196 (2011) 1–12.
[12] M. Li, S. Zhou, L. Cheng, F. Mo, L. Chen, S. Yu, J. Wei, 3D printed supercapacitor: Techniques, materials, designs, and applications, Adv. Funct. Mater. 33 (2022) 2208034.
[13] H. Gao, K. Lian, Proton-conducting polymer electrolytes and their applications in solid supercapacitors: A review, RSC Adv. 4 (2014) 33091–33113.
[14] Z. Xu, X. Chu, Y. Wang, H. Zhang, W. Yang, Three-dimensional polymer networks for solid-state electrochemical energy storage, Chem. Eng. J. 391 (2020) 123548.
[15] J. Liang, C. Jiang, W. Wu, Printed flexible supercapacitor: Ink formulation, printable electrode materials and applications, Appl. Phys. Rev. 8 (2021) 021319.
[16] H. Sun, Y. Zhang, J. Zhang, X. Sun, H. Peng, Energy harvesting and storage in 1D devices, Nat. Rev. Mater. 2 (2017) 17023.
[17] J. Diao, J. Yuan, A. Ding, J. Zheng, Z. Lu, Flexible supercapacitor based on inkjet-printed graphene@polyaniline nanocomposites with ultrahigh capacitance, Macromol. Mater. Eng. 303 (2018) 1800092.
[18] S. Sollami Delekta, M.-M. Laurila, M. Mäntysalo, J. Li, Drying-mediated self-assembly of graphene for inkjet printing of high-rate micro-supercapacitors, Nano-Micro Lett. 12 (2020) 40.
[19] L. Fan, N. Zhang, K. Sun, Flexible patterned micro-electrochemical capacitors based on PEDOT, Chem. Commun. 50 (2014) 6789–6792.
[20] J. Ma, S. Zheng, Y. Cao, Y. Zhu, P. Das, H. Wang, Y. Liu, J. Wang, L. Chi, S. Liu, Z. S. Wu, Aqueous MXene/PH1000 hybrid inks for inkjet-printing micro-supercapacitors with unprecedented volumetric capacitance and modular self-powered microelectronics, Adv. Energy Mater. 11 (2021) 2100746.
[21] Z. Li, V. Ruiz, V. Mishukova, Q. Wan, H. Liu, H. Xue, Y. Gao, G. Cao, Y. Li, X. Zhuang, J. Weissenrieder, S. Cheng, J. Li, Inkjet printed disposable high-rate on-paper microsupercapacitors, Adv. Funct. Mater. 32 (2021) 2108773.
[22] W. Liu, C. Lu, H. Li, R. Y. Tay, L. Sun, X. Wang, W. L. Chow, X. Wang, B. K. Tay, Z. Chen, J. Yan, K. Feng, G. Lui, R. Tjandra, L. Rasenthiram, G. Chiu, A. Yu, Paper-based all-solid-state flexible micro-supercapacitors with ultra-high rate and rapid frequency response capabilities, J. Mater. Chem. A 4 (2016) 3754–3764.
[23] Z. Wang, Q. E. Zhang, S. Long, Y. Luo, P. Yu, Z. Tan, J. Bai, B. Qu, Y. Yang, J. Shi, H. Zhou, Z. Y. Xiao, W. Hong, H. Bai, Three-dimensional printing of polyaniline/reduced graphene oxide composite for high-performance planar supercapacitor, ACS Appl. Mater. Interfaces 10 (2018) 10437–10444.
[24] Y. Liu, B. Zhang, Q. Xu, Y. Hou, S. Seyedin, S. Qin, G. G. Wallace, S. Beirne, J. M. Razal, J. Chen, Development of graphene oxide/polyaniline inks for high performance flexible microsupercapacitors via extrusion printing, Adv. Funct. Mater. 28 (2018) 1706592.
[25] L. Li, J. Meng, X. Bao, Y. Huang, X. P. Yan, H. L. Qian, C. Zhang, T. Liu, Direct-ink-write 3D printing of programmable micro-supercapacitors from MXene-regulating conducting polymer inks, Adv. Energy Mater. 13 (2023) 2203683.

[26] R. Xing, Y. Xia, R. Huang, W. Qi, R. Su, Z. He, Three-dimensional printing of black phosphorous/polypyrrole electrode for energy storage using thermoresponsive ink, Chem. Commun. 56 (2020) 3115–3118.

[27] Y. Gu, Y. Zhang, Y. Shi, L. Zhang, X. Xu, 3D all printing of polypyrrole nanotubes for high mass loading flexible supercapacitor, ChemistrySelect 4 (2019) 10902–10906.

[28] Z. Qi, J. Ye, W. Chen, J. Biener, E. B. Duoss, C. M. Spadaccini, M. A. Worsley, C. Zhu, 3D-printed, superelastic polypyrrole–graphene electrodes with ultrahigh areal capacitance for electrochemical energy storage, Adv. Mater. Technol. 3 (2018) 1800053.

[29] T. Cheng, F. Wang, Y.-Z. Zhang, L. Li, S.-Y. Gao, X.-L. Yang, S. Wang, P.-F. Chen, W.-Y. Lai, 3D printable conductive polymer hydrogels with ultra-high conductivity and superior stretchability for free-standing elastic all-gel supercapacitors, Chem. Eng. J. 450 (2022) 138311.

[30] J. Yang, Q. Cao, X. Tang, J. Du, T. Yu, X. Xu, D. Cai, C. Guan, W. Huang, 3D-Printed highly stretchable conducting polymer electrodes for flexible supercapacitors, J. Mater. Chem. A 9 (2021) 19649–19658.

[31] C. Y. Foo, H. N. Lim, M. A. Mahdi, M. H. Wahid, N. M. Huang, Three-dimensional printed electrode and its novel applications in electronic devices, Sci. Rep. 8 (2018) 7399.

[32] J. V. Vaghasiya, C. C. Mayorga-Martinez, M. Pumera, Smart energy bricks: Ti_3C_2@polymer electrochemical energy storage inside bricks by 3D printing, Adv. Funct. Mater. 31 (2021) 2106990.

[33] L. Christophe, L. B. Jean, B. Thierry, Challenges and prospects of 3D micro-supercapacitors for powering the internet of things, Energy Environ. Sci. 12 (2019) 96–115.

[34] H. Zhao, Y. Lei, 3D nanostructures for the next generation of high-performance nanodevices for electrochemical energy conversion and storage, Adv. Energy Mater. 10 (2020) 2001460.

[35] J. S. Chae, S. K. Park, K. C. Roh, H. S. Park, Electrode materials for biomedical patchable and implantable energy storage devices, Energy Storage Mater. 24 (2020) 113–128.

[36] Y. S. Choi, R. T. Yin, A. Pfenniger, J. Koo, R. Avila, K. Benjamin Lee, S. W. Chen, G. Lee, G. Li, Y. Qiao, A. Murillo-Berlioz, A. Kiss, S. Han, S. M. Lee, C. Li, Z. Xie, Y. Y. Chen, A. Burrell, B. Geist, H. Jeong, J. Kim, H. J. Yoon, A. Banks, S. K. Kang, Z. J. Zhang, C. R. Haney, A. V. Sahakian, D. Johnson, T. Efimova, Y. Huang, G. D. Trachiotis, B. P. Knight, R. K. Arora, I. R. Efimov, J. A. Rogers, Fully implantable and bioresorbable cardiac pacemakers without leads or batteries, Nat. Biotechnol. 39 (2021) 1228–1238.

[37] A. Khodabandehlo, A. Noori, M. S. Rahmanifar, M. F. El-Kady, R. B. Kaner, M. F. Mousavi, Laser-scribed graphene–polyaniline microsupercapacitor for internet-of-things applications, Adv. Funct. Mater. 32 (2022) 2204555.

[38] K. Shrestha, S. Sharma, G. B. Pradhan, T. Bhatta, S. M. S. Rana, S. Lee, S. Seonu, Y. Shin, J. Y. Park, A triboelectric driven rectification free self-charging supercapacitor for smart IoT applications, Nano Energy 102 (2022) 107713.

[39] J. Chen, Y. Huang, N. N. Zhang, H. Y. Zou, R. Y. Liu, C. Y. Tao, X. Fan, Z. L. Wang, Micro-cable structured textile for simultaneously harvesting solar and mechanical energy, Nat. Energy 1 (2016) 16138.

[40] W. Wang, L. Xu, L. Zhang, A. Zhang, J. Zhang, Self-powered integrated sensing system with in-plane micro-supercapacitors for wearable electronics, Small (2023) 2207723.

[41] J. Luo, F. R. Fan, T. Jiang, Z. Wang, W. Tang, C. Zhang, M. Liu, G. Cao, Z. L. Wang, Integration of micro-supercapacitors with triboelectric nanogenerators for a flexible self-charging power unit, Nano Res. 8 (2015) 3934–3943.

[42] N. Obidin, F. Tasnim, C. Dagdeviren, The future of neuroimplantable devices: A materials science and regulatory perspective, Adv. Mater. 32 (2020) 1901482.

[43] J. Deng, X. Sun, H. Peng, Power supplies for cardiovascular implantable electronic devices, EcoMat 5 (2023) 12343.

[44] G. Chen, Y. Li, M. Bick, J. Chen, Power supplies for cardiovascular implantable electronic devices, Chem. Rev. 120 (2020) 3668–3720.

[45] Y. Zou, A. Libanori, J. Xu, A. Nashalian, J. Chen, Triboelectric nanogenerator enabled smart shoes for wearable electricity generation, Research 2020 (2020) 1–20.

[46] K. Meng, S. Zhao, Y. Zhou, Y. Wu, S. Zhang, Q. He, X. Wang, Z. Zhou, W. Fan, X. Tan, J. Yang, J. Chen, A wireless textile-based sensor system for self-powered personalized health care, Matter 2 (2020) 896–907.

[47] X. Chu, G. Chen, X. Xiao, Z. Wang, T. Yang, Z. Xu, H. Huang, Y. Wang, C. Yan, N. Chen, H. Zhang, W. Yang, J. Chen, Air-stable conductive polymer ink for printed wearable micro-supercapacitors, Small 17 (2021) 2100956.

[48] J. Lv, L. Yin, X. Chen, I. Jeerapan, C. A. Silva, Y. Li, M. Le, Z. Lin, L. Wang, A. Trifonov, S. Xu, S. Cosnier, J. Wang, Wearable biosupercapacitor: Harvesting and storing energy from sweat, Adv. Funct. Mater. 31 (2021) 2102915.

[49] Y. Zou, V. Raveendran, J. Chen, Wearable triboelectric nanogenerators for biomechanical energy harvesting, Nano Energy 77 (2020) 105303.

16 3D-Printed Conducting Polymers for Metal-Ion Batteries

Nora Chelfouh, Manon Faral, Tristan Perodeau, Mickaël Dollé, and Audrey Laventure

16.1 INTRODUCTION

16.1.1 CONTEXT

The field of additive manufacturing holds numerous promises for a myriad of technological applications. One of these technological applications is another extensively investigated field, energy storage, more specifically related to batteries in the context of this chapter. While researchers working either in additive manufacturing or batteries have made important advances in their respective fields in recent years, at the time of writing this chapter, multiple challenges remain to be tackled, both at the fundamental and applied research forefront. However, some of the challenges that each field faces are complementary, and working at their interface could bring valuable insights for both.

The topic of this chapter is specifically dedicated to the 3D printing of conducting polymers for metal-ion batteries, and most of the examples are related to Li-ion batteries. Since this topic has attracted much attention in recent years, which has led to multiple publications, this chapter presents an overview of the most recent advances made at the intersection of the material and additive manufacturing front. These advances are covered for the electrodes, more specifically formulation considerations and to illustrate how conducting polymers can be included in 3D-printed electrodes to overcome some shortcomings in 3D-printed electrodes that do not include conducting polymers. These advances are also covered for electrolytes, that is, solid polymer electrolytes (blended with a salt for the conduction of ions) and hybrid electrolytes that are formulated using the polymer acting as an ion conductor. For additional details, the interested reader is invited to consult exhaustive reviews related to the manufacturing aspect of 3D printed batteries [1], the comparison of pros and cons for different types of materials used with different additive manufacturing techniques [2–4], and future perspectives [5] and applications [6] for 3D-printed batteries.

16.1.2 GENERAL OVERVIEW OF THE TOPICS COVERED

While significant research efforts have already been dedicated to the 3D printing of electrodes (and to a lesser extent electrolytes, *vide infra*) these studies can be

DOI: 10.1201/9781003415985-16

classified into different subcategories. The first one involves the use of 3D printing to demonstrate a proof-of-concept battery where one or multiple components are 3D printed, followed by the assembly of the other components using other techniques. A few others explore the possibility of creating a 3D array, with the aim of increasing the active surface and the contact between the electrode and the electrolyte. Finally, some studies have focused on interconnected systems via specific architecture designs to facilitate ion transport within the battery. These works are conducted with the goal of designing electrodes with higher areal capacity and higher power density than what is currently possible to achieve with conventional (tape-casting or doctor-blade coating) methods. For instance, the Ragone plot in Figure 16.1 shows that 3D scaffolds have the potential to achieve both high power and energy density compared to interdigitated and fiber architectures.

All these studies demonstrate that research conducted at the interface of additive manufacturing and batteries is useful to advance knowledge and/or technological development, but the last two address specific challenges in metal-ion batteries, which are the interfaces between the electrode and the electrolyte, where the charge transfer reaction at the heart of metal-ion battery operation occurs. This reaction can also be limited by mass transport issues that may be hindered in thicker electrodes. While several questions remain unanswered in the context of the formulation (e.g., what are the properties needed to ensure its printability?), additive manufacturing clearly stands out as an emerging method of choice to design architectures that may

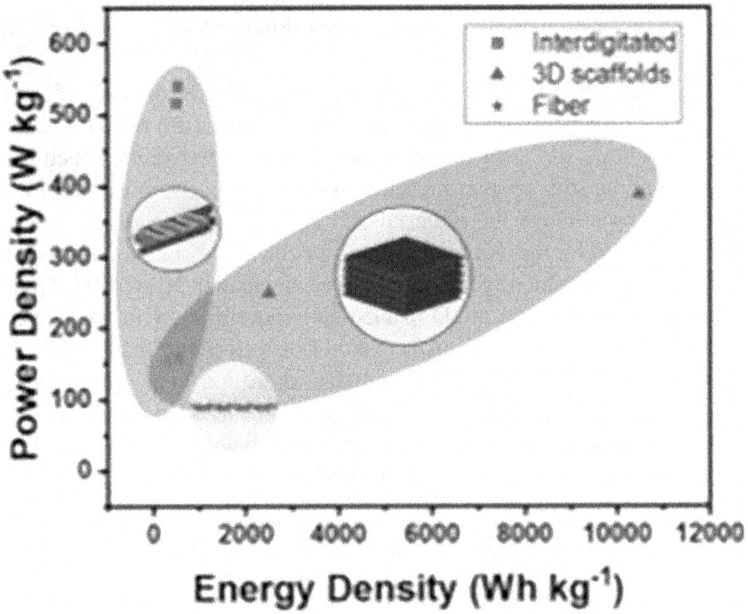

FIGURE 16.1 Ragone plots of 3D printed Li-ion batteries for three types of architectures: interdigitated, 3D scaffolds, and fiber. Adapted with permission [7]. Copyright 2019 John Wiley and Sons.

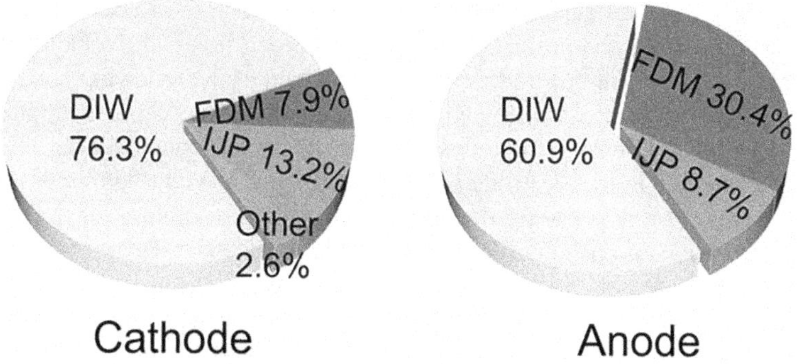

FIGURE 16.2 Proportions of the different additive manufacturing techniques used to prepare cathode and anode electrodes in batteries. DIW stands for direct-ink writing, FDM stands for fused deposition modeling, and IJP stands for inkjet printing. The proportions are calculated using the citations referred to in [1]. Adapted with permission [1]. Copyright 2021, Elsevier.

solve some of the mass transport and interface problems. Additive manufacturing is also promising as a tool that could be used to process materials in designs that are not achievable using conventional methods, designs that could potentially help to solve some electrochemical-related questions.

Additive manufacturing comprises a variety of different processing techniques, which all have specific requirements to ensure an appropriate 3D printability of the formulations. The reader is referred to this extensive review on polymer additive manufacturing [8] for more details on each technique. A rough categorization can be made between *material extrusion techniques*, including fused deposition modeling (FDM) and direct-ink writing (DIW); *powder bed fusion* (selective laser sintering (SLS)); and *vat-photopolymerization* (stereolithography (SLA), digital light processing (DLP)). As illustrated in Figure 16.2, as of 2021, most of the electrodes reported in the literature were 3D printed using extrusion-based techniques (FDM and DIW) [9]. Thus, most of the examples discussed in the next sections along with rheological consideration of the formulation will focus on this technique. Yet other specific examples will also be briefly presented for SLA and DLP, which have emerged in the field of 3D-printed metal-ion batteries.

16.2 3D PRINTING THE COMPONENTS OF A BATTERY

16.2.1 Electrodes

16.2.1.1 Formulation Considerations

A metal-ion battery is a secondary battery that is composed of two electrodes: a positive electrode (often imprecisely referred to as the "cathode") and a negative electrode (often imprecisely referred to as the "anode"). These electrodes enable the redox reactions that allow electrical energy storage by converting it to chemical energy. Positive and negative electrodes must fulfill specific requirements to ensure redox reactions at the interface between the electrode and the electrolyte. As illustrated in Figure 16.3, the positive or negative electrode in metal-ion batteries is, in most cases, a composite electrode, containing a blend of the following components: active material, polymeric binder, and carbon additive. Each of these compounds has a key role to play to enable the redox reactions at the interface between the active material contained in the electrode and the electrolyte. Indeed, the role of the electrode is a dual one: i) to ensure the ionic conductivity for lithium ions involved in the redox reaction and ii) to ensure the electronic conductivity to allow electrons to circulate in the external circuit.

The conventional way to process such electrodes at a laboratory or industrial scale generally involves a typical coating process such as tape casting. In this process, the coated dispersion (also referred to as a slurry) must meet several rheological considerations to be coated efficiently. The main advantage related to the processing of the electrode with additive manufacturing lies in the architectural freedom it provides. Indeed, conventional tape-casting methods do not offer the possibility of designing architectures that minimize or eliminate the ionic and electronic mass transport hindrance at the electrode/electrolyte interface. Yet each category of additive manufacturing techniques, as mentioned in the introduction of this chapter, has its own specific rheological requirements that the formulation (slurry) must meet to be printable. To date, most of the studies conducted on electrode fabrication via 3D printing techniques have been focused on extrusion printing techniques, that is, FDM and DIW. This choice is motivated by the fact that the extrusion process is the technique that exhibits the largest flexibility in terms of rheological considerations that a slurry must meet, paving the way towards printing a variety of materials, making it perfectly suitable for different kinds of active material and additives which the composite electrodes are made of.

However, the rheological considerations that formulations must meet to be extruded as electrodes while presenting an appropriate print fidelity, that is, the extent to which the resulting 3D printed architecture meets the features programmed via computer-aided design (CAD), remain different from those needed for conventional

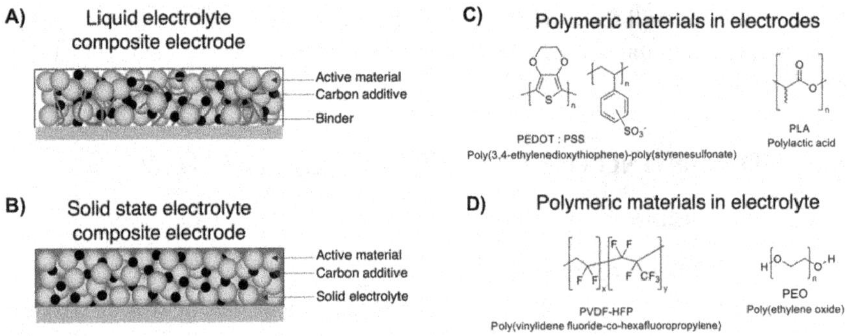

FIGURE 16.3 Schematic illustration of A) liquid electrolyte combined with a composite electrode and B) solid-state electrolyte combined to a composite electrode. Chemical structures of examples of polymers involved in C) composite electrode and D) solid-state electrolyte.

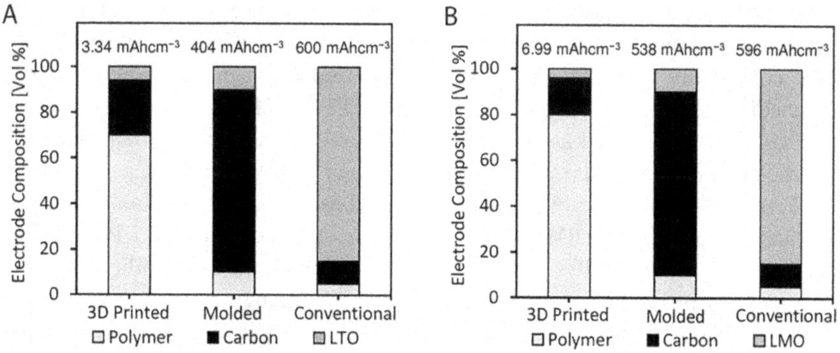

FIGURE 16.4 A) Anode and B) cathode composition comparison. 3D-printed (fused deposition modeling) and molded samples both have carbon:active material (LTO or LMO) ratios of 80:20. Volumetric discharge capacities are shown above each type of electrode type. Adapted with permission [10]. Copyright 2018, American Chemical Society.

tape-casting techniques. These differences are reported in Figure 16.4, where the material volumetric ratios are compared for a typical suspension used in tape casting and 3D-printed FDM processing (an additional comparison is added with the molding technique). In this case, the polymer matrix used is polylactic acid (PLA), which is commonly used in FDM and has been reported to act as an efficient polymer matrix for FDM [10], provided that appropriate plasticizers are added to the filament formulation [11]. Its ratio, compared to the carbon additive and the active material (LTO [$Li_4Ti_5O_{12}$] in the case of the anode and LMO [$LaMnO_3$] in the case of the cathode) is several times larger (70–80% v/v) than for the conventional process (5% v/v). This major difference originates from the rheological requirements difference between the two techniques, where a higher range of viscosity [12] is needed to ensure an appropriate printability for FDM compared to the one needed for the tape casting method. Unfortunately, the need to use such a large amount of polymer can lead to a decrease in the electrochemical performance. Indeed, in this specific comparison, the capacity reached in 3D-printed materials is lower (10% of the capacity measured for an electrode prepared using tape-casting) than conventional cast material due to a higher proportion of isolating phase covering active material particles and thus hindering their efficient percolation.

16.2.1.2 Use of Conducting Polymers as Fillers to Increase the Performance of Electrodes

To achieve better electrochemical performance, a few studies have explored the addition of conductive fillers into the composite electrode with the goal of enhancing the electrical contact between the filler and the active material. Multi-walled carbon nanotubes are one of the additives that can be used to enhance the electrical conductivity in the composite electrode [13]. Of note, this strategy is not only used in electrodes prepared using extrusion techniques but also with other additive manufacturing techniques, such as SLS, for which it has been reported that the addition of graphite to polyurethane leads to highly porous electrodes [14].

Another way to improve electrical conductivity is the addition of electronic conducting polymers in the formulation to be 3D printed into an electrode. One of the conducting polymers that has been reported for such purposes is poly(3,4-ethylenedioxythiophene): polystyrene sulfonate (PEDOT:PSS). Indeed, its high electronic conductivity as well as its ease of processability (suspension) make it a useful and efficient additive to consider for additive manufacturing processing of electrodes [15]. For instance, the use of PEDOT:PSS has been reported in printed composite (positive) electrodes via DIW. More specifically, a blend of PEDOT:PSS and carboxymethylcellulose (CMC) has been used to enhance the printability as well as the conducting properties of composite electrodes. Rheology measurements revealed that adding PEDOT:PSS increased the apparent viscosity by one order of magnitude compared to formulations that did not contain PEDOT:PSS while preserving the same steep viscosity decrease as a function of the shear rate (shear-thinning behavior). Oscillatory frequency sweep measurements have also been performed to determine the viscoelastic behavior of the formulation. These measurements revealed that the storage modulus can remain higher than the loss modulus, indicating that the formulation exhibits an elastic dominant region, which is essential for 3D printing stacked architecture, that is, that a layer can support the addition of successive layers without spreading undesirably or affecting the print fidelity. Moreover, it was shown that the nanofibril microstructure of the PEDOT:PSS plays a significant role in the resulting electrode microstructure. Indeed, the interactions between the positively charged functional groups of the CMC and the negatively charged functional groups of the polystyrene sulfonate components of the PEDOT:PSS enable the PEDOT chains to interact together via stacking. This type of interaction creates a 3D-connected network, leading to a higher viscosity, promoting the printability of the formulation and increased electronic conduction pathways, ultimately driving the electrochemical performance of the electrodes. The creation of such a network could pave the way towards the understanding of the microstructure that is needed to ensure that the mass transport is efficient, even for thick (and extra-thick) electrodes.

PEDOT:PSS has also been reported as a conductive agent in composite electrodes for negative electrode applications, processed via DLP. This technique is based on the use of a photocurable resin. The resin is composed of the active material (silicon nanoparticles) and a polymeric matrix that contains polyethylene glycol (PEG) for ensuring the mechanical properties of the resulting electrode, as well as PEDOT:PSS to promote the electronic conductivity of the printed architecture. It is important to highlight that DLP offers many advantages compared to extrusion-based techniques. The resulting architectures can be printed with a higher printing resolution than the one that is achievable via extrusion [16] while using a formulation that is less viscous. However, one of the main disadvantages is the need to use a photoinitiator to initiate the photopolymerization of the photocurable resin. This photoinitiator must be selected while keeping in mind its chemical compatibility with the other materials used in the resin formulation. Another potential limitation stems from the numerous components entering the composition of the electrode could also settle at the bottom of the vat and/or contribute to the scattering of the irradiation, which is detrimental to the quality of the print fidelity [17].

In another report [18], the electrode prepared via DLP resulted in a honeycomb gel-like architecture. This gel-like property can be interesting for materials such as silicon due to their volumetric expansion during electrochemical cycling. Yet the active material mass loading in such electrodes remains lower than what is produced in conventional electrodes: in this case, only 31.7 wt% of silicon nanoparticles are present in the composite electrode. Electrochemical characterizations using galvanostatic cycling comparing an electrode processed by a conventional method and an electrode processed by DLP were conducted. Conventional electrodes exhibit high charge and discharge capacity (2560 and 3146 mA.h/g, respectively) on the first cycle, but this performance decreases quickly over time (until 155 mA.h/g on the 125th cycle). As a comparison, DLP-processed electrodes reveal a lower charge/discharge capacity (1539 and 1783 mA.h/g, respectively) on the first cycle, but the decrease in capacity over time was less important, reaching 1105 mA.h/g on the 125th cycle. The DLP electrode's architecture as well as its gel-like constitution promoted better cycling stability due to its better capability of enduring the mechanical stress caused by the extreme volumetric fluctuations of Si particles during the cycle. This case is an interesting example where matching the appropriate additive manufacturing technique to a specific design can contribute to solving the challenges encountered when processing a material using a conventional technique.

The use of electronic conducting polymers (PEDOT:PSS) in two different printing techniques, DIW and DLP, succeeded in enhancing the electronic conductivity, which improved the overall electrochemical performance of the composited electrodes as well as maintaining suitable mechanical properties. While PEDOT:PSS meets the different criteria for these purposes, several challenges still remain, as the increase of active material loading is still necessary for 3D-printed electrodes to compete with the performance of electrodes processed in a conventional way. This observation highlights that another type of electronic conductive polymer could be designed to be used in electrode formulation to prevent electronic contact loss with the active material.

16.2.2 Electrolytes

16.2.2.1 Formulation Considerations

Polymers in metal-ion batteries can also be found in the electrolyte component of the system. So far, we have focused on composite electrodes for metal-ion batteries that contain liquid electrolytes. In such cases, the liquid electrolyte acts as an ion carrier media that needs to infiltrate the porous microstructure of the electrode. This infiltration ensures an efficient contact at the interface between the electrolyte and the composite electrode to enable an efficient charge transfer reaction. Liquid electrolyte batteries have been the focus of most of the research efforts in the field and are by far the dominant technology on the market. These past few years, another type of electrolyte has grown in interest owing to its potential to achieve higher energy densities. These electrolytes are called solid-state electrolytes. As suggested by the name, a solid-state electrolyte is an electrolyte that is in a solid-state phase, more commonly made from ceramic or ion-conductive polymer.

Using solid-state electrolytes in metal-ion batteries offers several advantages, mostly related to safety issues: without any volatile or flammable solvent (found in liquid electrolytes), the risk of thermal runaway or leakage is drastically reduced. One of the major solid-state electrolyte weaknesses comes from the poor interfacial contact between the electrolyte and the electrode. In contrast to liquid electrolytes (Figure 16.3), solid-state electrolytes cannot infiltrate the porous part of the electrode to take part in redox reactions. The only way to promote redox reactions would be to increase the contact area between the electrode and the electrolyte.

Considering the specific requirements of solid-state electrolytes, the architectural freedom that is allowed via the use of 3D printing could be a game changer for all-solid-state batteries (ASSBs)—the reader is referred to these reviews for more details on this specific topic [19, 20]. Indeed, the electrode-electrolyte interface can be designed in a way that can promote shorter ion diffusion in the electrolyte as well as shorter electronic pathways in the 3D composite electrode. As illustrated in Figure 16.5, Cobb *et al.* have developed a taxonomy to classify the different architectures that could promote contact between the electrode and the electrolyte, spanning interdigitated electrodes to structured and hybrid ones. The design freedom allowed by additive manufacturing can help lower the interfacial resistance between the solid electrolyte and the composite electrode and thus contribute to the improvement of the performance of 3D-printed batteries compared to batteries prepared by conventional methods.

Combining the promises of additive manufacturing and those of AASBs is still a hot topic of research as of writing this chapter, and the state of the art is yet to be established. Indeed, up to now, most of the research has focused on the preparation and characterization of 3D-printed electrodes in comparison to 3D-printed

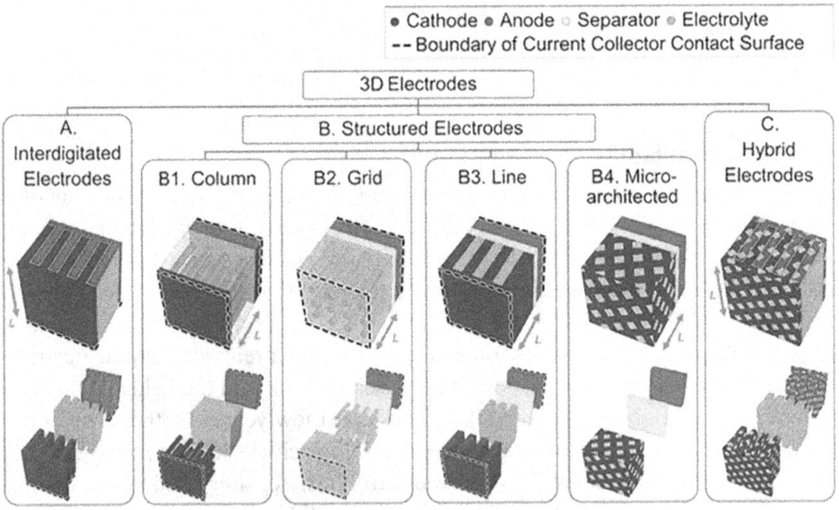

FIGURE 16.5 Taxonomy proposed by Cobb *et al.* to classify 3D-printed architectures. Adapted with permission [21]. Copyright 2023, American Chemical Society.

electrolytes. This observation can be explained by the recent use of solid polymer electrolytes in batteries. For instance, 3D printing an electrolyte involves the electrolyte in its solid state, leading into the realm of ASSB, which is already less well investigated and mastered than those involving a liquid electrolyte.

The development of 3D printable electrolytes for ASSB, as illustrated in Figure 16.5, involves the optimization of the contact surface with the electrode via the creation of customizable shapes and architectures, opening the possibility of creating a preferential orientation using ordered microchannels to increase the conductivity. It is important to mention that the electrolytes used in ASSBs can be divided into three major categories: polymer-based electrolytes, ceramic-based electrolytes, and hybrid electrolytes (here defined as a blended composition of solid polymer electrolytes and ceramics) [22]. This section will focus on polymer and hybrid electrolytes. Polymer electrolytes are based on a polymer blended with salt to act as an ion conductor. The hybrid electrolyte uses a polymer (also blended with salt) as a matrix containing ceramic particles, both parts acting synergistically as ion conductors (a topic still open to debate in the battery community). However, it is important to note that the additive manufacturing of ceramic electrolytes is also an active area of research. For instance, ceramic electrolytes, such as $Li_7La_3Zr_2O_{12}$ have been 3D printed via direct-ink writing using a sacrificial polymer as a binder (that is calcinated before the sintering of the ceramic) [19], and different types of 3D-printed ceramic architectures were successfully designed to enhance the contact between the electrolyte and the electrode.

16.2.2.2 Hybrid Electrolytes

Currently, studies conducted on the additive manufacturing of hybrid electrolytes are the most common, as the presence of both the polymer and the ceramic components enables the achievement of an ink formulation possessing the appropriate rheological properties. For instance, the polymer acts as a polymer matrix providing viscoelastic behavior to the formulation (and/or the filament if fused deposition modeling is used), while the ceramic particles contribute to its shear-thinning and thixotropic behavior. Both components also play a complementary role in improving the electrolyte's ionic conductivity, mechanical stability, and thermal properties.

An example of a hybrid architected electrolyte designed to prepare a 3D bicontinuous ordered ceramic and polymer microchannels [23] is illustrated in Figure 16.6A. The electrolyte relies on a combination of ceramic ion conductors ($Li_{1.4}Al_{0.4}Ge_{1.6}(PO_4)_3$, also referred to as LAGP) with a polymer. In this example, a polymer template is photopolymerized and later filled with LAGP powder. The polymer template is calcinated upon heating, followed by the sintering of the LAGP. Then, epoxy or polypropylene is used to fill the gaps between the sintered LAGP domains. While this example does not illustrate the 3D printing of a ceramic-polymer blend (detailed in [24]), it was specifically selected to demonstrate that 3D printing can be used to design an architecture where the size, orientation, and location of specific domains can be controlled.

16.2.2.3 Solid Polymer Electrolytes

In addition to hybrid electrolytes, polymer electrolytes have also been developed for 3D-printed batteries. In the context of Li-ion batteries, these solid-state electrolytes

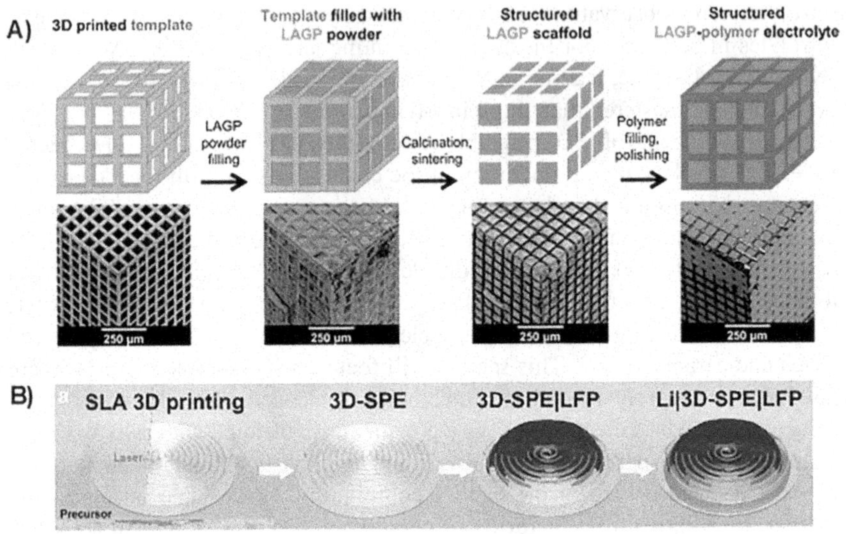

FIGURE 16.6 A) Schematic illustration of the process used to prepare 3D bicontinuous ordered ceramic and polymer microchannels. Corresponding scanning electron microscopy (SEM) images are displayed. Adapted with permission [23]. Copyright 2018, RSC Publishing. B) Schematic illustration of the process used to prepare a solid polymer electrolyte using stereolithography. Adapted with permission [25]. Copyright 2020, American Chemical Society.

are prepared from polymers blended with lithium salts (LiTFSI) to achieve high ionic conductivities. Different additive manufacturing methods have been reported to 3D print these types of blends, such as stereolithography. For instance, Figure 16.6B illustrates an example in which a blend of poly(ethylene glycol) diacrylate, succinonitrile, and LiTFSI is 3D printed via stereolithography into a spiral architecture [25], maximizing the contact with the LFP composite electrode that is subsequently doctor-bladed on top of the architecture. The resulting mechanical properties of an SLA-prepared electrolyte can also be controlled via the curing degree, which is interesting for applications requiring a gel-like electrolyte [26]. Boyer *et al.* also reported important advances in 3D-printed electrolytes using digital light processing by taking advantage of the polymerization-induced microphase separation to form bicontinuous nanoscopic domains to promote ionic conduction [27]. This study proves that merging photopolymerization, self-assembly, and additive manufacturing techniques is promising for controlling the architecture of the electrolyte both at the nano/microscopic and macroscopic level. This advanced technology would also have the potential to optimize the electrode-electrolyte interfaces.

Finally, solid polymer electrolyte properties are suitable for extrusion-based techniques, for which the benefits of a solvent-free process can be taken advantage of. Among the examples related in the literature, FDM has been successfully used to 3D print an SPE electrolyte composed of a PEO/LiTFSI blend with a resulting conductivity of 2.18×10^{-3} S.cm^{-1} at 90 °C [28]. Another example showed that PLA, which is also a common polymer matrix used for the 3D printing of electrodes when blended with PEO, N-butyl-N-methylpyrrolidinium bis(trifluoromethane-sulfonyl)imide (Pyr_{14}TFSI), and LiTFSI could be used to prepare a SPE with a conductivity around 0.2 $mS.cm^{-1}$ at 60 °C (at an optimized PLA/PEO blend ratio) [29]. Finally, elevated temperature direct-ink writing (also referred to as hot melt extrusion) was performed using a copolymer matrix, that is, poly(vinylidene-fluoride-co-hexafluoropropylene) (PVDF-co-HFP) blended with LiTSFI. As reported in the section related to 3D-printed electrodes, additional additives can be added to optimize the rheological properties of the formulation, such as TiO_2 nanoparticle fillers. 3D printing at elevated temperatures allowed for better print fidelity compared to FDM, leading to new possibilities in the context of conformal printing [30].

16.2.3 A Few Words on Assessing the Printability of Ink Components and Evaluating the Properties of 3D-Printed Battery Components

16.2.3.1 Printability and Print Fidelity

As discussed throughout the sections related to the 3D printing of electrodes or batteries, rheological considerations are important aspects of formulations. Whether the formulation is a filament, a melt from a pellet-based blend, or a slurry, it needs to fall within the appropriate range of viscosity that fits the printability requirements of the additive manufacturing technique that has been selected for the processing of the component. The authors of this chapter thus wanted to emphasize the importance of systematically assessing the printability of the formulation to be printed using appropriate rheological measurements to avoid using the time- and resource-consuming Edisonian approach of the 3D printing trial-and-error process. Many detailed reviews have been published on this topic, including two targeted formulations designed for extrusion, more specifically for direct-ink writing, which will be highlighted herein. In their review, Cooke *et al.* detail the rheological tests that need to be conducted to evaluate if a gel-like formulation possesses the appropriate shear-thinning behavior, that is, decrease of the viscosity as a function of the shear rate in the case of non-Newtonian fluids and thixotropic behavior, that is, the time that is necessary for the formulation to recover its initial viscosity value (at a specific shear rate) [31]. Printing an ink that does not present an important thixotropic behavior, meaning that the recovery time of the formulation viscosity to its initial value is quick, is of the utmost importance to avoid the spreading of the formulation upon the addition of an additional layer (in the case of an extrusion process).

Another review by Rau *et al.* is quite relevant in the context of assessing the printability of a formulation [32]. The rheology roadmap they propose enables one to predict on the sole basis of rheological tests if the formulation can be considered a "good" or a "bad" ink. This classification is built on the three following questions: i) does the flow ink rate match the nozzle translation speed, which is important to obtain a continuous print; ii) does the ink solidify before excessive spreading occurs, which is important to ensure an appropriate print fidelity; and iii) does the ink have strength/stiffness to support subsequent layers, which is important to realize grids and micro-architected electrodes (Figure 16.5)? Print fidelity was referred to a few times in the previous sections. It is important to mention that the comparison between the dimensions of the actual 3D-printed object and the dimensions of the CAD file is a comparison that can be quantified [33]. Depending on the size of the features, imaging techniques such as stereomicroscopy coupled to image processing software and/or profilometry can be used to measure the dimensions of the 3D-printed architecture. Depending on the resulting print fidelity indicators (e.g., uniformity factor, printing coefficient), a systematic adjustment of the CAD can be applied to obtain an architecture with the desired dimensions, which is important to achieve proof-of-concept, lab scale–specific architecture but also for industrial considerations.

16.2.3.2 Characterization of the Components' Microstructure

In the same way as with rheological considerations in the pre-printing process, the post-printing characterization steps are also important to take into consideration, especially considering that the microstructure of the resulting electrode and electrolyte does have an impact on the performance of the devices. Cutting-edge techniques such as X-ray nano-computed tomography [34, 35] used to map the electrode microstructure are specifically of interest in this context. Indeed, such advanced techniques could be coupled to computational studies where a specific architecture of an electrode (or electrolyte) may be designed so it matches its microstructure to optimize the electrochemical properties and performance. On that note, it is also important to highlight that a customized electrochemical setup often needs to be designed and/or customized to evaluate the performance of specific architecture and thus needs to be accounted for in the experimental design when 3D-printed batteries' components are studied.

16.3 CONCLUSION: RECENT ADVANCES AND FUTURE DIRECTIONS

Most of the advances have been made in relation to the 3D printing of electrode components. Similar ongoing research efforts are currently focusing on electrolyte components. Of course, the development of novel conducting polymers, and more generally novel materials, will contribute to enhancing the performance of 3D-printed metal-ion batteries, but it is the synergy between the development of novel materials, novel architectures, novel processing methods, and novel (*in situ* and *in operando*) characterization methods—possibly coupled to computational advances—that will certainly lead to the most important advances. Examples of research conducted at the interface of the additive manufacturing and battery fields include studies taking advantage of the shearing forces experienced by the formulation to 3D print composite electrolytes, such as PVDF blended with boron nitride, which showcases anisotropic thermal conductivity due to the alignment of the boron nitride nanosheets [36]. It has been shown that such anisotropy could contribute to the reduction of hot-spot formation and thus increase the useful lifetime of the electrolyte. Another example involves co-axial extrusion, where the current collector, the electrodes, and the electrolyte are simultaneously extruded [37], and even multi-axis coextrusion [38], where solid polymer electrolyte-coated carbon fibers are coextruded with the cathode formulation, altogether paving the way towards batteries that can be included in niche applications, such as smart textiles [39].

Such research works contribute to the advancement of both the battery and the additive manufacturing fields, in addition to providing important insights in terms of characterization tools that need to be developed to further our understanding of the mechanisms underlying the printability of these components. Overall, these studies allow for the establishment of processing-microstructure-performance relationships, leading to a more rational design of 3D-printed components for metal-ion batteries.

ACKNOWLEDGMENTS

T.P. thanks the Natural Sciences and Engineering Research Council of Canada (NSERC) for a Canada Graduate Scholarship—master's program. A.L. and M.D. thank the NSERC and the Fonds de Recherche du Québec—Nature et Technologie for funding. A.L. also acknowledges the Canada Research Chairs program for funding.

REFERENCES

[1] Lyu, Z.; Lim, G. J. H.; Koh, J. J.; Li, Y.; Ma, Y.; Ding, J.; Wang, J.; Hu, Z.; Wang, J.; Chen, W.; Chen, Y. Design and Manufacture of 3D-Printed Batteries. *Joule* **2021**, *5* (1), 89–114.

[2] Pang, Y.; Cao, Y.; Chu, Y.; Liu, M.; Snyder, K.; MacKenzie, D.; Cao, C. Additive Manufacturing of Batteries. *Adv Funct Mater* **2020**, *30* (1), 1906244.

[3] Pinto, R. S.; Gonçalves, R.; Lanceros-Méndez, S.; Costa, C. M. Three-Dimensional Printing for Solid-State Batteries. In *Solid State Batteries Volume 2: Materials and Advanced Devices*; ACS Symposium Series; American Chemical Society, Washington, DC, 2022; Vol. 1414, pp 331–350.

[4] Pei, M.; Shi, H.; Yao, F.; Liang, S.; Xu, Z.; Pei, X.; Wang, S.; Hu, Y. 3D Printing of Advanced Lithium Batteries: A Designing Strategy of Electrode/Electrolyte Architectures. *J Mater Chem A Mater* **2021**, *9* (45), 25237–25257.

[5] Narita, K.; Saccone, M. A.; Sun, Y.; Greer, J. R. Additive Manufacturing of 3D Batteries: A Perspective. *J Mater Res* **2022**, *37* (9), 1535–1546.

[6] Zheng, M.; Sun, X. 3D Printed Batteries: A Critical Overview of Progress and Future Outlooks. In *Handbook of Energy Materials*; Gupta, R., Ed.; Springer Nature Singapore: Singapore, 2022; pp 1–33.

[7] Cheng, M.; Deivanayagam, R.; Shahbazian-Yassar, R. 3D Printing of Electrochemical Energy Storage Devices: A Review of Printing Techniques and Electrode/Electrolyte Architectures. *Batter Supercaps* **2020**, *3* (2), 130–146.

[8] Ligon, S. C.; Liska, R.; Stampfl, J.; Gurr, M.; Mülhaupt, R. Polymers for 3D Printing and Customized Additive Manufacturing. *Chem Rev* **2017**, *117* (15), 10212–10290.

[9] Maurel, A.; Grugeon, S.; Armand, M.; Fleutot, B.; Courty, M.; Prashantha, K.; Davoisne, C.; Tortajada, H.; Panier, S.; Dupont, L. Overview on Lithium-Ion Battery 3D-Printing By Means of Material Extrusion. *ECS Trans* **2020**, *98* (13), 3.

[10] Reyes, C.; Somogyi, R.; Niu, S.; Cruz, M. A.; Yang, F.; Catenacci, M. J.; Rhodes, C. P.; Wiley, B. J. Three-Dimensional Printing of a Complete Lithium Ion Battery with Fused Filament Fabrication. *ACS Appl Energy Mater* **2018**, *1* (10), 5268–5279.

[11] Maurel, A.; Courty, M.; Fleutot, B.; Tortajada, H.; Prashantha, K.; Armand, M.; Grugeon, S.; Panier, S.; Dupont, L. Highly Loaded Graphite–Polylactic Acid Composite-Based Filaments for Lithium-Ion Battery Three-Dimensional Printing. *Chemistry of Materials* **2018**, *30* (21), 7484–7493.

[12] Zhai, Y.; Wang, Z.; Kwon, K.-S.; Cai, S.; Lipomi, D. J.; Ng, T. N. Printing Multi-Material Organic Haptic Actuators. *Advanced Materials* **2021**, *33* (19), 2002541.

[13] Gupta, V.; Alam, F.; Verma, P.; Kannan, A. M.; Kumar, S. Additive Manufacturing Enabled, Microarchitected, Hierarchically Porous Polylactic-Acid/Lithium Iron Phosphate/Carbon Nanotube Nanocomposite Electrodes for High Performance Li-Ion Batteries. *J Power Sources* **2021**, *494*, 229625.
[14] Lahtinen, E.; Kukkonen, E.; Jokivartio, J.; Parkkonen, J.; Virkajärvi, J.; Kivijärvi, L.; Ahlskog, M.; Haukka, M. Preparation of Highly Porous Carbonous Electrodes by Selective Laser Sintering. *ACS Appl Energy Mater* **2019**, *2* (2), 1314–1318.
[15] Bao, P.; Lu, Y.; Tao, P.; Liu, B.; Li, J.; Cui, X. 3D Printing PEDOT-CMC-Based High Areal Capacity Electrodes for Li-Ion Batteries. *Ionics (Kiel)* **2021**, *27* (7), 2857–2865.
[16] Walker, D. A.; Hedrick, J. L.; Mirkin, C. A. Rapid, Large-Volume, Thermally Controlled 3D Printing Using a Mobile Liquid Interface. *Science (1979)* **2019**, *366* (6463), 360–364.
[17] Shen, F.; Dixit, M. B.; Zaman, W.; Hortance, N.; Rogers, B.; Hatzell, K. B. Composite Electrode Ink Formulation for All Solid-State Batteries. *J Electrochem Soc* **2019**, *166* (14), A3182.
[18] Ye, X.; Wang, C.; Wang, L.; Lu, B.; Gao, F.; Shao, D. DLP Printing of a Flexible Micropattern Si/PEDOT:PSS/PEG Electrode for Lithium-Ion Batteries. *Chemical Communications* **2022**, *58* (55), 7642–7645.
[19] McOwen, D. W.; Xu, S.; Gong, Y.; Wen, Y.; Godbey, G. L.; Gritton, J. E.; Hamann, T. R.; Dai, J.; Hitz, G. T.; Hu, L.; Wachsman, E. D. 3D-Printing Electrolytes for Solid-State Batteries. *Advan Mat* **2018**, *30* (18), 1707132.
[20] Cheng, M.; Jiang, Y. 3D-Printed Solid-State Electrolytes for Electrochemical Energy Storage Devices. *J Mater Res* **2021**, *36* (22), 4547–4564.
[21] Hung, C.-H.; Huynh, P.; Teo, K.; Cobb, C. L. Are Three-Dimensional Batteries Beneficial? Analyzing Historical Data to Elucidate Performance Advantages. *ACS Energy Lett* **2023**, *8* (1), 296–305.
[22] Keller, M.; Varzi, A.; Passerini, S. Hybrid Electrolytes for Lithium Metal Batteries. *J Power Sources* **2018**, *392*, 206–225.
[23] Zekoll, S.; Marriner-Edwards, C.; Hekselman, A. K. O.; Kasemchainan, J.; Kuss, C.; Armstrong, D. E. J.; Cai, D.; Wallace, R. J.; Richter, F. H.; Thijssen, J. H. J.; Bruce, P. G. Hybrid Electrolytes with 3D Bicontinuous Ordered Ceramic and Polymer Microchannels for All-Solid-State Batteries. *Energy Environ Sci* **2018**, *11* (1), 185–201.
[24] Jian, S.; Cao, Y.; Feng, W.; Yin, G.; Zhao, Y.; Lai, Y.; Zhang, T.; Ling, X.; Wu, H.; Bi, H.; Dong, Y. Recent Progress in Solid Polymer Electrolytes with Various Dimensional Fillers: A Review. *Mater Today Sustain* **2022**, *20*, 100224.
[25] He, Y.; Chen, S.; Nie, L.; Sun, Z.; Wu, X.; Liu, W. Stereolithography Three-Dimensional Printing Solid Polymer Electrolytes for All-Solid-State Lithium Metal Batteries. *Nano Lett* **2020**, *20* (10), 7136–7143.
[26] Gambe, Y.; Kobayashi, H.; Iwase, K.; Stauss, S.; Honma, I. A Photo-Curable Gel Electrolyte Ink for 3D-Printable Quasi-Solid-State Lithium-Ion Batteries. *Dalton Transactions* **2021**, *50* (45), 16504–16508.
[27] Melodia, D.; Bhadra, A.; Lee, K.; Kuchel, R.; Kundu, D.; Corrigan, N.; Boyer, C. 3D Printed Solid Polymer Electrolytes with Bicontinuous Nanoscopic Domains for Ionic Liquid Conduction and Energy Storage. *Small* **2023**, 2206639.
[28] Maurel, A.; Armand, M.; Grugeon, S.; Fleutot, B.; Davoisne, C.; Tortajada, H.; Courty, M.; Panier, S.; Dupont, L. Poly(Ethylene Oxide)–LiTFSI Solid Polymer Electrolyte Filaments for Fused Deposition Modeling Three-Dimensional Printing. *J Electrochem Soc* **2020**, *167* (7), 070536.
[29] Vinegrad, A.; Ragones, H.; Jayakody, N.; Ardel, G.; Goor, M.; Kamir, Y.; Dorfman, M. M.; Gladkikh, A.; Burstein, L.; Horowitz, Y.; Greenbaum, S.; Golodnitsky, D. Plasticized 3D-Printed Polymer Electrolytes for Lithium-Ion Batteries. *J Electrochem Soc* **2021**, *168* (11); 110549.

[30] Cheng, M.; Jiang, Y.; Yao, W.; Yuan, Y.; Deivanayagam, R.; Foroozan, T.; Huang, Z.; Song, B.; Rojaee, R.; Shokuhfar, T.; Pan, Y.; Lu, J.; Shahbazian-Yassar, R. Elevated-Temperature 3D Printing of Hybrid Solid-State Electrolyte for Li-Ion Batteries. *Advan Mater.* **2018**, *30* (39), 1800615.

[31] Cooke, M. E.; Rosenzweig, D. H. The Rheology of Direct and Suspended Extrusion Bioprinting. *APL Bioeng* **2021**, *5* (1), 011502.

[32] Rau, D. A.; Bortner, M. J.; Williams, C. B. A Rheology Roadmap for Evaluating the Printability of Material Extrusion Inks. *Addit Manuf* **2023**, *75*, 103745.

[33] Schwab, A.; Levato, R.; D'Este, M.; Piluso, S.; Eglin, D.; Malda, J. Printability and Shape Fidelity of Bioinks in 3D Bioprinting. *Chem Rev* **2020**, *120* (19), 11028–11055.

[34] Lu, X.; Bertei, A.; Finegan, D. P.; Tan, C.; Daemi, S. R.; Weaving, J. S.; O'Regan, K. B.; Heenan, T. M. M.; Hinds, G.; Kendrick, E.; Brett, D. J. L.; Shearing, P. R. 3D Microstructure Design of Lithium-Ion Battery Electrodes Assisted by X-Ray Nano-Computed Tomography and Modelling. *Nat Commun* **2020**, *11* (1), 2079.

[35] Nagda, V.; Kulachenko, A.; Lindström, S. B. Image-Based 3D Characterization and Reconstruction of Heterogeneous Battery Electrode Microstructure. *Comput Mater Sci* **2023**, *223*, 112139.

[36] Rasul, M. G.; Cheng, M.; Jiang, Y.; Pan, Y.; Shahbazian-Yassar, R. Direct Ink Printing of PVdF Composite Polymer Electrolytes with Aligned BN Nanosheets for Lithium-Metal Batteries. *ACS Nanoscience Au* **2022**, *2* (4), 297–306.

[37] Ragones, H.; Vinegrad, A.; Ardel, G.; Goor, M.; Kamir, Y.; Dorfman, M. M.; Gladkikh, A.; Golodnitsky, D. On the Road to a Multi-Coaxial-Cable Battery: Development of a Novel 3D-Printed Composite Solid Electrolyte. *J Electrochem Soc* **2020**, *167* (7), 070503.

[38] Thakur, A.; Dong, X. Additive Manufacturing of 3D Structural Battery Composites with Coextrusion Deposition of Continuous Carbon Fibers. *Manuf Lett* **2020**, *26*, 42–47.

[39] Praveen, S.; Santhoshkumar, P.; Joe, Y. C.; Senthil, C.; Lee, C. W. 3D-Printed Architecture of Li-Ion Batteries and Its Applications to Smart Wearable Electronic Devices. *Appl Mater Today* **2020**, *20*, 100688.

17 Nanocomposites of Conducting Polymers for 3D-Printed Metal–Sulfur Batteries

Wei Ni and Ling-Ying Shi

17.1 INTRODUCTION

Conducting polymers (CPs), also called electrically conductive polymers or semiconducting polymers, have been adapted for additive manufacturing (AM) techniques, including three-dimensional (3D) printing, and thus for customized energy storage devices, including metal–sulfur batteries [1–4]. By the formation of polymer composites or nanocomposites, these mostly brittle insoluble/infusible conducting polymers become 3D printable and show controllable electrode geometries, robust strength, and high electronic/ionic conductivity and competitive electrochemical performance, promising for 3D wearable electronic devices [2, 5].

Various conducting polymers, such as poly(3,4-ethylene-dioxythiophene) (PEDOT), polyaniline (PANI), polypyrrole (PPy), and polythiophene (PTh), have been applied in metal–sulfur batteries [6–8]. Poly(3,4-ethylenedioxythiophene):poly styrene sulfonate (PEDOT:PSS) is a kind of thiophene-based block copolymer with favorable aqueous solubility and superior printability and is thus an ideal and currently widely investigated/used material candidate for 3D printing of microstructures with high aspect ratio and energy storage devices with high resolution [2, 9]. With tunable Young's modulus, flexibility, solubility, and high electrical conductivity (up to 28 S cm^{-1} in the hydrogel state and over 155 S cm^{-1} in the dry state), to date, it is probably the single conductive polymer most utilized for 3D printing without additional additives, including adhesives. To further enhance the conductivity of conducting polymer nanocomposites, additional conductive additives such as carbon, graphene, or metals may be incorporated into these matrices [10]. Electrically conductive polymer composites are another kind of candidate for 3D printing; while these composites by combining carbons (or metals) and resins are not focused on, for more information, please refer to recent reviews [1, 4, 11].

Rechargeable metal–sulfur batteries (RMSBs, metal = Li, Na, K, Mg, Ca, Al, etc.) are a new class of promising and competitive next-generation electrochemical energy systems due to their high energy density, low cost, and hypotoxicity compared to currently commercial lithium-ion batteries (LIBs), partly resulting from the high theoretical specific capacity (1675 mAh g^{-1}) and abundance in the Earth's crust for the

DOI: 10.1201/9781003415985-17

sulfur-based cathode materials as well as the high theoretical capacity and low redox potentials of metal anodes (e.g., Li = 3860 mAh g^{-1}, Na = 1166 mAh g^{-1}, Mg = 2205 mAh g^{-1}, Al = 2979 mAh g^{-1}) [12–14]. These room-temperature and environmentally friendly metal–sulfur (metal–S) batteries deliver high theoretical gravimetric energy density, such as 2600 Wh kg^{-1} for Li–S batteries [15] and 1274 Wh kg^{-1} for Na–S batteries [16], and the general data on metal–S batteries are presented in Figure 17.1 and Table 17.1 [13, 14, 17]. The incorporation of sulfur cathodes with conductive

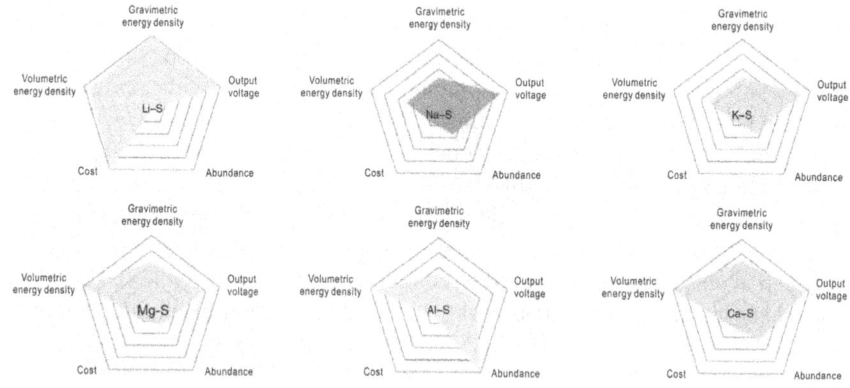

FIGURE 17.1 Comparison of the characteristics, including gravimetric and volumetric energy densities, output voltage, abundance, and cost of typical metal–sulfur batteries (e.g., Li–S, Na–S, K–S, Mg–S, Al–S, and Ca–S batteries). Figure 17.1 adapted from reference [14].

TABLE 17.1
Comparison of Theoretical Parameters of Metal–Sulfur Batteries [13, 14]

Types	Redox Chemistry	Gibbs Free Energies ($\Delta_r G$, kJ mol^{-1})	Thermo-dynamic *emf* (V)	Theoretical Cell Voltage (E^0, V)	Gravimetric Energy Density (Wh kg^{-1})	Volumetric Energy Density (Wh L^{-1})
Li–S	$2Li + S \leftrightarrow Li_2S$	–432.57	2.6	2.24	2612	2955 (2723)
Na–S	$2Na + S \leftrightarrow Na_2S$	–357.77	2.3	1.85	1270 (1262)	1545 (1680)
K–S	$2K + S \leftrightarrow K_2S$	–362.73	2.5	1.88	916 (914)	952 (1149)
Mg–S	$Mg + S \leftrightarrow MgS$	–341.44	1.8	1.77	1685 (1682)	3221 (3195)
Al–S	$2Al + 3S \leftrightarrow Al_2S_3$	–713.20	1.1	1.23	1319 (1185)	2981 (2747)
Ca–S	$Ca + S \leftrightarrow CaS$	–477.4	2.4	2.47	1838 (1802)	3202 (3244)

Note: Some data may vary from different sources/methods, for example, theoretical gravimetric and volumetric energy densities.

polymers is one of the most effective approaches to solve the critical problems (e.g., low conductivity, shuttle effect, sulfur utilization ratio, volume change) encountered in conventional metal–S batteries and has received significant attention in recent years [6, 7].

17.2 TECHNIQUES

For 3D printing, direct 3D printing (e.g., direct ink writing, fused deposition modeling, stereolithography, extrusion-based AM, and electrohydrodynamic printing) and the post-modification thereof are typical approaches used for next-generation electrochemical energy storage devices (EESDs), including advanced metal–S batteries with rational, regulated, and elaborate architectures [2, 18–22]; with appropriate formulae and printing technologies, commercial-level high-mass loading electrodes and high-performance metal–S batteries could be implemented [19, 20, 23, 24].

The design, synthesis, and engineering of 3D printing materials are of great importance to the electrochemical properties and the applications thereof, including EESDs [21, 25–27]. Freeze-drying of the printed electrode endows it with much better flexibility and avoids fracture compared to common drying in an oven [28]. Functional components including conductive agents (e.g., CNTs, carbon black, graphene) [10, 29–31], sulfur hosts (e.g., multishelled Ni_2P microspheres [32], MXenes [33]), electrocatalysts [31], binders (e.g., cellulose nanofibrils [29], PVDF, sodium alginate), and their combinations are often incorporated into inks for enhanced mechanical and electrochemical performances. Micro-batteries are playing a key role in 3D-printed miniaturized energy storage devices (MESDs) [34], which will be another significant trend for the development and application of next-generation MESDs and 3D-printed batteries, including metal–S batteries. The electrochemical mechanisms and materials design are vital to the insight and development of metal–S batteries with specific properties (Figure 17.2).

17.3 APPLICATIONS

17.3.1 Sulfur-Based Cathodes

Conductive polymers with high electronic conductivity and polarity could be ideal organic carriers for sulfur in the cathode, since the poor binding ability between traditional inert carbon surface and polar Li_2S/Li_2S_2 results in deteriorated electrical contact or even detachment [36]. Core–shell structured micro-/nanospheres can be a promising candidate for 3D printing of metal–S batteries [37, 38]. For example, via the surface coating of monodisperse hollow sulfur nanospheres with conductive polymers in a facile and scalable aqueous polymerization system, core–shell nanocomposites could be fabricated with controllable coating thickness and high sulfur content, which plays a distinct role in improving the electrochemical performance of nanostructured sulfur composite cathodes [37, 39].

Functional additives for efficient adsorption and/or (electrocatalytic) conversion of polysulfides in metal–S batteries are currently intensively investigated to solve some of the major obstacles (e.g., shuttle effect, sluggish transformation) hindering

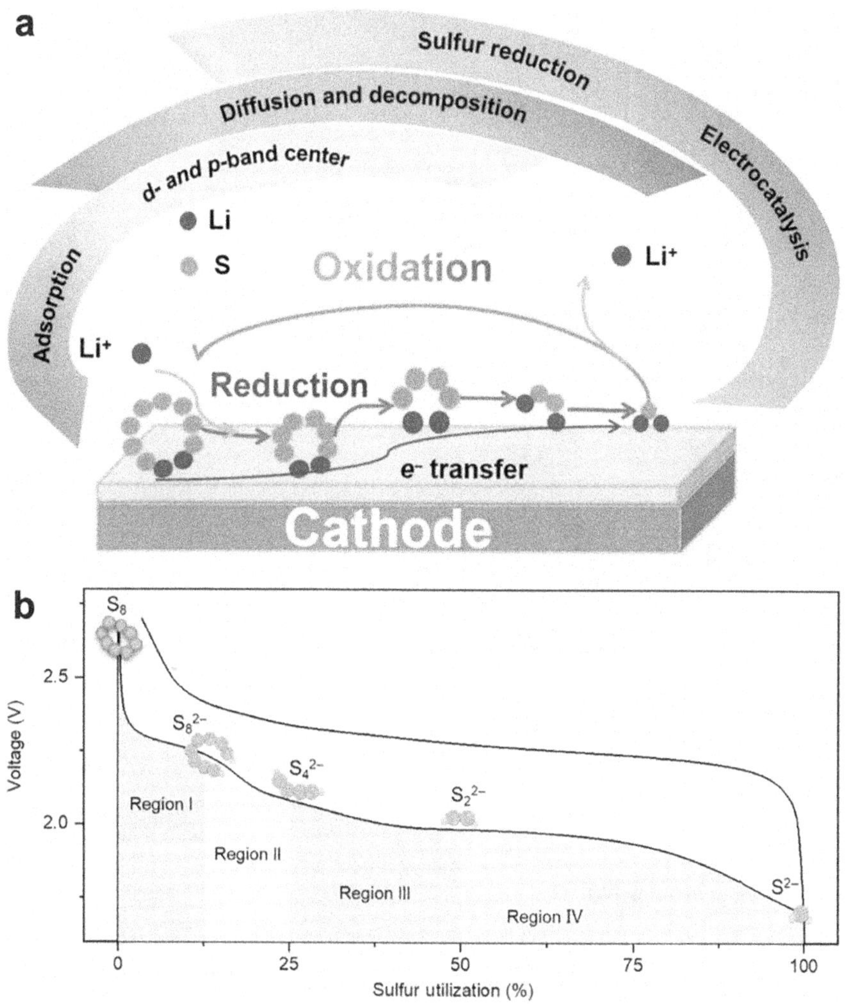

FIGURE 17.2 (a) Schematic illustration of theoretical models on mechanism investigation and materials design of sulfur host for typical metal–S batteries. (b) Typical discharge–charge profiles and corresponding phase changes of sulfur in a typical metal–S battery. Figure 17.2 adapted from reference [14, 35]. Copyright the authors, some rights reserved; exclusive licensee UESTC and John Wiley & Sons Australia, Ltd. Distributed under a Creative Commons Attribution License 4.0 (CC BY) https://creativecommons.org/licenses/by/4.0/.

them from practical and widespread application [35]. Some polar materials beyond heteroatom-doped carbon materials such as metal–organic frameworks (MOFs), (single-atom) electrocatalysts (CoFe alloy [31]), and metallic compounds (e.g., metal boride LaB_6) have been applied in 3D-printed metal–S batteries [40, 41]. For example, with a sulfur loading of 2.7 mg cm^{-2}, the printed Al-MOF/S composite cathode (shown in Figure 17.3) demonstrates a high initial capacity of 1171 mAh g^{-1} and an

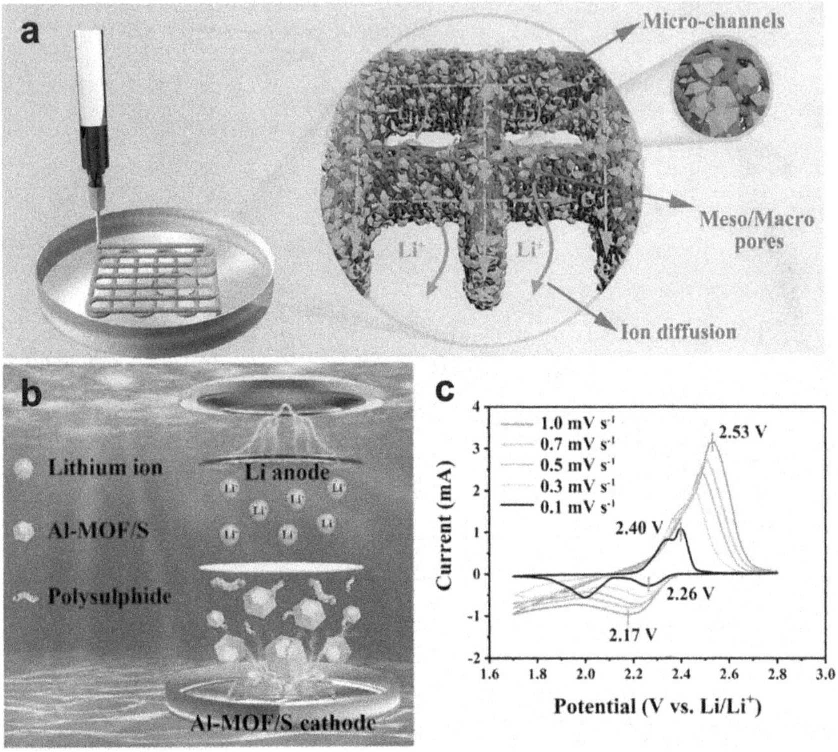

FIGURE 17.3 (a and b) Schematic illustration of the electron/ion transport in a 3DP Al-MOF-160/S cathode as well as the interaction between Al-MOF and lithium polysulfides. (c) CV curves of the 3DP Al-MOF-160/S electrode at different scan rates (from 0.1 to 1.0 mV s^{-1}). Adapted with permission [40]. Copyright (2023), The Royal Society of Chemistry.

acceptable capacity of 504 mAh g^{-1} after a long-term cycling (300 cycles at 0.1 C). A much higher sulfur loading, up to ~16 mg cm^{-2}, can be realized with a fair capacity of 531 mAh g^{-1} [40]. The high performance can be attributed to the 3D-printed hierarchically micro-/nanostructured cathode with enhanced electrolyte penetration, facilitated electron/ion transport, and improved sulfur utilization and reaction kinetics, as well as the suppressed polysulfide shuttling and buffered electrode volume change. With the incorporation of an electrocatalyst such as LaB_6 (shown in Figure 17.4), the electrochemical kinetics of sulfur cathode could be expedited, and the rate capability and areal capacity along with capacity retention are therefore improved [41]. Besides, compared to conventional polymeric PVDF additives, polyvinylidene fluoride-hexafluoropropylene (PVDF-HFP) seems to show better electrochemical performance regardless of the price [28].

Conducting polymer-based soft fiber electronics have attracted increasing attention in various applications in information interfaces, healthcare and medicine, and energy conversion and storage, in which different techniques (e.g., continuous spinning, electrospinning, melt spinning), compositions (e.g., forming various composites), and structures (e.g., 2D/3D textiles and integration) are referred to [3]. Hofmann

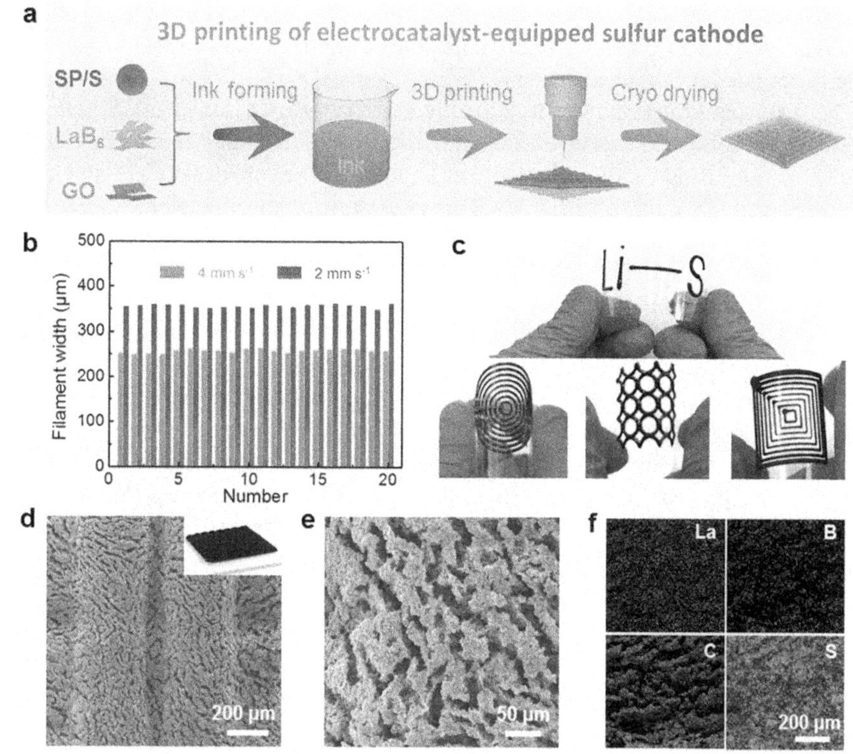

FIGURE 17.4 (a) Schematic illustration of the fabrication procedure of a 3DP-LaB_6/SP@S electrode. (b) The width distribution of 3D-printed LaB_6/SP@S filaments under two different printing speeds. (c) Photographs of diverse printing patterns on PET matrix using LaB_6/SP@S composite ink. (d and e) SEM images of the as-printed 3DP-LaB_6/SP@S electrodes (inset: macroscopic appearance) and (f) the corresponding elemental mappings. Adapted with permission [41]. Copyright (2020), Elsevier Ltd.

et al. exploited melt processing along with templated polymerization for elastic all-polymer conducting fibers (i.e., PEDOT:Nafion fibers, a conjugated polymer doped by polymeric counter ion) and the 3D prints thereof [42]. Different from the conventional harmful/unstable dopant-containing solution processing techniques, the method used here involves specific melt extrusion/modification featured by fiber spinning or fused filament printing (i.e., fused filament fabrication (FFF), a typical 3D printing technique), which makes it processable in bulk with complex structures and may serve as a promising candidate for wearable electronics or electrochemical energy storage devices/modules.

Wearable electronics and integrated energy storage units/subsystems have attracted ever-increasing interest. To meet the high requirements of high flexibility and high energy density, 3D Li–S battery is a preferred candidate. For example, a 3D-printed high-loading Li–S battery using a computer-aided design (CAD) technique can deliver a high specific capacity of 505 mAh g^{-1} at 0.2 C over 500 cycles with an active material loading up to 10.2 mg cm^{-2} [43]. Typical 3D printing techniques such as direct ink writing (DIW) and fused deposition modeling (FDM, also known as FFF) implement both a flexible/customized cathode and battery case (e.g., wearable Li–S bracelet battery).

To date, liquid organic electrolytes are commonly used for metal–S batteries; however, all-solid-state batteries, including those using inorganic solid-state electrolytes, are promising candidates. Wang et al. fabricated a low-density glass-ceramic solid electrolyte (Li_3PS_4–$2LiBH_4$, $\rho \approx 1.5$ g cm^{-3}) with relatively high bulk ionic conductivity (6.0 mS cm^{-1} at 25 °C), showing an impressive high-temperature performance, that is, a high discharge capacity of ~1145 mAh g^{-1} (at a current density of 167.5 mA g^{-1}) and high capacity retention (77.5%, fading rate ~0.028% per cycle) over 800 cycles at 60 °C [15]. However, for high-performance metal–S batteries, the conductivity of these solid electrolytes should be further enhanced for high rate or high power applications. With regard to 3D printability, gel-type and polymer solid-state electrolytes seem to possess great potential for future applications.

As a specific sulfur source, the sulfur copolymer (e.g., elemental sulfur reacting with 1,3-diisopropenylbenzene, DIB, via inverse vulcanization process [36, 44]) may be incorporated with a conductive agent such as ideal 2D graphene to form an effective 3D printing ink to construct diverse complex cathode architectures (Figure 17.5). Due to the suppressed dissolution of polysulfides and the high electrical conductivity of the whole cathode, the 3D-printed Li–S battery demonstrates a high reversible capacity of 813 mAh g^{-1} and good cycling performance [45]. To further enhance the conductivity,

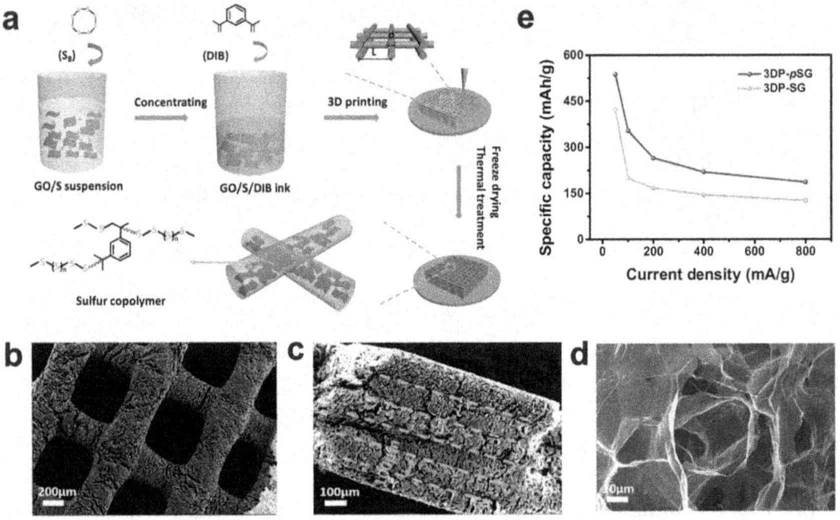

FIGURE 17.5 (a) Schematic illustration of the 3D printing of sulfur copolymer-graphene (3DP-*p*SG) layer-by-layer architectures followed by freeze drying and thermal treatment. (b-d) SEM images of the as-printed 3DP-*p*SG architectures (top view, lateral view, and magnified image). (e) Rate capabilities of the 3DP-*p*SG compared with conventional 3DP-SG (physical mixture of elemental sulfur and graphene) at different current densities of 50–800 mA g^{-1}. Adapted with permission [45]. Copyright (2017), WILEY-VCH Verlag GmbH & Co. KGaA, Weinheim.

conducting polymers with conductive backbones are used to copolymerize with elemental sulfur, such as poly(*m*-aminothiophenol) (PMAT), PANI, PPy, and PTh [36]. In some aspects such as electrochemical impedance, specific capacity, or rate performance, the sulfur copolymer can partially surpass the sulfur particles, although the overall performance needs further improvement compared to state-of-the-art Li–S batteries.

17.3.2 Other Electrodes

In addition to the classic S and Li_2S cathodes [29], some other sulfur compounds such as SeS_2 are also candidates as sulfur sources, which ultimately work in the form of polysulfides in the cell. For example, Shen et al. fabricated a 3D-printed cellular SeS_2 cathode with the assistance of ketjenblack (KB), polyvinylidene fluoride (PVDF), and carbon nanotubes (CNTs) [30]. The as-prepared cathode demonstrates hierarchically porous structures and high mass loading up to 7.9 mg cm^{-2} and can deliver a high initial discharge capacity (9.5 mAh cm^{-2}) and high coulombic efficiency (CE, 96% after 80 cycles) in Li–SeS_2 batteries (Figure 17.6). However, the mass ratio of active materials (49

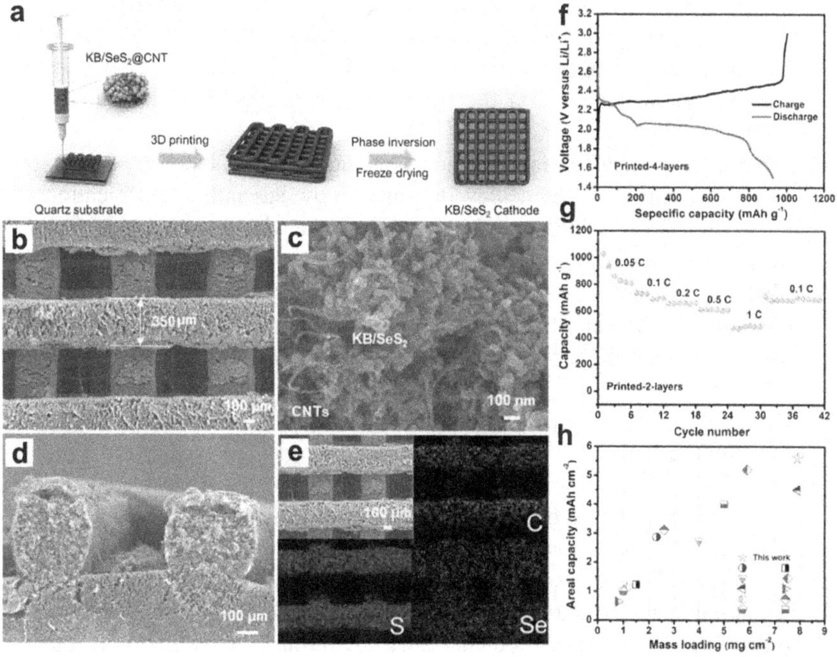

FIGURE 17.6 (a) Schematic illustration of extrusion-based 3D printing of SeS_2/ketjenblack composite (3DP-KB/SeS_2) cellular architectures. (b–d) SEM images of the as-printed 3D-KB/SeS_2 cellular cathode (top view at different magnifications and cross-sectional view) and (e) the corresponding elemental mapping. (f and g) Typical galvanostatic charge-discharge (GCD) curves and rate performance of 3DP-KB/SeS_2 cellular cathodes. (h) Comparison of the areal capacities for Li–Se_xS_y batteries of different mass loading to other recently published works. Adapted with permission [30]. Copyright (2020), Elsevier Ltd.

wt.%), rate, and cycling performance (1 C and above) may be further improved for efficient practical application. Besides the mentioned cathodes, the anodes/protective interlayer [46], separators [47], and packaging can also be 3D printed for high-performance metal–S batteries. Furthermore, it should be mentioned that these selenides or even tellurides are quite different to elemental sulfur or sulfur copolymers [48] and should not be confused with typical sulfide LiS_2; however, alkali metal–Se/Te batteries such as Li–Se, Li–Te, and Na–Se batteries are quite similar to Li–S batteries.

17.4 CONCLUSION AND PROSPECTIVES

Although some pioneering research has been conducted and considerable progress has been made in the application of conductive polymers in 3D-printed sulfur nanocomposite cathodes for metal–sulfur batteries, some critical issues, including the synthesis, structure, and functionality of sulfur nanocomposites, processing properties, printing techniques, electrochemical behaviors, security, and cost-effectiveness should be addressed in the future for ultimate practical application.

a) The inks or filaments for electrodes, especially cathodes, are key components for printed metal–sulfur batteries. They should be systematically optimized via tuning the active materials, conducting polymers, and/or other binders/additives, not only on the chemical compositions and electrochemical behaviors but also on the microstructures, rheology behaviors, and processabilities.
b) 3D-printed energy storage devices, including Li–S batteries based on nanocomposites, can be achieved, although the streamlined fabrication of integrated devices into practical application environments needs further exploration, not only on the batteries themselves but also on the interfaces [49] and the synergistic effects of subunits. Full battery systems including electrochemical stability, reversibility, and lifetime of all cell components, including cathode, anode, interlayer/separator, and electrolyte, should be taken into account.
c) Direct ink writing is usually utilized for 3D printing of conducting polymers for metal–S batteries. Although for practical application, the initial/major obstacles such as high areal mass loading of cathodes are now solved due to progressive achievements, the configuration with anodes for smart metal–S batteries has not been intensively investigated. Most works are based on the conventional assembly in the form of a coin-cell or pouch battery, which limits the potential and performance of 3D-printed cathodes. Emerging smart energy storage devices such as fibrous, membrane, and microbatteries will be a promising trend soon.
d) Fused filament fabrication or extrusion-based printing is another promising 3D printing technique for bulk processing of energy storage devices, including metal–S batteries with customized/complex structures or favorable integration.
e) Although post-Li batteries such as Na–S batteries have lower voltage output or capacity, these batteries are showing extraordinary long-term cycling

performance with negligible capacity fading if electrodes/cells are appropriately designed and constructed [16].

f) The utilization of 4D printing based on classic 3D printing techniques with smart materials (e.g., active polymers, hydrogel resins) responsive to external stimulus will be an interesting exploration of advanced metal–S batteries with adequate battery shape and optimized physical/electrochemical behavior thereof. These printed batteries with customization, flexibility, stretchability, and/or self-healing characteristics will be promising in wearable electronics and Internet of Things (IoT) applications featuring specific physical properties as well as high energy/power energy densities.

ACKNOWLEDGMENTS

This work was supported by the National Natural Science Foundation of China (Grant No. 51403193).

REFERENCES

[1] Y. Yan, Y. Jiang, E.L.L. Ng, Y. Zhang, C. Owh, F. Wang, Q. Song, T. Feng, B. Zhang, P. Li, X.J. Loh, S.Y. Chan, B.Q.Y. Chan, Progress and opportunities in additive manufacturing of electrically conductive polymer composites, Mater. Today Adv. 17 (2023) 100333.

[2] M. Criado-Gonzalez, A. Dominguez-Alfaro, N. Lopez-Larrea, N. Alegret, D. Mecerreyes, Additive manufacturing of conducting polymers: Recent advances, challenges, and opportunities, ACS Appl. Polym. Mater. 3 (2021) 2865–2883.

[3] F. Sun, H. Jiang, H. Wang, Y. Zhong, Y. Xu, Y. Xing, M. Yu, L.-W. Feng, Z. Tang, J. Liu, H. Sun, H. Wang, G. Wang, M. Zhu, Soft fiber electronics based on semiconducting polymer, Chem. Rev. 123 (2023) 4693–4763.

[4] K.R. Ryan, M.P. Down, N.J. Hurst, E.M. Keefe, C.E. Banks, Additive manufacturing (3D printing) of electrically conductive polymers and polymer nanocomposites and their applications, eScience 2 (2022) 365–381.

[5] S. Park, W. Shou, L. Makatura, W. Matusik, K. Fu, 3D printing of polymer composites: Materials, processes, and applications, Matter 5 (2022) 43–76.

[6] J. Wang, W. Zhang, H. Wei, X. Zhai, F. Wang, Y. Zhou, F. Tao, P. Zhai, W. Liu, Y. Liu, Recent advances and perspectives in conductive-polymer-based composites as cathode materials for high-performance lithium–sulfur batteries, Sustainable Energy Fuels 6 (2022) 2901–2923.

[7] X. Chen, C. Zhao, K. Yang, S. Sun, J. Bi, N. Zhu, Q. Cai, J. Wang, W. Yan, Conducting polymers meet lithium–sulfur batteries: Progress, challenges, and perspectives, Energy Environ. Mater. 6 (2023) e12483.

[8] X. Hong, Y. Liu, Y. Li, X. Wang, J. Fu, X. Wang, Application progress of polyaniline, polypyrrole and polythiophene in lithium-sulfur batteries, Polymers 12 (2020) 331.

[9] H. Yuk, B. Lu, S. Lin, K. Qu, J. Xu, J. Luo, X. Zhao, 3D printing of conducting polymers, Nat. Commun. 11 (2020) 1604.

[10] W. Ni, L.-Y. Shi, Graphene–sulfur nanocomposites as cathode materials and separators for lithium–sulfur batteries, in: R.K. Gupta, T.A. Nguyen, H. Song, G. Yasin (Eds.) Lithium-Sulfur Batteries, Elsevier, Cambridge, 2022, pp. 289–314.

[11] P. Keane, Electrically Conductive Polymer Composites for 3D Printing, https://3dprinting.com/3d-printing-use-cases/conductive-materials-for-3d-printing/ (last accessed on 1/21/2024)

[12] M. Salama, Rosy, R. Attias, R. Yemini, Y. Gofer, D. Aurbach, M. Noked, Metal–sulfur batteries: Overview and research methods, ACS Energy Lett. 4 (2019) 436–446.
[13] D. Meggiolaro, M. Agostini, S. Brutti, Aprotic sulfur–metal batteries: Lithium and beyond, ACS Energy Lett. 8 (2023) 1300–1312.
[14] H. Ye, Y. Li, Room-temperature metal–sulfur batteries: What can we learn from lithium–sulfur? InfoMat 4 (2022) e12291.
[15] D. Wang, L.-J. Jhang, R. Kou, M. Liao, S. Zheng, H. Jiang, P. Shi, G.-X. Li, K. Meng, D. Wang, Realizing high-capacity all-solid-state lithium-sulfur batteries using a low-density inorganic solid-state electrolyte, Nat. Commun. 14 (2023) 1895.
[16] E. Zhang, X. Hu, L. Meng, M. Qiu, J. Chen, Y. Liu, G. Liu, Z. Zhuang, X. Zheng, L. Zheng, Y. Wang, W. Tang, Z. Lu, J. Zhang, Z. Wen, D. Wang, Y. Li, Single-atom yttrium engineering Janus electrode for rechargeable Na–S batteries, J. Am. Chem. Soc. 144 (2022) 18995–19007.
[17] X. Yu, A. Manthiram, A progress report on metal–sulfur batteries, Adv. Funct. Mater. 30 (2020) 2004084.
[18] X. Xu, Y.H. Tan, J. Ding, C. Guan, 3D printing of next-generation electrochemical energy storage devices: From multiscale to multimaterial, Energy Environ. Mater. 5 (2022) 427–438.
[19] H. Yang, Z. Feng, X. Teng, L. Guan, H. Hu, M. Wu, Three-dimensional printing of high-mass loading electrodes for energy storage applications, InfoMat 3 (2021) 631–647.
[20] Q. Zhang, J. Zhou, Z. Chen, C. Xu, W. Tang, G. Yang, C. Lai, Q. Xu, J. Yang, C. Peng, Direct ink writing of moldable electrochemical energy storage devices: Ongoing progress, challenges, and prospects, Adv. Eng. Mater. 23 (2021) 2100068.
[21] S. Mubarak, D. Dhamodharan, H.-S. Byun, Recent advances in 3D printed electrode materials for electrochemical energy storage devices, J. Energy Chem. 81 (2023) 272–312.
[22] M. Pei, H. Shi, F. Yao, S. Liang, Z. Xu, X. Pei, S. Wang, Y. Hu, 3D printing of advanced lithium batteries: A designing strategy of electrode/electrolyte architectures, J. Mater. Chem. A 9 (2021) 25237–25257.
[23] Y. Ye, F. Wu, S. Xu, W. Qu, L. Li, R. Chen, Designing realizable and scalable techniques for practical lithium sulfur batteries: A perspective, J. Phys. Chem. Lett. 9 (2018) 1398–1414.
[24] L. Zeng, P. Li, Y. Yao, B. Niu, S. Niu, B. Xu, Recent progresses of 3D printing technologies for structural energy storage devices, Materials Today Nano 12 (2020) 100094.
[25] H. Zhou, H. Yang, S. Yao, L. Jiang, N. Sun, H. Pang, Synthesis of 3D printing materials and their electrochemical applications, Chin. Chem. Lett. 33 (2022) 3681–3694.
[26] C.M. Costa, R. Gonçalves, S. Lanceros-Méndez, Recent advances and future challenges in printed batteries, Energy Storage Mater. 28 (2020) 216–234.
[27] J. Yan, S. Huang, Y.V. Lim, T. Xu, D. Kong, X. Li, H.Y. Yang, Y. Wang, Direct-ink writing 3D printed energy storage devices: From material selectivity, design and optimization strategies to diverse applications, Mater. Today 54 (2022) 110–152.
[28] X. Gao, Q. Sun, X. Yang, J. Liang, A. Koo, W. Li, J. Liang, J. Wang, R. Li, F.B. Holness, A.D. Price, S. Yang, T.-K. Sham, X. Sun, Toward a remarkable Li-S battery via 3D printing, Nano Energy 56 (2019) 595–603.
[29] L. Xue, L. Zeng, W. Kang, H. Chen, Y. Hu, Y. Li, W. Chen, T. Lei, Y. Yan, C. Yang, A. Hu, X. Wang, J. Xiong, C. Zhang, 3D printed Li–S batteries with in situ decorated Li_2S/C cathode: Interface engineering induced loading-insensitivity for scaled areal performance, Adv. Energy Mater. 11 (2021) 2100420.
[30] C. Shen, T. Wang, X. Xu, X. Tian, 3D printed cellular cathodes with hierarchical pores and high mass loading for $Li–SeS_2$ battery, Electrochim. Acta 349 (2020) 136331.
[31] Z. Shi, Z. Sun, J. Cai, Z. Fan, J. Jin, M. Wang, J. Sun, Boosting dual-directional polysulfide electrocatalysis via bimetallic alloying for printable Li–S batteries, Adv. Funct. Mater. 31 (2021) 2006798.

[32] F. Zhang, Z. Li, T. Cao, K. Qin, Q. Xu, H. Liu, Y. Xia, Multishelled Ni_2P microspheres as multifunctional sulfur host 3D-printed cathode materials ensuring high areal capacity of lithium–sulfur batteries, ACS Sustainable Chem. Eng. 9 (2021) 6097–6106.

[33] C. Wei, M. Tian, M. Wang, Z. Shi, L. Yu, S. Li, Z. Fan, R. Yang, J. Sun, Universal *in situ* crafted MO_x-MXene heterostructures as heavy and multifunctional hosts for 3D-printed Li–S batteries, ACS Nano 14 (2020) 16073–16084.

[34] W. Ni, L.-Y. Shi, Microbatteries for advanced applications, in: R. Gupta (Ed.) Handbook of Energy Materials, Springer Nature Singapore, Singapore, 2022, pp. 1–25.

[35] S. Feng, Z.-H. Fu, X. Chen, Q. Zhang, A review on theoretical models for lithium–sulfur battery cathodes, InfoMat 4 (2022) e12304.

[36] Q. Zhang, Q. Huang, S.-M. Hao, S. Deng, Q. He, Z. Lin, Y. Yang, Polymers in lithium–sulfur batteries, Adv. Sci. 9 (2022) 2103798.

[37] W. Li, Q. Zhang, G. Zheng, Z.W. Seh, H. Yao, Y. Cui, Understanding the role of different conductive polymers in improving the nanostructured sulfur cathode performance, Nano Lett. 13 (2013) 5534–5540.

[38] T.-G. Jeong, Y.-S. Lee, B.W. Cho, Y.-T. Kim, H.-G. Jung, K.Y. Chung, Improved performance of dual-conducting polymer-coated sulfur composite with high sulfur utilization for lithium-sulfur batteries, J. Alloys Compd. 742 (2018) 868–876.

[39] G. Ma, Z. Wen, J. Jin, Y. Lu, X. Wu, M. Wu, C. Chen, Hollow polyaniline sphere@sulfur composites for prolonged cycling stability of lithium–sulfur batteries, J. Mater. Chem. A 2 (2014) 10350–10354.

[40] W. Xi, J. Zhang, Y. Zhang, R. Wang, Y. Gong, B. He, H. Wang, J. Jin, A 3D-printed Al metal–organic framework/S cathode with efficient adsorption and redox conversion of polysulfides in lithium–sulfur batteries, J. Mater. Chem. A 11 (2023) 7679–7689.

[41] J. Cai, Z. Fan, J. Jin, Z. Shi, S. Dou, J. Sun, Z. Liu, Expediting the electrochemical kinetics of 3D-printed sulfur cathodes for Li–S batteries with high rate capability and areal capacity, Nano Energy 75 (2020) 104970.

[42] A.I. Hofmann, I. Östergren, Y. Kim, S. Fauth, M. Craighero, M.-H. Yoon, A. Lund, C. Müller, All-polymer conducting fibers and 3D prints via melt processing and templated polymerization, ACS Appl. Mater. Interfaces 12 (2020) 8713–8721.

[43] C. Chen, J. Jiang, W. He, W. Lei, Q. Hao, X. Zhang, 3D printed high-loading lithium-sulfur battery toward wearable energy storage, Adv. Funct. Mater. 30 (2020) 1909469.

[44] W.J. Chung, J.J. Griebel, E.T. Kim, H. Yoon, A.G. Simmonds, H.J. Ji, P.T. Dirlam, R.S. Glass, J.J. Wie, N.A. Nguyen, B.W. Guralnick, J. Park, Á. Somogyi, P. Theato, M.E. Mackay, Y.-E. Sung, K. Char, J. Pyun, The use of elemental sulfur as an alternative feedstock for polymeric materials, Nat. Chem. 5 (2013) 518–524.

[45] K. Shen, H.L. Mei, B. Li, J.W. Ding, S.B. Yang, 3D printing sulfur copolymer-graphene architectures for Li-S batteries, Adv. Energy Mater. 8 (2018) 1701527.

[46] C. Wei, M. Tian, Z. Fan, L. Yu, Y. Song, X. Yang, Z. Shi, M. Wang, R. Yang, J. Sun, Concurrent realization of dendrite-free anode and high-loading cathode *via* 3D printed *N*-Ti_3C_2 MXene framework toward advanced Li–S full batteries, Energy Storage Mater. 41 (2021) 141–151.

[47] Y. Liu, Y. Qiao, Y. Zhang, Z. Yang, T. Gao, D. Kirsch, B. Liu, J. Song, B. Yang, L. Hu, 3D printed separator for the thermal management of high-performance Li metal anodes, Energy Storage Mater. 12 (2018) 197–203.

[48] W. Ni, X. Li, L.-Y. Shi, J. Ma, Research progress on ZnSe and ZnTe anodes for rechargeable batteries, Nanoscale 14 (2022) 9609–9635.

[49] Z. Ji, L. Feng, Z. Zhu, X. Fu, W. Yang, Y. Wang, Polymeric interface engineering in lithium-sulfur batteries, Chem. Eng. J. 455 (2023) 140462.

18 3D-Printed Conducting Polymers for Metal-Air Batteries

Allen Davis and Ram K. Gupta

18.1 INTRODUCTION

In 2018, global energy demands for battery-sourced electricity hovered around 184 GWh, with an over 5× increase to 971 GWh expected by 2025 [1]. Truly, electronics have insinuated themselves into daily life, but little heed is given to their power sources. Batteries are responsible for powering many of the world's disconnected devices. Cars, cell phones, and personal computers are examples of devices that require powerful batteries to operate. Since the development of the first voltaic piles, batteries have advanced to power all aspects of society. Though similar in principle and design to the voltaic pile, modern batteries are composed of vastly more complex materials. The best example of this is the comparison of the voltaic pile to modern lithium ion (Li-ion) batteries. The voltaic pile bore a straightforward design, with a cell consisting of alternating layers of copper and zinc, separated by electrolyte-soaked cloth. During operation, this battery generates low voltage, with a single cell producing around 0.18 V. Meanwhile, a rechargeable Li-ion battery is composed of a lithium-based compound at the cathode facing a porous carbon electrode. In between, there is a complex electrolyte composed of lithium hexafluorophosphate dissolved in a 50/50 ethylene carbonate/dimethyl carbonate solution [2]. While much more complex, a single Li-ion cell has a nominal voltage of 3.7 V. This represents a more than 20-fold increase, which, alongside rechargeability, shows the massive advances in modern battery technology. These advances primarily lie in the materials used, as the general construction of batteries has hardly changed since their first iterations.

The specific materials used in a battery depend on the type of battery in question, as well as its availability for application. As seen in Figure 18.1, only a handful of elements are even compatible with traditional battery applications [3]. First and foremost, only a few dozen elements on the periodic table fall within the ranks of non-toxic and affordable. Heavy metals can range from long-term environmental hazards to acute neurotoxins, making them suboptimal for energy applications. Meanwhile, certain metals are extremely rare, leading to their application being very narrow and expensive. Finally, some elements are simply not conducive to battery application, either due to inactivity, legality, or some other unique elemental property.

In terms of modern battery application, primary cell alkaline batteries are the most general, representing the bulk of non-rechargeable batteries seen in daily use.

DOI: 10.1201/9781003415985-18

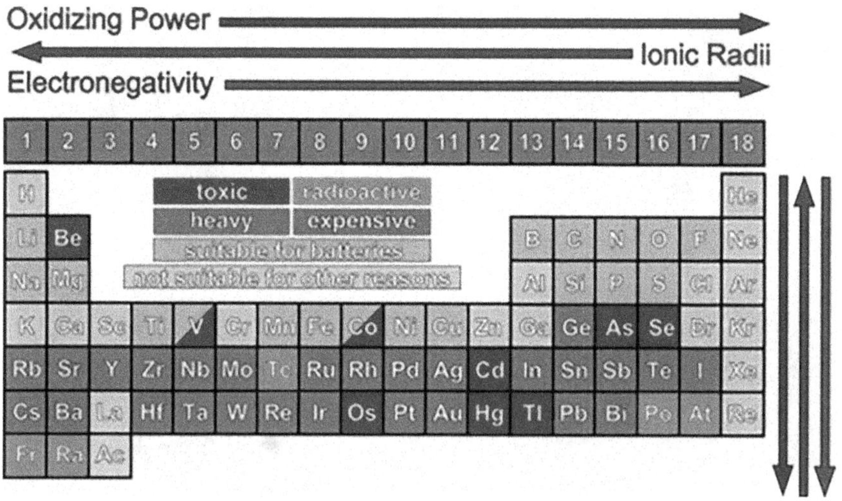

FIGURE 18.1 Graphic demonstrating the suitability of different elements on the periodic table for battery applications. Adapted with permission [3]. Copyright 2013, American Chemical Society.

These batteries consist of a metal and a metal oxide at opposite terminals, with a basic electrolyte in between. When applied to a load, the components undergo redox, eventually draining the cell. As a result of the wastefulness of such one-use batteries, rechargeable batteries like the Li-ion battery have exploded in popularity. The convenience of a rechargeable battery is mirrored in its environmental friendliness, as they produce much less waste when compared to disposable batteries. In the push towards environmentally compatible energy storage, alternative types of materials and batteries have emerged. Of these new developments, conducting polymers and metal air batteries have proved especially interesting.

Conducting polymers are a unique class of polymer that demonstrate a high conductivity unusual among standard polymers. In terms of energy storage, conducting polymers are attractive as either an electrolyte, or as a component of a device's electrodes. This is due to the organic nature of these polymers, which grants additional benefits beyond what other options may produce. These benefits include flexibility, high electrochemical surface area (ECSA), and low equivalent series resistance [4]. These benefits make conducting polymers attractive for use as a material binder [5], electrolyte [6], and/or synthetic precursor [7]. Additionally, the polymeric nature of these materials makes them suitable for a wider variety of processing methods. Additive manufacturing, for example, can improve the capabilities of conducting polymers beyond their normal expectations. This can occur in the form of unique synthesis schemes [8], improved conductivity [9], better elasticity [10], and expanded surface areas [11]. Additionally, additive manufacturing can allow for the design of unique shapes, as seen in Figure 18.2 [12].

Given this, metal-air batteries are one technology that is poised to benefit the most from these developments. Metal-air batteries have a storied history, with their

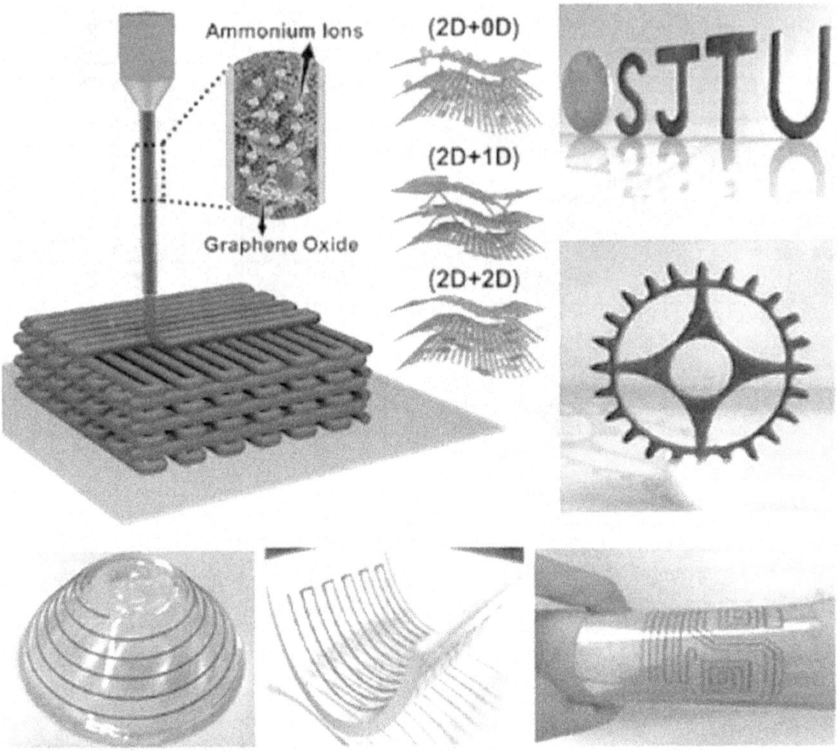

FIGURE 18.2 Demonstration of selectively printed shapes using a conductive graphene-based gel. Adapted with permission [12]. Copyright 2018, American Chemical Society.

discovery first in the mid-19th century [13]. Despite their age, the wide application of this style of battery is limited due to stringent restrictions on its use. First and foremost, the energy-generating reaction in this type of cell is generally inefficient, reducing the battery's effective energy density. Meanwhile, the requirement of steady access to oxygen means that the cells must be exposed to open air or otherwise incorporated as part of a flow battery. This same restriction also limits the size of the battery, meaning that functional batteries must consist of either button-sized cells or large 2D planar batteries.

Considering all this, this chapter first explores the basics of metal-air batteries alongside their general chemistries and components. Next to be explored are the different varieties of conducting polymers, their electrochemical properties, and importance in energy storage. Next, the principals of 3D printing are explored, with special consideration given to the requirements needed for conducting polymer applications. Additionally, the unique features that 3D printing endows to energy storage technology are explored, with a focus on geometry. Finally, the potential application of these three concepts is explored, alongside future prospects and endeavors.

18.2 METAL-AIR BATTERIES

18.2.1 Basics of Metal-Air Batteries

Metal-air batteries are a distinct form of battery that relies on a chemical reaction between metal and air. While the name is somewhat self-explanatory, the nature of the reaction can greatly vary depending on the materials used in their design. Nevertheless, metal-air batteries operate with the help of three active components. These components are the cathode, electrolyte, and anode. At the cathode, oxygen is allowed to diffuse into the system, where it is promptly reduced. The now-reduced ions then travel through the electrolyte until they react with the metal anode. This reaction generates electricity, which can then be put towards powering a device. A generalized depiction of a metal air battery during operation is visible in Figure 18.3 [14].

Metal-air batteries can be distinguished by a variety of factors, such as the type of electrolyte, the metal used at the anode, and the rechargeability of the cell. The electrolyte used within a metal-air battery is strongly dependent on the metals to be used. Nonaqueous electrolytes are suitable for alkali metal-based batteries, such as lithium, sodium, and potassium. Meanwhile, aqueous electrolytes are more suitable for other metals, such as iron, zinc, and aluminum. In terms of the metal-air redox reaction, different metals demonstrate different theoretical energy densities, making them

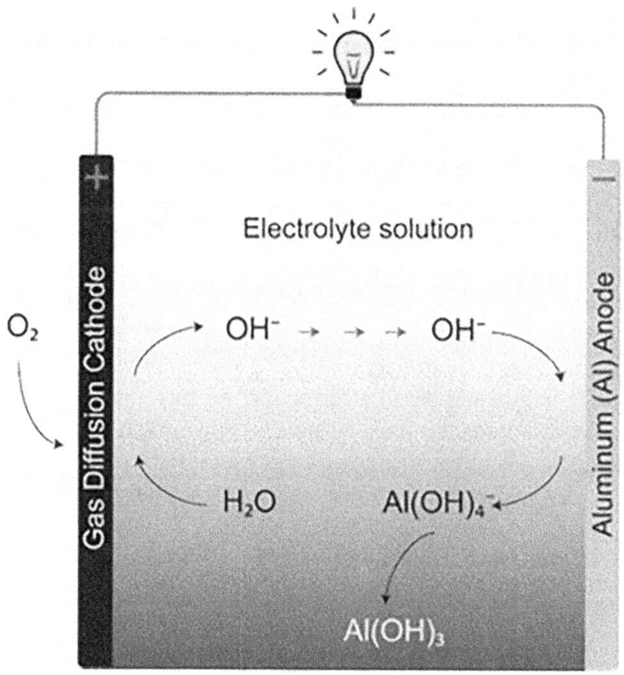

FIGURE 18.3 Educational diagram depicting the operation of an aluminum-air battery. Adapted with permission [14]. Copyright 2023, American Chemical Society.

more attractive for certain applications. Lithium air batteries, for example, have one of the highest theoretical energy densities, around 11,429 Wh/Kg. The high energy density of lithium makes it ubiquitous in the energy industry, especially when applied to Li-ion batteries. Contrastingly, iron and zinc have vastly lower energy densities, yet they are in greater abundance and can be easily recycled, making them popular for metal-air batteries as well. The reusability of a metal-air battery also acts as a method to categorize them. Most metal air batteries are non-rechargeable, leaving nothing but their respective metal oxide at the anode of the device. However, some of these batteries are rechargeable, allowing for their overall lifespan to be expanded. While the varied categories are somewhat different when compared to each other, they are all vastly dependent on the chemistries that occur in the battery itself.

18.2.2 Chemistries and Electrochemical Properties

The generative force in a metal-air battery is known as the metal-air redox reaction. Redox is short for reduction-oxidation and refers to the process that the two primary reactants undergo during the reaction. In a redox reaction, one agent is oxidized, wherein it loses some electrons. These electrons then join with the material that was reduced, completing the reaction. As mentioned previously, oxygen is reduced in a metal air battery, which in turn oxidizes the metal in the battery. The actual chemical reactions that take place in a metal air battery vary drastically based on the metal used in the anode, with the only commonality being the use of oxygen as an oxidizing agent. Zinc-air batteries are a popular option due to their stable voltage and low cost. One advantage of zinc-air batteries is the potential rechargeability that can be attained by tuning the specifics of the cell. Otherwise, zinc suffers from a low theoretical energy density, meaning that individual cells are less efficient than other battery options. Iron-air batteries demonstrate similar benefits and flaws when compared to zinc-air batteries, barring one important exception. As a side product of the reaction, hydrogen is generated, leading to special considerations that must be taken to prevent unwanted side effects during operation. Compared to zinc and iron-air batteries, aluminum air batteries demonstrate a sixfold energy density increase. The aluminum-air battery reaction generates aluminum hydroxide as a product, which can be mechanically recycled for eventual reuse, much like the metal oxides generated in other batteries. Alkali metal-air batteries are unique in that they are usually incompatible with water-based electrolytes. As such, their chemistries rely on organic-based ionic liquids, such as 1-butyl-3-methylimidazolium hexafluorophosphate. Such ionic liquids offer other advantages aside from compatibility, as they can support higher energy densities as well, making them a popular subject for energy research.

18.2.3 Electrodes, Electrolytes, and Membranes

A metal-air battery is only as efficient as the sum of its parts, which is to say that the components used in its construction are of the upmost importance. While only the electrodes and electrolyte are active participants in the metal-air reaction, there are other important components that need to be considered. These components include the battery casing, membrane seal, and any potential separators. As far as the electrodes are concerned, the cathode and the anode are designed with contrasting functions in mind. A metal-air battery cathode is designed to have a porous architecture to allow airflow into the device. Additionally, the cathode must exhibit electrocatalytic activity towards the oxygen reduction reaction (ORR). These properties can either extend from a single material or, more often, a composite of two complementary compounds. The anode of the device can be considerably simpler in design. Many metal-air batteries use a solid piece of the desired metal, but other designs do exist. These designs can include metals that have been doped, composited, or otherwise modified to increase activity. Meanwhile, the electrolyte can vary drastically depending on the application of the battery. Aqueous electrolytes such as alkali salts and hydroxides are low cost and environmentally compatible, while ionic liquids are more expensive but usable with a wider range of materials. As an additional consideration, the electrolyte can also be presented as either a free fluid or contained in an absorbent material such as gel or paper. This choice is again influenced by the design goals and limitation of the metal-air battery itself. When selecting a casing for a metal-air battery, thought must be given to oxygen availability. Lack of adequate oxygen is one of the largest limitations faced in expanding this technology. In larger batteries this can be circumvented by ensuring direct oxygen low to the cathode. In smaller coin cell batteries, small holes are often added to the surface that allow for passive diffusion. With these holes on the surface, one could foresee the issue of electrolyte spillage, but this issue is avoided with the use of an air-permeable membrane. These membranes are specifically tuned to allow for maximum oxygen diffusion without compromising the electrolyte's function. One final consideration is the potential application of a separator within the battery. The advantage of a separator is that it allows for the use of two separate electrolytes, which can extend battery life and reduce the odds of a short circuit [15].

18.2.4 Role of Catalysts

Of all the components in a metal-air battery, the cathode stands to benefit the most from further research. This is because the efficiency of the cathode is intrinsically linked to the output of the battery. The ORR is extremely sluggish as compared to the oxidation that takes place at the anode. As a result, oxygen reduction acts as a rate limiting step that must be overcome to increase battery efficiency. This shows the importance of the electrocatalyst tuning in the design of a metal air battery. When designing a good electrocatalyst for a metal-air battery, a few factors should be kept in mind. A prime concern is what materials are selected to construct the electrocatalyst and their functions within. Noble metals such as platinum often demonstrate optimal catalytic efficiency but are costly and rare. Contrastingly, carbon-based materials are much lower in cost but require extensive modifications to improve their catalytic activity. Metal oxide-based electrocatalysts lie in the middle of the two extremes, offering varied efficiency at varied costs. While each of these selections can be used on their own, they often benefit from compositing or co-doping to improve the efficiency of the catalyst. Another consideration in electrocatalyst design is the functionality of the material. In primary cell metal-air batteries, a monofunctional

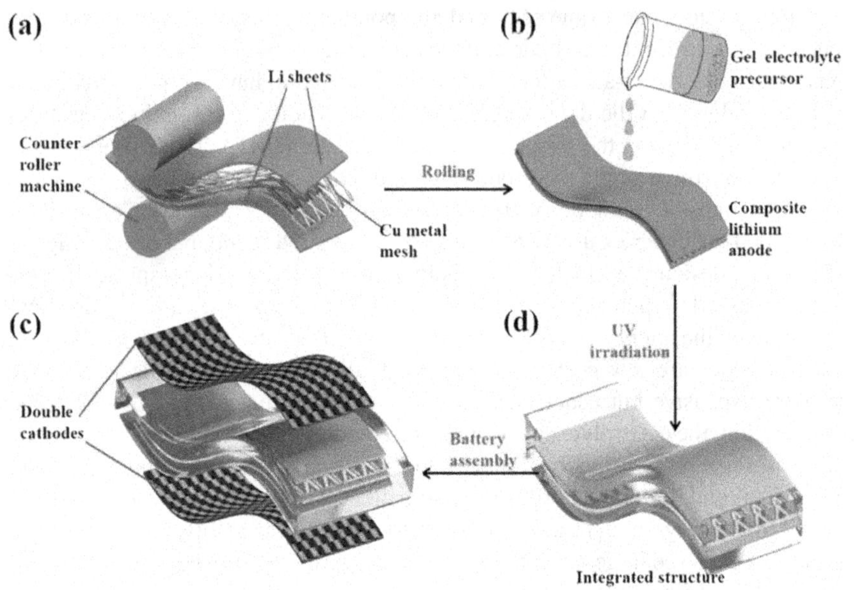

FIGURE 18.4 The creation and assembly of a flexible metal-air battery over four steps. (a) The formation of the anode. (b) The application of the precursor electrolyte. (c) The application of the flexible cathodes. (d) The integration of the gel electrolyte.

Adapted with permission [16]. Copyright 2021, American Chemical Society.

electrocatalyst is preferable, as the battery will be disposed of or recycled after use. In rechargeable secondary cell batteries, an OER-ORR bifunctional electrocatalyst should be used to facilitate both the charge and discharge of the battery. Other considerations for electrocatalyst design include electrolyte compatibility, architectural concerns, and long-term stability. For flexible metal-air batteries, extra thought must be given to the electrocatalyst's construction. For such a battery, too much compression may limit oxygen accessibility, while expansion may damage or perforate the battery. Thus, a flexible catalyst must be designed and tuned to the specific purpose in mind. An example of a flexible electrocatalyst when applied to the construction of a battery is shown in Figure 18.4 [16].

18.3 CONDUCTING POLYMERS

18.3.1 Fundamentals and Types of Conducting Polymers

The history of conducting polymers begins in the 20th century, where they were first confirmed to be synthesized in the mid-1900s. While the first official synthesis is up for debate, Dr. Henry Letheby is generally accepted to be the first to describe such a compound [17]. The material in question was polyaniline, and its discovery paved the way for a wide variety of conducting polymers. Conducting polymers are generally classified by their chain structure. The three classifications of a conducting polymer chain

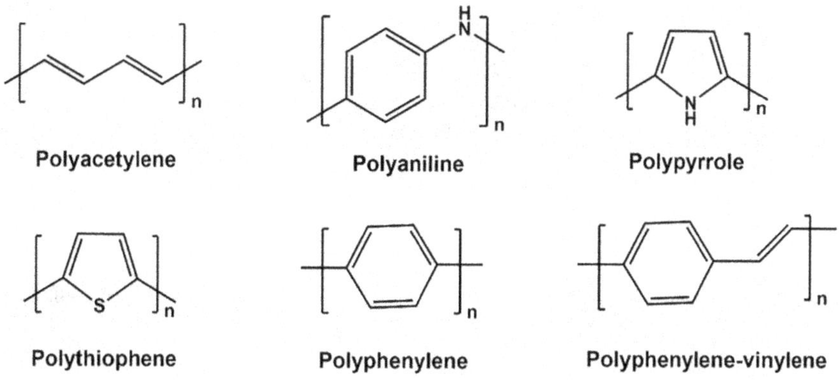

FIGURE 18.5 Examples of the different types of conducting polymer based on their structure and heteroatom content.

structure are those connected by double bonds, by aromatic chains, and by an alternating chain of the two. Polyacetylene is the archetypical example of a non-aromatic conducting polymer, with a structure likened to a completely conjugated polyethylene. Aromatically linked chains are a bit more complicated; this is because they can be further subdivided by the presence of heteroatoms in and around the aromatic ring. Polyphenylene is a self-explanatory example of non-heteroatomic aromatic chained conducting polymers. Meanwhile, the previously mentioned polyaniline, alongside polypyrrole, represent nitrogen-inclusive polyaromatics. Polythiophene, meanwhile, is remarkably similar in structure to polypyrrole, swapping out the secondary amine of the latter for a sulfur group. As for the final category, polyphenylene-vinylene combines the structures of polyphenylene with polyacetylene, creating a complex conjugation structure. As an additional consideration, the physical structure and properties of a conducting polymer can vary drastically depending on its synthesis method. Conducting polymer gels and films can change the flexural properties of a conducting polymer [18]. Meanwhile, powdered conducting polymers can be applied to a nonconductive matrix to increase the conductivity of the bulk composite [19].

18.3.2 Electrochemical Properties of Conducting Polymers

There are several unique electrochemical properties present in conducting polymers. First and foremost is the unique property for an organic molecule to transmit electricity. Through the repeating pi bonds that make up the polymer chains, electrons can flow. The conductivity of these materials often begins in the range of semiconductors but can often be improved through the doping process. In standard semiconducting materials, tiny amounts of dopant are added to the semiconductor, increasing conductivity. In this, dopants are divided into N-type dopants and P-type dopants. N-type dopants, such as phosphorus, function as electron donors, adding extra electrons to the electronic structure. Meanwhile, P-type dopants, such as boron, are electron acceptors, creating vacancies in the electronic framework of the material. These

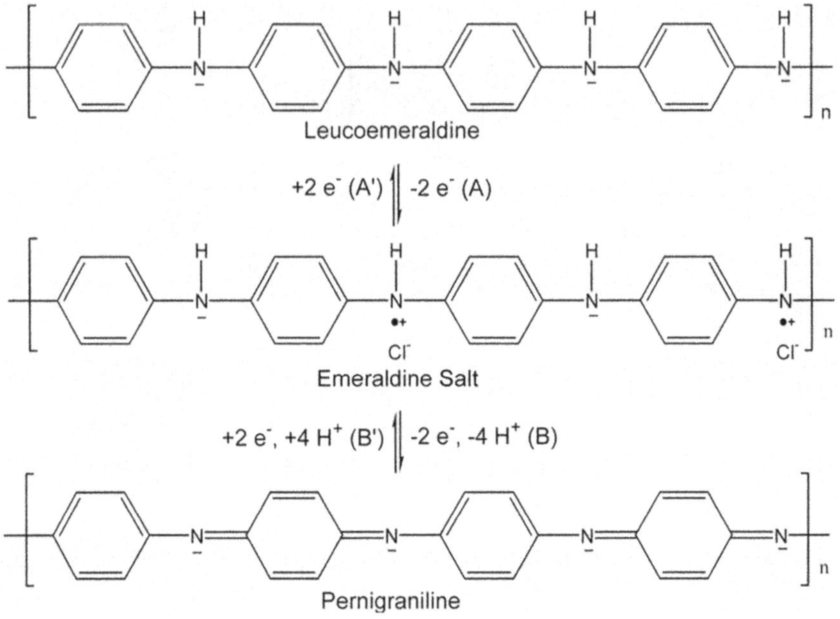

FIGURE 18.6 Demonstration of the different forms of PANI, alongside the electrochemical transitions between them. Adapted with permission [22]. Copyright 2014, American Chemical Society.

irregularities encourage electrons to flow through the material, like how high-pressure and low-pressure atmospheres create the wind. While the result is similar, the doping process for conducting polymers is drastically different. Instead of having atoms inundated in a matrix, conducting polymers are directly doped using a doping ion. In this, electrons are either added or removed via a redox reaction, creating a form of charge carrier known as a polaron [20]. The positive polarons achieved from P-doping are generally preferred due to their stability, though the overall material can be tuned using both methods. In fact, tunability is one of the greatest strengths seen in conducting polymers. Polyaniline (PANI), for example, can have different structures based off its current oxidation state. Luekoemraldine form PANI occurs when the material is fully reduced, wherein it acts as an insulator. The fully oxidized pernigraniline form also acts as an insulator, making both forms unsuitable for conductor applications. Emraldine form PANI acts as the middle ground between these two forms, being partially oxidized and reduced. In this state electrons can flow freely through the material as a result of the change to the PANI's electronic structure, as seen in Figure 18.6 [21, 22]. Other factors that can affect the electrochemical properties of a conducting polymer include temperature, compositing, and compression/extension [23].

18.3.3 Importance in Energy Storage Devices

Modern energy storage devices stand to massively benefit from the application of conducting polymers. In terms of grid-scale energy storage, conducting polymers have found use as a photocatalyst for solar panels [24]. Without exploring the mechanisms of a photovoltaic cell in superfluous detail, conducting polymer photocatalysts help to improve the efficiency of electronic charge transfer. Like the photosynthesis seen in plants, the translation of light into energy occurs in a sequence. Conducting polymers can assist in this sequence by reducing the band gap between electronic states at the interface and/or improving the charge transfer process [25]. Conducing polymer stability also plays a role in this process, as a stable catalyst can improve the active lifetime of a photovoltaic cell. In addition to its use in solar cells, the pseudocapacitive nature of conducting polymers is beneficial to supercapacitor and battery designs. For supercapacitor applications, the redox potential of the material acts as chemical energy storage [26]. In terms of battery operation, conducting polymers are attractive as an electrolyte matrix material [27]. For such an application, the conducting polymer works best as a hydrogel, wherein ions can move through the open pore structure. The greatest benefit that conducting polymers contribute towards energy storage is their high potential for flexibility. The traditional metals used in a battery often permanently deform under strain. Meanwhile, liquid electrolytes are incompressible, limiting the range of movement available for a battery or supercapacitor. Conducting polymers, meanwhile, can be designed to flex, stretch, and compress. Flexible energy storage devices are useful for applications in the medical field, where they can be applied to portable health monitors and other such devices.

18.4 3D PRINTING TECHNOLOGY

18.4.1 Principles and Technology of 3D Printing

Additive manufacturing, more commonly known as 3D printing, is a manufacturing process that is growing increasingly popular across a variety of fields. The popularity first began in the mid-2010s, where expiring patents allowed for the rapid improvement of the technology. These improvements allowed 3D printers to shift from an expensive yet efficient prototyping device to a miniaturized manufactory with prices comparable to other home appliances. The basic principles of 3D printing technology first involve the creation of a stereolithography (.STL) file. The. STL file creates a triangulated 3D model to the dimensions labeled in the program. This file is then uploaded to the printer, after which 3D printing can begin [28]. There are many distinct types of 3D printer, with each family having unique permutations of its own [29]. Of the many different branches of 3D printer, three varieties often function as a baseline for design. Fused deposition modeling (FDM) printers are the archetypal 3D printer due to their ubiquity. FDM printers commonly consist of three different motors on the X,Y, and Z axes alongside an extruder. The extruder melts a thermoplastic filament and, in conjunction with the three motors, deposits it on a build plate. Successive layers of polymer are stacked and fused upon one another until the desired object is printed. Stereolithography (SLA) printers, from which. STL files get their name, take a vastly different approach to 3D object production. These printers lower the build plate into a vat of photopolymer resin within the printer. Opposing the build plate, a laser fires, tracing out a 2D slice of the desired object. The build plate then raises, and the process repeats until the desired object is complete. Digital light processing (DLP) is like this method; however, it uses a light shown on a high-resolution screen to print the object. Powder bed fusion (PBF) printers are the final base concept for 3D printers, and they are also the most niche. In this process, a thin layer of powder is spread over the build plate, then hit with a high energy beam. The build plate is then lowered, and the process again repeats to create the object. PBF printers are energy intensive and expensive to obtain, making them more apt for industrial applications. As previously mentioned, there are a wide range of other printing methods that can be expanded from these three. For example, some printers combine certain aspects of FDM and SLA printers, spraying a photopolymer ink that is promptly cured by a laser at the deposition site. Two other methods that should be mentioned are prudent for the printing of conducting polymers. The first of these methods is inkjet printing. Like traditional office printers, the inkjet printing method sprays a non-viscous ink from a nozzle onto a surface. Unlike office printers, these inks are cured and stacked to create a 3D object. Another popular method is direct ink writing (DIW), where, much like a pen, ink is allowed to flow from nozzle, creating the desired object over successive layers. An example of a few of these methods can be seen in Figure 18.7 [30]. Furthermore, plastics, resins, and inks are only the tip of the iceberg when considering the material technology for printing. Clay 3D printing involves the extrusion of clay through a syringe, which can then be thermally cured to create an object. Meanwhile, selective laser sintering (SLS) printers allow for the printing of metals. However, as with all 3D printers, certain requirements must be met to even consider a material for 3D printing.

18.4.2 Requirements for Materials

The specific property requirements of a material often vary from printer to printer, with rheological, chemical, and physical properties all needing to be considered for

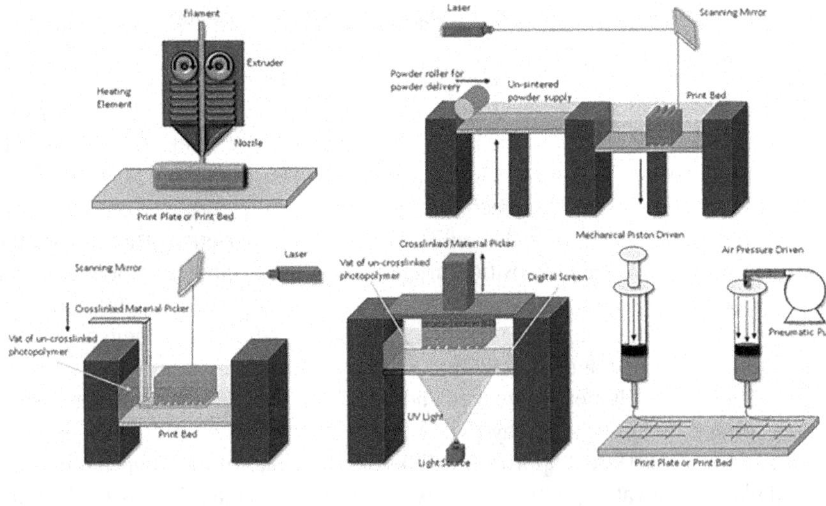

FIGURE 18.7 Examples of different additive manufacturing methods, as observed clockwise from the top right, FDM, SLS, DIW, DLP, and SLA. Adapted with permission [30]. Copyright 2023, American Chemical Society.

the method selected. For example, consider the selection of material for FDM printing. At the very least, a thermoplastic material must be selected, as other materials would burn at the nozzle and be unable to flow. Next to be considered is the melting temperature, both for production and application. A material with a low melting temperature will be easier to process, while a high-temperature material will be more difficult yet available for wider application. Toughness and rigidity are also key factors, with extremely flexible materials interfering with the extrusion process and extremely rigid materials snapping during the unspooling process. With these aspects in mind, a prime example of this selection is polylactic acid (PLA) for FDM printing. PLA is a biobased thermoplastic that demonstrates a low melting temperature and high degree of stiffness, making it a popular selection for 3D printing [31]. Regardless, PLA would be next to useless if applied to SLA printing, where photoset resins are required. With processability being such a major concern, conducting polymers find themselves in an interesting spot. In general, conducting polymers are not considered thermoplastic, making FDM and PBF printing infeasible. Meanwhile, very few conducting polymers make use of photopolymerization, making SLA a niche choice. This leads back to the ink-based 3D printing methods mentioned prior. The fickle nature of conducting polymer processing can be somewhat curtailed when the polymer is made into ink. The advantage of ink processing is twofold, with the solvent improving the rheological properties of the ink while later being able to evaporate away. In terms of rheology, such inks must demonstrate shear-thinning behavior. Shear-thinning is a property that allows for a material to flow when pressed, then coagulate when released [32]. This property is useful because the lack of flow when still affords the object more stability during printing. The high solvent content of inks requires additional post-processing when creating an object. One solution to this issue is to place the object within a heated environment. The added heat increases the rate of evaporation, leading to a faster cure. In inkjet printing, the electronically active nature of conducting polymers can come into play. One style of inkjet printing uses polarizers in the nozzle mechanism, allowing for the ink to be more forcefully ejected due to the electronic interaction within [33]. This trait allows for faster printing and higher resolutions when compared to the direct ink writing method. Further material modifications can be achieved using additives. Such additives can enhance the properties of conducting polymer. These enhancements can range from minor improvements to the polymer's characteristics, to enabling new methods of 3D printing altogether [8].

18.4.3 Advantages and Challenges of 3D Printing

As with every manufacturing process, there are clear advantages and challenges when employing 3D printing. One of the major advantages endowed by 3D printing is the formation of complex shapes that may be unobtainable through other methods. This is because unlike other methods, 3D printing does not require the use of a mold. The lack of a mold allows for intersecting parts, such as chains, struts, and spongiform structures, as seen in Figure 18.8 [34].

Complex formations do face limitations, as overhanging structures require the use of supports. These supports extend from the build plate to the model itself, helping to maintain the structural integrity of the piece. Regrettably, these supports require extra material and extra effort to remove and create waste that must be disposed of. Thankfully, such waste can often be recycled, reducing the environmental impact. In fact, 3D printing can serve as an environmentally friendly option for production, as many polymers can be sourced from green options such as lactic acid and vegetable oil [30]. Another advantage of 3D printing is the rapid creation of protypes, alongside the speed at which said protypes can be modified. Using 3D modeling software, mockups can be quickly created, modified, and transferred into an. STL file. Prints can then be evaluated for their properties, and the cycle can be repeated. The costs saved through additive manufacturing extend to the printer itself, as prices have dropped considerably since the device's inception. FDM and SLA printers are some of the few devices that allow for at home plastics processing. Furthermore, the small footprints of these devices make them optimal for both small-scale benchtop printing and macro-scale printing with multiple devices in sequence. Additive manufacturing stands as an extremely cost-effective method for creating 3D objects; however, many of its benefits come alongside significant drawbacks. For example, the small size of benchtop 3D printers disallows the creation of large objects. As such, the size of any given project must be thoroughly considered due to the limited area of the build plate. 3D printers also tend to have issues regarding tolerance and resolution. 3D printers not only require a flat surface to operate on but may require tuning for different projects. Delamination is also an issue that can occur during 3D printing, especially in conjunction with other print failures. Delamination occurs when the stacked layers of polymer fail to adhere to one another [35]. This issue can arise both during printing

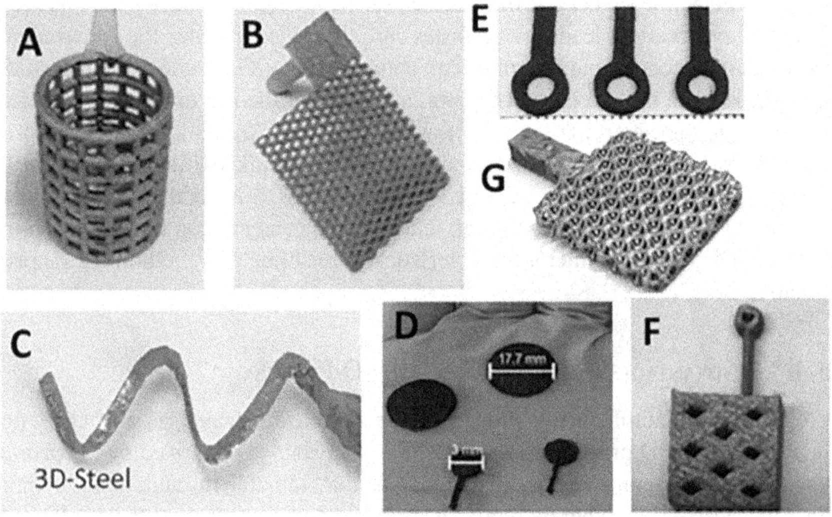

FIGURE 18.8 Examples of complex 3D-printed architectures for electrochemical applications: (A) basket architecture, (B) mesh architecture, (C) ribbon architecture, (D) cyclic architecture, (E) hollow circle architecture, (F) square monolith, (G) square mesh architecture. Adapted with permission [34]. Copyright 2020, American Chemical Society.

and after, often leading to structural issues with the final product. Material selection can also be an issue, as some polymers and their composites require special consideration when used. FDM printers must consider if any additives or fillers are in the filament. Many consumer-grade FDM printers use brass nozzles for their extruders, for which additives like carbon fiber or iron can cause considerable damage. Meanwhile, SLA printers can emit toxic fumes during operation. These fumes require filtration and/or ventilation while the printer is operating, adding to safety concerns [36].

18.5 3D-PRINTED CONDUCTING POLYMERS IN METAL-AIR BATTERIES

18.5.1 Conducting Polymers

While the topics of additive manufacturing, conducting polymers, and metal-air batteries are vastly researched fields, the topic of 3D printed conducting polymers applied to metal-air batteries is not. Indeed, it seems that despite the popularity of the three subjects discussed previously, little research has been done pertaining to their combined use. Nevertheless, work on 3D-printed conducting polymers, 3D-printed metal-air batteries, and conducting polymers in metal-air batteries is rather common. In response to this reality, focus will be placed more on the potential implications of adding the missing element to an already published work.

For example, Liu et al. explored the creation of a fully 3D-printed solid-state zinc-air battery [37]. In this battery, every part of the device was created via 3D printing, from the frame to the electrode. To achieve this, multiple inks were developed for each portion of the battery. The electrocatalyst ink was created by completely dissolving polyvinylidene fluoride (PVDF) in a dimethylformamide (DMF) solution. After complete dissolution, cobalt oxide, graphene, and Super P, a proprietary form of carbon black, were added and homogenized. The zinc anode ink was made in a similar fashion, with the only changes being the swapping of cobalt oxide with zinc and a slight change in ratio. The gel-based electrolyte was synthesized by dissolving polyvinyl alcohol and tetraethylammoniumhydroxide (TEAOH) in distilled water at elevated temperature. After dissolution, a potassium hydroxide (KOH) solution was added and stirred to create the electrolyte slurry. As for the battery's frame, a commercial silicone rubber adhesive/sealant material was used. The printing of the solid-state zinc-air battery occurred over four distinct steps. In step 1, the inks, electrolyte slurry, and rubber were loaded into four separate syringes to facilitate the DIW printing process. In the following step, the electrode inks were printed to a specified pattern on a PET sheet, vacuum dried, then returned to the printer. Step 3 involved printing the rubber frame, which was promptly followed by step 4, wherein the electrolyte was printed, and the battery was frozen to cure the materials. After the freezing process, the battery was allowed to return to room temperature for testing. These print steps, the principles of operation, and the resulting batteries are visible in Figure 18.9 [37].

There are many benefits that can be gleaned from this manufacturing process. First and foremost, this process is highly applicable to a manufacturing line style of mass production. Figure 18.9d demonstrates that at least 16 batteries can be

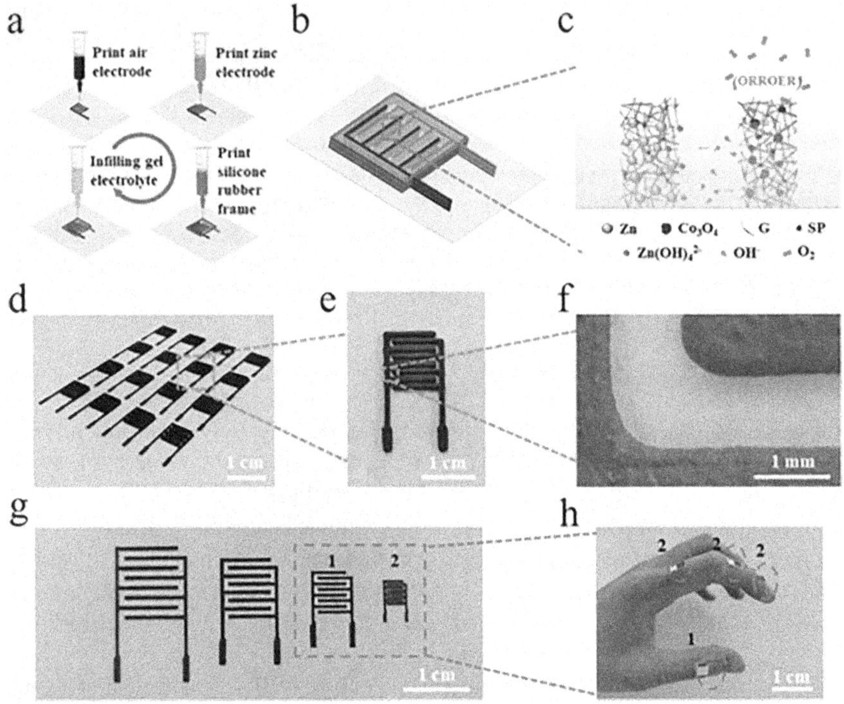

FIGURE 18.9 (a) Simplified fabrication steps for a single battery, (b) schematic of the battery's configuration, (c) diagram of the working principal of the battery, (d) demonstration of print scalability, (e) depiction of a single battery alongside its interdigital structure, (f) microscope image demonstrating print resolution, (g) demonstration of size tunability, (f) demonstration of different-sized batteries adhered to a human hand. Adapted with permission [37]. Copyright 2023, American Chemical Society.

manufactured on one build plate. As such, multiple printers can run both parallel and in series to massively upscale the general production of the batteries. Furthermore, it was demonstrated that these devices could be manufactured to specified sizes and thicknesses, further expanding their potential applications. As for the construction of the battery itself, graphene acts as a conductive substrate for the electrode, whereon cobalt oxide bifunctional catalyst and zinc metal resides. The addition of Super P serves to further enhance the conductivity of the substrate, improving battery efficiency. In terms of electrochemical properties, the most optimal interdigital battery demonstrated an open circuit voltage of ~ 1.37 V, with an areal capacity of 42.2 mAh/cm^2. This battery also demonstrated a high areal specific energy of 77.2 mWh/cm^2 alongside a good cycle stability up to 50 hours. While this work demonstrates the benefits of 3D printing in metal air batteries, further studies could explore the application of conducting polymers to such a device. Using the design of this battery as a baseline, conducting polymers could be applied to the construction of either the electrodes or the electrolyte. In terms of electrode application, conducting polymers have been used natively as well as sacrificially. In one study, Isci et al. explored the use of a thienothiophene-triphenylbenzene copolymer for use in ORR [38]. This material demonstrated a high porosity alongside an ORR onset potential of ~ 0.9 V. While not directly applied to a metal-air battery, such a material showed great promise for future research. More apt to this specific study, however, are sacrificial conducting polymers for catalyst development. In one example, Liu et al. carbonized polypyrrole as part of the cathode of a zinc-air battery [39]. The advantage of carbonizing polypyrrole over other cyclic polymers is the presence of nitrogen within its chain. Nitrogen's unique valance state makes it attractive as an N-type dopant for carbonaceous materials [40, 41].

18.5.2 Composites of Conducting Polymers

There are many advantages to designing a conducting polymer composite framework over more traditional mono-material designs. First and foremost, conducting polymers on their own lack readily available active sites. To rectify this, a wide range of materials have been observed to improve catalytic activity. Metals and their oxidized counterparts are popular options for this application due to their plethora of active site availability. Additionally, these materials are low in cost, abundant, and highly stable, making their use preferred [42]. Nevertheless, metal oxides on their own suffer from a lack of available surface area while also lacking structural stability when applied without a substrate. As such, these materials stand to complement each other when combined, enhancing the overall benefits by reducing the drawbacks.

Towards the goal of exploring conducting polymer composites, Chen et al. devised a Pd-Cu nanoparticle ORR electrocatalyst with N-doped graphene acting as a substrate. In recent years, graphene has grown in popularity for application in the energy field. This popularity stems from the massive surface area and high conductivity that is observed in its structure. Regrettably, native graphene faces many issues that hinder its general application. Price, harsh synthesis methods, and a tendency for agglomeration are all issues faced for graphene in energy applications. While the first two issues are outside the scope of the in-situ application of graphene in energy, agglomeration does have a practical solution. By oxidizing and subsequently reducing graphene, it can be exfoliated, exposing individual flake layers. The resulting material, known as reduced graphene oxide (rGO), can also be synthesized into a hydrogel structure, allowing the material to maintain its exfoliated structure in a 3D environment. The catalyst was created over the course of three steps, two to create the substrate and one to add the nanoparticles. The initial substrate consisted of a graphene oxide GO-PANI co-gel, with the PANI synthesized on the surface of the GO. After synthesis and washing, the composite gels were treated to lyophilization to create a GO-PANI aerogel. The GO-PANI aerogels were subsequently pyrolyzed, creating a more homogenous nitrogen doped carbon aerogel (N-rGO). To create the final catalyst, the N-rGO aerogel was dissolved in DI water, after which palladium chloride and copper chloride were added and the whole solution was ultrasonicated to ensure even dispersion. Sodium carbonate and hydrazine hydrate were then added to catalyze gelation, whereafter the solution was placed into an autoclave and hydrothermally treated. The resulting material was then washed and lyophilized, creating the Pd_3Cu_1/N-rGO aerogel catalyst. Structural characterization determined that the aerogel demonstrated a high surface area alongside vast interconnecting networks. Transmission electron microscopy determined that the embedded nanoparticles also demonstrated promising crystal face availability, which is directly correlated to catalytic activity [43]. The enhanced catalytic availability was promptly observed during electrochemical testing, where it was compared to commercial Pt/C. During testing, the limiting current density of 164.39 mW/cm^2 was over twice that of Pt/C, indicating a marked improvement in that regard. Additionally, the catalyst's current stability was maintained around 82.79% after 2+ hours of testing.

While unused in this study, DIW printing is potentially applicable as a manufacturing method for this device. Multiple studies have demonstrated a synergetic interaction between graphene and 3D printing. One of these synergies is a further increase in functional surface area when compared to non-printed electrodes. For example, Chandrasekaran et al. devised a MoS_2/graphene composite aerogel for electrocatalytic application [44]. In this study, a lattice network was printed via direct ink writing and compared to an aerogel monolith of the same material. In this study, the ECSA of the monolithic electrode was higher than that of the lattice, being ~1725 and ~3100 cm^2, respectively. While the initial ECSA of the monolith seemed higher, prolonged testing demonstrated a >60% loss in ECSA in the monolith when compared to the lattice. This is due to the buildup of gas within the electrocatalyst, which inhibits how much of the material is available for reaction. This property allows for the device to function at higher currents without majorly impacting efficiency. While this technique was used for an HER electrocatalyst, the same benefits can be applied to OER catalysts, which is attractive for secondary cell applications [45]. Another benefit of 3D printing composites is the ability to make a conductive material from a nonconductive matrix. For all intents and purposes, PLA is a nonconductive polymer. However, when composited with graphene, PLA can be used as part of an electrode, as explored by Palenzuela et al. [46]. The creation of this composite was rather simple, with the G/PLA filament being created, then chemically activated for use as a chemical sensor electrode. The deactivating agent used was DMF, which served to partially dissolve the PLA, exposing the graphene within. The increase in conductivity was observed through cyclic voltammetry, wherein a two order of magnitude increase in conductivity was observed post-activation.

18.5.3 Flexible 3D-Printed Conducting Polymers for Metal-Air Batteries

Of all the potential applications available for metal-air batteries, wearable electronics is the most popular. This fact is most evident in the application of hearing aids, in which zinc-air batteries are a popular option. However, wearable technology represents a larger market than just hearing aids alone and as such comes with more stringent requirements. These requirements include light weight, resistance to wear and tear, biocompatibility, and flexibility. As with many batteries, flexibility is a difficult obstacle to overcome due to inherent electrode and electrolyte limitations. In response to this issue, two distinct solutions are primarily being explored, each with different advantages. The first solution involves the creation of the battery on a thin but flexible material such as paper [47], carbon cloth [48], or plastic film [49]. An excellent example of this style of battery was explored in a study by Li et al [50]. In this report, an all solid-state, bifunctional, zinc-air battery was developed using hydroxide@polydopamine core–shell nanosheet arrays. Polydopamine is a bio-inspired polymer that is generally considered non-conductive. However, research has shown that the conductivity of polydopamine can be massively increased by thinly coating it on the surface of an electrode [51]. To create the electrocatalyst, a pretreated carbon cloth substrate was placed into an autoclave alongside cobalt nitrate, aluminum nitrate, ammonium fluoride, and hexamethylenetetramine and allowed to hydrothermally react. This process coated the carbon cloth in a cobalt/aluminum layered double hydroxide nanostructure (CoAl-LDH). The CoAl-LDH sample was promptly washed, then immersed in a dopamine solution, after which Tris-HCl was added to induce polymerization. After one final cycle of washing, the polydopamine coated CoAl-LDH was subjected to calcination, creating a material dubbed Co-CoOχ/N-C. The remainder of the battery was composed of a PVA gel electrolyte and a zinc foil anode. During battery testing, a power density of 20.7 mW/cm^2 was observed, which exceeded that of a more traditional catalyst. Furthermore, the battery maintained a stable voltage of ~1.32 V at up to 135° of deformation, demonstrating suitability for flexural applications.

While the prior paper demonstrated a large capacity for folding, its general design inhibits its ability to stretch. This is where the second method gains ground, in which material geometry is modified to allow for multidirectional stretching. Such a concept was explored by Liu et al, wherein a gel derived electrocatalyst was designed with adaptability in mind [52]. To synthesize this material, 4′-(4-boronatophenyl)-2,2′:6′,2″-terpyridine was suspended alongside melamine in a KOH DMF–H_2O solution, then heated to create a homogenous solution. After this, guanosine was added under slight heating, resulting in a clarified solution. This solution was then promptly cooled to produce an opaque, melamine-doped, guanosine-based hydrogel. To increase the material's catalytic activity, a 1:1 molar ratio of potassium ferricyanide and nickel nitrate was added to the top of the hydrogel and allowed to naturally percolate for 12 hours. The resulting gel was subsequently pyrolyzed to create the catalytically active material designated NiFe-M-GSMG. To create the desired flexible electrocatalyst, the NiFe-M-GSMG material was dispersed in an ethanol-Nafion solution to create a catalytic ink. This ink was then applied to a carbon nanotube paper and attached to a pre-punctured and stretched dielectric elastomer. The rest of the battery consisted of a guanosine-based gel electrolyte and a zinc paste on silver nanowire anode. In testing, a high power density of 159.0 mW/cm^2 alongside a current retention of 87.5% after 20 hours were observed. More impressive however, were the battery's material properties and environmental resistance. Testing determined that the battery could maintain a consistent voltage after 10,000 cycles of omnidirectional stretching. Furthermore, the battery was able to maintain its elastic properties at −60 °C, where it was stretched to 1000% of its original dimension while still maintaining voltage. In addition to the flexibility, the battery was also observed to be waterproof. In this test, it was demonstrated to function for at least 7 hours while immersed in water. With these properties in mind, such a battery could be extremely useful when applied to amphibious electronics, where its flexibility and waterproof nature would prove beneficial.

While these prior papers explored examples of flexible metal-air batteries, there are a few directions where the application of conducting polymers and additive manufacturing could serve to benefit them. In the first paper, polydopamine was explored as a sacrificial source of nitrogen-doped carbon. In this regard, other conducting polymers such as PANI or polyindole could be substituted to modify the catalytic effect. Polyindole may be extra attractive for this specific application due to its structural similarity to polydopamine. Meanwhile, the inking process used in the second paper could easily be modified to fit the DIW method for 3D printing. In fact, 3D printing is immensely popular for flexible electronics due to the metamaterial properties that can be acquired through specific complex shape printing [53, 54].

18.6 CONCLUSION

The potential impact of 3D-printed conducting polymers for use in metal-air batteries is not to be understated. While the lack of research including all three of these terms is considerable, this simply translates to a massive research opportunity. In one way or another, each of these cutting-edge concepts is at the forefront of research, with hundreds of articles on each topic [34, 55, 56]. Metal-air batteries act as a potential solution to some of the woes faced in portable energy storage technologies. The high energy density of Li-ion batteries is equally matched by the rising prices of lithium. Metal-air batteries retaliate against this issue, demonstrating a high energy density and low cost due to the nature of the materials used in their construction. Concurrently, conducting polymers are growing ever more popular for energy applications due to a plethora of unique properties. Some conducting polymers can form gels that function as an electrolyte or membrane, while others act as a precursor for further synthesis. Additionally, conducting polymers are one of the few materials available for 3D printing, making them extra popular. The rapid proliferation of 3D printing technology in modern society reflects their potential as more than a mere novelty. Additive manufacturing has found applications in almost every field imaginable, from healthcare an aerospace, to even more esoteric topics such as food and textiles [57, 58, 59]. Altogether, the greatest hurdle faced in the application of these subjects is the newness of some technologies, especially in the case of 3D printers. Conducting polymers have seen great use as a sacrificial component for electrocatalyst construction or as a component of hydrogel construction. However, the DIW method has only seen wide use for conducting polymer printing within the last 5 years, restricting the available research. Likewise, the non-sacrificial application of conducting polymers in metal-air batteries has hardly seen exploration either. This is because other materials that are either cheaper, easier to manufacture, or more effective are selected instead. Reflecting on this issue, potential solutions can be found in the use of more niche applications for conducting polymers. Polymer-air batteries were a popular concept for research in the mid-2010s, wherein redox active polymers served as the anode of the device [60]. However, this technology fell out of favor due to substandard results and lower stability. Perhaps, however, this cousin of the metal-air battery could be improved with the application of 3D printing, reigniting this field of research. In terms of 3D printing capability, a variety of methods have been developed for non-metal-air battery applications that could translate to these devices. Some studies explore complex 3D-printed frameworks that maximize the available reactive surface area [61]. Meanwhile, other studies explore the printing of flexible self-healing gels that could be applied for electrolyte applications [62].

As a summary of this information, current research on 3D printed conducting polymers in metal-air batteries is scarce, with a vast amount of potential research being underutilized. However, certain avenues of research run parallel to this concept, implying a wide range of potential applications towards the development of a 3D-printed conducting polymer metal-air battery. The newness of this specific concept all but ensures that massive amounts of data could be gleaned from further development on this topic. This subject faces many challenges, however, as certain components in its design have unique requirements. The physical requirements for oxygen in metal-air battery construction inhibit its dimensions, while conducting polymers require special treatment to activate them. 3D printing meanwhile, suffers issues with scale, losing out to traditional manufacturing methods as demand increases. Nevertheless, the outlook for this technology is both promising and astounding. In the medical field, environmentally friendly flexible metal-air batteries will be useful for wearable health monitors. Meanwhile, the high potential energy density of some metals may allow future metal-air batteries to dethrone lithium-ion batteries as the default high-energy battery. In closing, while this subject of research deserves more effort towards its proliferation, the potential applications available to be discovered could change the way humankind thinks about energy storage.

REFERENCES

[1] Y. Zhao, O. Pohl, A.I. Bhatt, G.E. Collis, P.J. Mahon, T. Rüther, A.F. Hollenkamp, A review on battery market trends, second-life reuse, and recycling, Sustain. Chem. 2 (2021) 167–205.

[2] Y. Wang, B. Liu, Q. Li, S. Cartmell, S. Ferrara, Z.D. Deng, J. Xiao, Lithium and lithium ion batteries for applications in microelectronic devices: A review, J. Power Sources. 286 (2015) 330–345.

[3] B.C. Melot, J.-M. Tarascon, design and preparation of materials for advanced electrochemical storage, Acc. Chem. Res. 46 (2013) 1226–1238.

[4] C.B.T.-H. of N. for I.A. Mustansar Hussain, ed., Chapter 41—Engineered nanomaterials for energy applications, in: Micro Nano Technol., Elsevier, 2018: pp. 751–767.

[5] T.M. Higgins, S.-H. Park, P.J. King, C. (John) Zhang, N. McEvoy, N.C. Berner, D. Daly, A. Shmeliov, U. Khan, G. Duesberg, V. Nicolosi, J.N. Coleman, A commercial conducting polymer as both binder and conductive additive for silicon nanoparticle-based lithium-ion battery negative electrodes, ACS Nano. 10 (2016) 3702–3713.

[6] D. Du, X. Hu, D. Zeng, Y. Zhang, Y. Sun, J. Li, H. Cheng, Water-insoluble side-chain-grafted single ion conducting polymer electrolyte for long-term stable lithium metal secondary batteries, ACS Appl. Energy Mater. 3 (2020) 1128–1138.

[7] C. Strietzel, K. Oka, M. Strømme, R. Emanuelsson, M. Sjödin, An alternative to carbon additives: The fabrication of conductive layers enabled by soluble conducting polymer precursors—A case study for organic batteries, ACS Appl. Mater. Interfaces. 13 (2021) 5349–5356.

[8] A.I. Hofmann, I. Östergren, Y. Kim, S. Fauth, M. Craighero, M.-H. Yoon, A. Lund, C. Müller, All-polymer conducting fibers and 3D prints via melt processing and templated polymerization, ACS Appl. Mater. Interfaces. 12 (2020) 8713–8721.

[9] I.M. Hill, V. Hernandez, B. Xu, J.A. Piceno, J. Misiaszek, A. Giglio, E. Junez, J. Chen, P.D. Ashby, R.S. Jordan, Y. Wang, Imparting high conductivity to 3D printed PEDOT:PSS, ACS Appl. Polym. Mater. 5 (2023) 3989–3998.
[10] Y. Chen, Z. Yu, Y. Ye, Y. Zhang, G. Li, F. Jiang, Superelastic, hygroscopic, and ionic conducting cellulose nanofibril monoliths by 3D printing, ACS Nano. 15 (2021) 1869–1879.
[11] K.R. Ryan, M.P. Down, N.J. Hurst, E.M. Keefe, C.E. Banks, Additive manufacturing (3D printing) of electrically conductive polymers and polymer nanocomposites and their applications, EScience. 2 (2022) 365–381.
[12] X. Tang, H. Zhou, Z. Cai, D. Cheng, P. He, P. Xie, D. Zhang, T. Fan, Generalized 3D Printing of Graphene-Based Mixed-Dimensional Hybrid Aerogels, ACS Nano. 12 (2018) 3502–3511.
[13] D. Ahuja, V. Kalpna, P.K. Varshney, Metal air battery: A sustainable and low cost material for energy storage, J. Phys. Conf. Ser. 1913 (2021) 12065.
[14] P.E. Olli, T. Romann, Educational metal–Air battery, J. Chem. Educ. 100 (2023) 259–266.
[15] H. Wang, Q. Xu, Materials design for rechargeable metal-air batteries, Matter 1 (2019) 565–595.
[16] J. Li, Z. Wang, L. Yang, Y. Liu, Y. Xing, S. Zhang, H. Xu, A flexible Li–air battery workable under harsh conditions based on an integrated structure: A composite lithium anode encased in a gel electrolyte, ACS Appl. Mater. Interfaces. 13 (2021) 18627–18637.
[17] S.C. Rasmussen, The early history of polyaniline: Discovery and origins, Substantia. 1 (2017) 99–109.
[18] P. Li, K. Sun, J. Ouyang, Stretchable and conductive polymer films prepared by solution blending, ACS Appl. Mater. Interfaces. 7 (2015) 18415–18423.
[19] P. Sengun, M.T. Kesim, M. Caglar, U. Savaci, S. Turan, İ. Sahin, E. Suvaci, Characterization of designed, transparent and conductive Al doped ZnO particles and their utilization in conductive polymer composites, Powder Technol. 374 (2020) 214–222.
[20] T.H. Le, Y. Kim, H. Yoon, Electrical and electrochemical properties of conducting polymers, Polymers (Basel). 9 (2017) 150.
[21] K. Namsheer, C.S. Rout, Conducting polymers: A comprehensive review on recent advances in synthesis, properties and applications, RSC Adv. 11 (2021) 5659–5697.
[22] M. Canales, J. Torras, G. Fabregat, A. Meneguzzi, C. Alemán, Polyaniline emeraldine salt in the amorphous solid state: Polaron versus bipolaron, J. Phys. Chem. B. 118 (2014) 11552–11562.
[23] Y. Liu, E. Asare, H. Porwal, E. Barbieri, S. Goutianos, J. Evans, M. Newton, J.J.C. Busfield, T. Peijs, H. Zhang, E. Bilotti, The effect of conductive network on positive temperature coefficient behaviour in conductive polymer composites, Compos. Part A Appl. Sci. Manuf. 139 (2020) 106074.
[24] L. Tao, J. Wang, Z. Luo, J. Ren, D. Yin, Fabrication of an S-scheme heterojunction photocatalyst MoS2/PANI with greatly enhanced photocatalytic performance, Langmuir. 39 (2023) 11426–11438.
[25] L. Li, Z. Zhang, C. Ding, J. Xu, Boosting charge separation and photocatalytic CO2 reduction of CsPbBr3 perovskite quantum dots by hybridizing with P3HT, Chem. Eng. J. 419 (2021) 129543.
[26] M. Moussa, M.F. El-Kady, D. Dubal, T.T. Tung, M.J. Nine, N. Mohamed, R.B. Kaner, D. Losic, Self-assembly and cross-linking of conducting polymers into 3D hydrogel electrodes for supercapacitor applications, ACS Appl. Energy Mater. 3 (2020) 923–932.
[27] L.C. Merrill, X.C. Chen, Y. Zhang, H.O. Ford, K. Lou, Y. Zhang, G. Yang, Y. Wang, Y. Wang, J.L. Schaefer, N.J. Dudney, Polymer–ceramic composite electrolytes for lithium batteries: A comparison between the single-ion-conducting polymer matrix and its counterpart, ACS Appl. Energy Mater. 3 (2020) 8871–8881.

[28] M. Szilvśi-Nagy, G. Mátyási, Analysis of STL files, Math. Comput. Model. 38 (2003) 945–960.
[29] N. Shahrubudin, T.C. Lee, R. Ramlan, An overview on 3D printing technology: Technological, materials, and applications, Procedia Manuf. 35 (2019) 1286–1296.
[30] G. Guggenbiller, S. Brooks, O. King, E. Constant, D. Merckle, A.C. Weems, 3D printing of green and renewable polymeric materials: Toward greener additive manufacturing, ACS Appl. Polym. Mater. 5 (2023) 3201–3229.
[31] E.H. Tümer, H.Y. Erbil, Extrusion-based 3D printing applications of PLA composites: A review, Coatings. 11 (2021).
[32] J.F. Ryder, J.M. Yeomans, Shear thinning in dilute polymer solutions, J. Chem. Phys. 125 (2006) 194906.
[33] M. Criado-Gonzalez, A. Dominguez-Alfaro, N. Lopez-Larrea, N. Alegret, D. Mecerreyes, Additive manufacturing of conducting polymers: Recent advances, challenges, and opportunities, ACS Appl. Polym. Mater. 3 (2021) 2865–2883.
[34] M.P. Browne, E. Redondo, M. Pumera, 3D printing for electrochemical energy applications, Chem. Rev. 120 (2020) 2783–2810.
[35] A. Katalagarianakis, E. Polyzos, D. Van Hemelrijck, L. Pyl, Mode I, mode II and mixed mode I-II delamination of carbon fibre-reinforced polyamide composites 3D-printed by material extrusion, Compos. Part A Appl. Sci. Manuf. 173 (2023) 107655.
[36] J.J. Tully, G.N. Meloni, A scientist's guide to buying a 3D printer: How to choose the right printer for your laboratory, Anal. Chem. 92 (2020) 14853–14860.
[37] G. Liu, Z. Ma, G. Li, W. Yu, P. Wang, C. Meng, S. Guo, All-printed 3D solid-state rechargeable zinc-air microbatteries, ACS Appl. Mater. Interfaces. 15 (2023) 13073–13085.
[38] R. Isci, T. Balkan, S. Tafazoli, B. Sütay, M.S. Eroglu, T. Ozturk, Thienothiophene and triphenylbenzene based electroactive conjugated porous polymer for oxygen reduction reaction (ORR), ACS Appl. Energy Mater. 5 (2022) 13284–13292.
[39] S. Liu, M. Wang, T. Qian, J. Liu, C. Yan, Selenium-doped carbon nanosheets with strong electron cloud delocalization for nondeposition of metal oxides on air cathode of zinc–air battery, ACS Appl. Mater. Interfaces. 11 (2019) 20056–20063.
[40] S. Guo, J. Wang, Y. Sun, L. Peng, C. Li, Interface engineering of Co3O4/CeO2 heterostructure in-situ embedded in Co/N-doped carbon nanofibers integrating oxygen vacancies as effective oxygen cathode catalyst for Li-O2 battery, Chem. Eng. J. 452 (2023) 139317.
[41] Y. Liu, F. Zhan, B. Wang, B. Xie, Q. Sun, H. Jiang, J. Li, X. Sun, Three-dimensional composite catalysts for Al–O2 batteries composed of CoMn2O4 nanoneedles supported on nitrogen-doped carbon nanotubes/graphene, ACS Appl. Mater. Interfaces. 11 (2019) 21526–21535.
[42] Z. Zhang, J. Liu, J. Gu, L. Su, L. Cheng, An overview of metal oxide materials as electrocatalysts and supports for polymer electrolyte fuel cells, Energy Environ. Sci. 7 (2014) 2535–2558.
[43] A. Trovarelli, J. Llorca, Ceria catalysts at nanoscale: How do crystal shapes shape catalysis? ACS Catal. 7 (2017) 4716–4735.
[44] S. Chandrasekaran, J. Feaster, J. Ynzunza, F. Li, X. Wang, A.J. Nelson, M.A. Worsley, three-dimensional printed MoS2/graphene aerogel electrodes for hydrogen evolution reactions, ACS Mater. Au. 2 (2022) 596–601.
[45] T. Zhao, Y. Wang, X. Chen, Y. Li, Z. Su, C. Zhao, Vertical growth of porous perovskite nanoarrays on nickel foam for efficient oxygen evolution reaction, ACS Sustain. Chem. Eng. 8 (2020) 4863–4870.
[46] C.L. Manzanares Palenzuela, F. Novotný, P. Krupička, Z. Sofer, M. Pumera, 3D-printed graphene/polylactic acid electrodes promise high sensitivity in electroanalysis, Anal. Chem. 90 (2018) 5753–5757.

[47] Y. Wang, H.Y.H. Kwok, W. Pan, Y. Zhang, H. Zhang, X. Lu, D.Y.C. Leung, Printing Al-air batteries on paper for powering disposable printed electronics, J. Power Sources. 450 (2020) 227685.

[48] P. Katsoufis, M. Katsaiti, C. Mourelas, T.S. Andrade, V. Dracopoulos, C. Politis, G. Avgouropoulos, P. Lianos, Study of a thin film aluminum-air battery, Energies. 13 (2020).

[49] L. Wang, J. Pan, Y. Zhang, X. Cheng, L. Liu, H. Peng, A Li–air battery with ultralong cycle life in ambient air, Adv. Mater. 30 (2018) 1704378.

[50] S. Li, W. Xie, Y. Song, M. Shao, Layered double hydroxide@polydopamine core–shell nanosheet arrays-derived bifunctional electrocatalyst for efficient, flexible, all-solid-state zinc–air battery, ACS Sustain. Chem. Eng. 8 (2020) 452–459.

[51] T. Eom, J. Lee, S. Lee, B. Ozlu, S. Kim, D.C. Martin, B.S. Shim, Highly conductive polydopamine coatings by direct electrochemical synthesis on Au, ACS Appl. Polym. Mater. 4 (2022) 5319–5329.

[52] J. Liu, M. Wang, C. Gu, J. Li, Y. Liang, H. Wang, Y. Cui, C.-S. Liu, Supramolecular gel-derived highly efficient bifunctional catalysts for omnidirectionally stretchable Zn–air batteries with extreme environmental adaptability, Adv. Sci. 9 (2022) 2200753.

[53] K.A. Deo, M.K. Jaiswal, S. Abasi, G. Lokhande, S. Bhunia, T.U. Nguyen, M. Namkoong, K. Darvesh, A. Guiseppi-Elie, L. Tian, A.K. Gaharwar, Nanoengineered ink for designing 3D printable flexible bioelectronics, ACS Nano. 16 (2022) 8798–8811.

[54] T.H. Vo, P.K. Lam, Y.-J. Sheng, H.-K. Tsao, Jammed microgels in deep eutectic solvents as a green and low-cost ink for 3D printing of reliable auxetic strain sensors, ACS Appl. Mater. Interfaces. 15 (2023) 33109–33118.

[55] F. Wu, C. Wang, K. Liao, Z. Shao, Air cathode design for light-assisted charging of metal–air batteries: Recent advances and perspectives, Energy & Fuels. 37 (2023) 8902–8918.

[56] J. Gamboa, S. Paulo-Mirasol, F. Estrany, J. Torras, Recent progress in biomedical sensors based on conducting polymer hydrogels, ACS Appl. Bio Mater. 6 (2023) 1720–1741.

[57] A. Jandyal, I. Chaturvedi, I. Wazir, A. Raina, M.I. Ul Haq, 3D printing—A review of processes, materials and applications in industry 4.0, Sustain. Oper. Comput. 3 (2022) 33–42.

[58] N. Nachal, J.A. Moses, P. Karthik, C. Anandharamakrishnan, Applications of 3D printing in food processing, Food Eng. Rev. 11 (2019) 123–141.

[59] Y.-Q. Xiao, C.-W. Kan, Review on development and application of 3D-printing technology in textile and fashion design, Coatings. 12 (2022).

[60] T. Kawai, K. Oyaizu, H. Nishide, High-density and robust charge storage with poly(-anthraquinone-substituted norbornene) for organic electrode-active materials in polymer–air secondary batteries, Macromolecules. 48 (2015) 2429–2434.

[61] V. Muñoz-Perales, M. van der Heijden, P.A. García-Salaberri, M. Vera, A. Forner-Cuenca, Engineering lung-inspired flow field geometries for electrochemical flow cells with stereolithography 3D printing, ACS Sustain. Chem. Eng. 11 (2023) 12243–12255.

[62] P.-C. Lai, Z.-F. Ren, S.-S. Yu, Thermally induced gelation of cellulose nanocrystals in deep eutectic solvents for 3D printable and self-healable ionogels, ACS Appl. Polym. Mater. 4 (2022) 9221–9230.

19 3D-Printed Conducting Polymers for Biomedical Applications

Emre Yılmazoğlu and Selcan Karakuş

19.1 INTRODUCTION

3D production is a manufacturing method that involves layering fluidized raw material, one layer at a time. This method enables the creation of both intermediate and final products. Thanks to advancements such as the diversification of raw materials and the availability of easily accessible production devices, 3D production has become applicable not only for industrial production but also for small-scale and even home production. Industrial sectors where 3D production is frequently utilized include the production of spare parts for automotive and other mechanical devices, construction, aviation and space exploration, electronics, medicine, aesthetics, and dentistry. The raw materials employed in 3D production vary extensively based on the specific requirements of the respective industries. These materials include polymers, metals and alloys, concrete, diverse composites, clay, sand, and food. When examining the consumption of raw materials in 3D production, it is observed that polymers are predominantly utilized. Due to the diverse range of properties that polymers possess; they can be employed in production using various methods. Depending on the specific procedure of the method, polymers may be in solid forms such as powder, pellets, filaments, or films, or they may be in liquid forms.

The first notable instance of 3D production with commercial implications was the stereolithography method, discovered by Hideo Kodama. Numerous studies were patented, presenting original approaches in this field after the stereoscopic shaping of liquid polymer resin. According to current standards, 3D production methods are categorized into seven categories: binder jetting (BJT), powder bed fusion (PBF), material extrusion (MEX), directed energy deposition (DED), material jetting (MJT), vat photopolymerization (VPP), and sheet lamination (SHL). Each of these methods is associated with a distinct production technique, and as a result, the raw materials used in one method may not be applicable in another. Consequently, not every method can be employed in the production of every product. The possibility of utilizing the conductivity of polymers became evident in the 1970s when Shirakawa obtained polyacetylene, a polymer capable of conducting electricity. Prior to this discovery, polymers were commonly regarded as insulating materials. However, this breakthrough marked the beginning of a new technological era, demonstrating that polymers could be utilized for electrical conduction and other intriguing properties. Poly(thiophene) (PTh), poly(pyrrole) (PPy), poly(styrene sulfonate) (PSS),

DOI: 10.1201/9781003415985-19

poly(3,4-ethylenedioxythiophene) (PEDOT), poly(aniline) (PANi), poly(phenylene vinylene) (PPV), poly(phenylene), and poly(paraphenylene) (PPP) are among the polymers known for their electrical conductivity. These polymers exhibit various other properties such as lightness, transparency, thermal resistance, chemical or mechanical strength, and low redox potential.

19.2 CONDUCTIVE POLYMERS

The utilization of conductive polymers, particularly at nanoscales, has significantly increased their application in advanced technology domains, facilitating the production of novel and superior products. When choosing from various conductive polymers, factors such as the degree of conductivity; structural properties enabling conductivity; reversibility of the doping process; control over chemical, electrical, and mechanical properties; cost; and sustainability play a significant role. As a result, numerous conductive polymers excel in different fields. Furthermore, apart from their electrical, chemical, and mechanical properties, various polymers possess distinguishing optical, magnetic, microwave-absorbing, or wetting properties. After doping, the polymeric material exhibits metal-like conductivity. When comparing the conductivity of conductive polymers with different nano-sized shapes, it has been observed that nanofibers demonstrate conductivity one to two orders of magnitude higher than nanotubes and nanowires. Nanocomposites, formed by incorporating metals, semiconductors, insulating polymers, and other nanomaterials such as carbon into conductive polymers, offer a wide range of materials suitable for electronic applications. These materials find utility in various electronic devices, including light-emitting diodes, memory devices, field-effect transistors, and photovoltaic devices. Conductive polymeric composites with different structures are capable of accommodating NH_3, H_2, CO, NO_2, H_2S, $CHCl_3$, Zn^{+2}, and Ni^{+2} ions. These composite materials find potential applications in areas such as environmental monitoring, healthcare, and food industry. Conductive polymers can be utilized as biosensors due to their ability to facilitate fast electron transfer and immobilization of biomolecules. For instance, poly(aniline)/poly(styrene) (PANi/PS) nanocomposites have shown functionality in detecting H_2O_2, which is a byproduct of various enzymatic reactions [1]. Enzyme-doped PANi/Fe_3O_4/carbon nanocomposites have the capacity to repair glucose. In instance, lipase-encapsulated conductive polymeric composites have successfully exploited these nanocomposites for enzymatic processes. Enzymatic processes can be aided by adding lipase enzymes to these composites. To demonstrate the potential of conductive polymeric materials in analytical chemistry, (polyacrylate/polyethylene) PA/PE composites have been used for amperometric triglyceride detection [2]. Applications for conductive polymers include DNA analysis and detection. DNA can be immobilized on certain materials, like PANi, making it easier to find and analyze DNA sequences. This creates opportunities for forensic analysis, diagnostics, and genetic study [3]. The use of conductive polymers in molecular biology and biotechnology has been expanded by the use of polyaniline doped with poly(methylvinyl ether-alt-maleic acid) for the detection of oligonucleotides [4]. Additionally, PANi-based nanonetworks have been used as biosensors for detecting a variety of bacteria, such as *Klebsiella pneumoniae, Pseudomonas aeruginosa* (*P. aeruginosa*), *Escherichia coli* (*E. coli*), and *Enterococcus faecalis*

(*E. faecalis*) [5]. The implications for applications in healthcare and the environment highlight the potential of PANi as a sensing material for bacterial detection and monitoring. Due to their low cost, simplicity in production, good biocompatibility, and chemical reactivity, PANi and PPy are especially preferred as conducting polymers. Due to these qualities, they are appealing choices for a variety of applications, including biosensors and bioelectronics. Their adaptability and compatibility with biological systems help explain why they are so well liked in a variety of scientific and technological domains.

By responding to the electrochemical reactions triggered by drugs, conductive polymeric composites play a significant role in facilitating the controlled release of pharmaceuticals. These composites, composed of polymers and active pharmaceutical ingredients, offer important advantages such as easy manufacturing, compatibility with metabolic processes in the body, adaptability, and the ability to enhance the chemical or mechanical effectiveness of the drug. Furthermore, chemically modified PPy nanoparticles exhibit remarkable versatility, extending beyond their applications in drug delivery to serving as affinity matrices for protein purification and separation.

Utilizing electrical stimulation to regulate various cellular activities, including intercellular communication, proliferation, migration, and differentiation, has demonstrated remarkable capabilities. Conductive polymers have emerged as the optimal choice in this regard when compared to other polymers, primarily due to their exceptional ability to respond to electrical impacts. Notably, several conductive polymer architectures, such as gelatin, poly(glycolic acid) (PGA)/gelatin, poly(lactic-co-glycolic acid) (PLGA), poly(lactide-co-ε-caprolactone) (PLCL), poly(lactic acid) (PLA)/PPy, poly(caprolactone) (PCL), and PANi, possess highly desirable qualities, including biocompatibility, biodegradability, and a large surface area. These attributes make them exceptionally well suited for a wide range of tissue engineering applications. The integration of conductive polymers into tissue engineering scaffolds or matrices allows for the creation of an electrically responsive environment. This electrical stimulation can be precisely controlled and harnessed to accelerate the regeneration and repair of damaged tissue. The conductive nature of these polymers enables the transmission of electrical impulses throughout the scaffold, closely resembling the natural electrical cues produced by the body. For instance, PANi, a conductive polymer, exhibits excellent electrical and biocompatibility characteristics. It has found extensive use in tissue engineering to promote cell adhesion, proliferation, and differentiation. Similarly, other conductive polymers such as PCL, PLGA, PLCL, and PLA/PPy offer a combination of mechanical strength, biodegradability, and electrical conductivity, making them well suited for scaffold construction. Furthermore, gelatin and PGA/gelatin composites demonstrate high biocompatibility and can be easily functionalized to exhibit conductive properties. These materials possess a wide surface area and high porosity, facilitating enhanced cellular adhesion, nutrient exchange, and tissue integration. The utilization of conductive polymer-based structures in tissue engineering holds significant potential for applications such as brain regeneration, cardiac tissue engineering, bone repair, and wound healing. Researchers are exploring novel strategies to enhance tissue development, functionality, and overall regenerative potential by leveraging the electrical responsiveness of these polymers.

For instance, a biocellulose-reinforced polyurethane composite has been utilized as a scaffold in bone tissue implants [6]. The application of a hydrogel scaffold coated with a conductive polymer nanomaterial on neural microelectrodes has proven effective in achieving controlled drug release, facilitated by low impedance and high charge capacity. Spectroscopic measurements of PEDOT nanotubes have been employed as biosensors capable of distinguishing between acute and chronic responses of brain cells [7]. PPy-based scaffolds have been employed as actuators in various electromechanical devices. The high electrical conductivity, flexural strength, and controllable doping-undoping mechanism of PPy make it a prominent material for actuation. Additionally, PPy can be utilized for temperature detection. PANi has been utilized as an artificial muscle, capable of changing volume through doping-undoping processes. PEDOT, on the other hand, has been employed as a transparent electrode in the fabrication of a focusable lens system.

19.3 3D PRINTING METHODS OF CONDUCTIVE POLYMERS

In the field of biomedicine, conductive polymers have found a variety of uses, including the creation of electrodes, electronic tissue components, wearable or directly applicable electronic devices, sensors, and recording apparatuses that can track metabolic rates, hormone levels, enzyme levels, and more (Figure 19.1). Biomedical devices have been transformed by the development of production processes from one-dimensional and two-dimensional technologies, such as fibers and films, to three-dimensional production, which has replaced rigid, heavy, and mechanically worn metallic components. The capacity of polymers to be precisely tuned for individual patients and certain application areas is a noteworthy benefit of employing polymers in this context. This personalization enables customized biomedical devices that enhance their usability and body compatibility. Depending on the chosen production technique, the synthesis of multi-component structures can also be seamlessly incorporated throughout the production stage, obviating the need for separate assembly stages. However, it is crucial to highlight that 3D creation of conductive polymers requires specialist understanding as these materials do not offer the same ease of production as thermoplastics and have different mechanical properties from other polymers. Therefore, strengthening the workability and mechanical properties of structures made of conductive polymers, as well as improving the conductivity of polymeric composites, are of utmost importance.

Conductive polymer inks can be precisely deposited during inkjet printing, allowing for the development of intricate patterns and designs. Layer-by-layer building of conductive polymer structures is possible using extrusion printing techniques like fused deposition modeling (FDM) or direct ink writing (DIW), providing flexibility and design freedom. By using electrical forces to regulate the deposition of conductive polymer solutions or suspensions, electrohydrodynamic printing makes it easier to fabricate small-scale objects. Complex three-dimensional structures can be created using light-based printing processes that use photopolymerization or photoablation to harden or remove areas of conductive polymer material, respectively. To improve the functionality, performance, and biocompatibility of biomedical devices, researchers and engineers are pushing the limits of conductive polymer applications

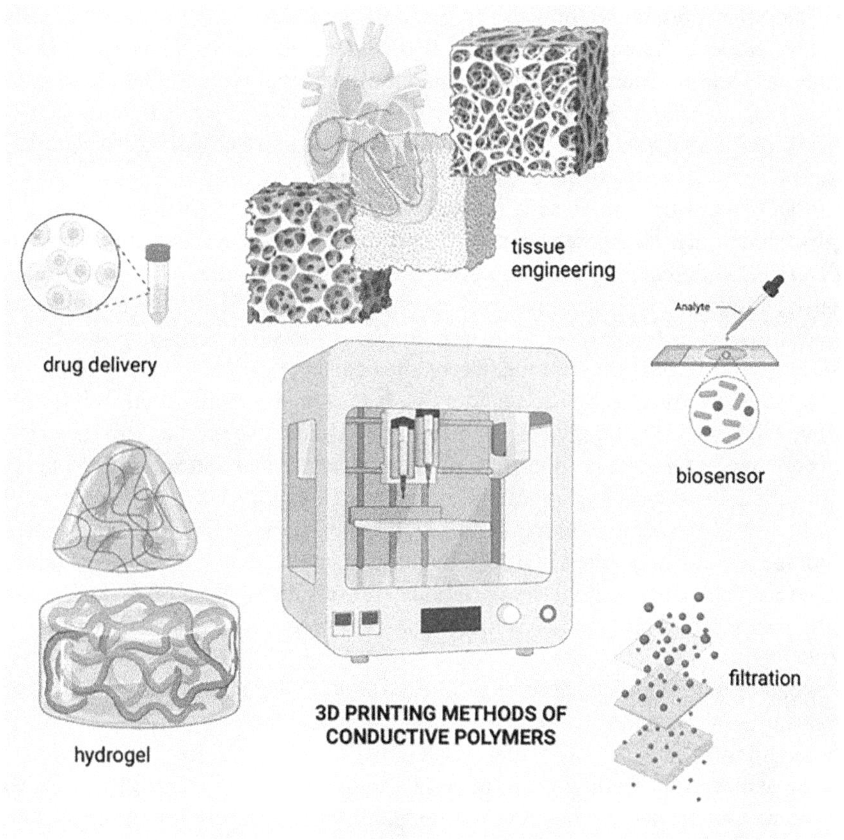

FIGURE 19.1 Schematic diagram of 3D printing methods for conductive polymers.

in biomedicine. Conducive polymer technology's continuing development offers enormous potential for the creation of ground-breaking remedies in healthcare, diagnostics, and customized medicine. Conductive polymers have the potential to completely transform the biomedical industry thanks to their flexibility to be customized to the unique needs of certain patients and application sectors and their use of cutting-edge 3D printing technologies. These polymers enable precise monitoring and therapy of metabolic rates, hormones, enzymes, and more using electrodes, electronic tissue components, wearable technology, and sensors. Utilizing conductive polymers to their potential will shape the future of biomedical applications, paving the way for revolutionary improvements in healthcare and raising people's quality of life all around the world.

19.3.1 Inkjet Printing

The technique used in 3D production is like conventional printers, where the material is sprayed onto the surface using a nozzle that delivers the material from a reservoir.

The necessary pressure to spray the droplets can be achieved through thermal, piezoelectric, or electromagnetic methods. It is crucial to maintain smooth spray performance, especially when working with fluid materials that typically have a viscosity of less than 100 mPas. While this characteristic enables the application of materials in the form of a solution, the limited solubility of conductive polymers is a significant disadvantage that restricts the use of this technique.

PEDOT is widely used as a conductive polymer in bioelectronics due to its high conductivity, transparent structure, and thermal and electrochemical stability. PEDOT:PSS or other polymeric mixtures are employed to create structures that offer adjustable electrical, mechanical, and biochemical properties. Applying this composite onto a polyamide or PET surface enhances the surface rigidity while maintaining the conductivity required for biomedical applications. By combining the PEDOT:PSS mixture with an ionic liquid such as 1-ethyl-3-methylimidazolium ethyl sulfate (EMIM:ES) in an inkjet system, a highly conductive product can be formed. This product can serve as a high-performance electronic and optical coating material or be utilized in the production of multilayer structures [8].

The use of the inkjet technique for placing the PEDOT:PSS mixture on carbon nanotube structures resulted in a 53% increase in conductivity compared to random placement. This approach allows for a homogeneous distribution of the composite's components across all layers, ensuring consistent properties throughout [9]. By utilizing the inkjet system, a triple mixture of graphene, PEDOT:PSS, and other polymers can be incorporated, harnessing the high structural strength and conductivity of graphene along with the flexibility and ease of application of the polymeric structure. The addition of metallic properties, such as silver nanoparticles, to the composite can be achieved by feeding these materials into the inkjet system. PPy is another biocompatible conductive polymer that exhibits superior electrical and mechanical properties when combined with a mixture of 9BA-4–9BA gemini acid surfactant, $FeCl_3$/iron porphyrin tetrasulfonate (FePTS) oxidant, and poly(vinyl alcohol) (PVA) stabilizer. When used in conjunction with ethanol in the inkjet system, the viscosity and surface tension decrease, resulting in a suitable bionanomaterial for 3D printing, albeit with slightly reduced conductivity [10]. In another application, where ethanol is employed to improve printing properties, PPy and collagen are combined and coated onto the surface of a polyarylate film using the inkjet system, creating a multi-layered polymeric structure [11]. While PANi is easily doped/dedoped through acid-base reactions, its stability is relatively low. However, by spraying aniline on ammonium persulfate with the inkjet technique, the resulting structure can be doped and gelatinized using phytic acid, allowing for the adjustment of electrochemical properties. The combined use of silver nanoparticles and PANi has also enabled the creation of a 3D printable conductive structure [12].

19.3.2 Extrusion Printing

In the FDM technique, polymeric fiber is fed into the device, melted at a hot end and then deposited onto the surface. In the DIW technique, the polymeric material is in a semi-molten state and is extruded from the end using pneumatic or mechanical means before being transferred to the surface. Conductive polymers utilized in these methods often require processing with other materials to attain suitable mechanical properties for 3D printing. For instance, the PEDOT:PSS mixture was shaped using DIW to produce a biogel suitable for neural stem cell production [13]. These techniques offer greater control over the shaping process compared to inkjet printing, making them particularly advantageous for fabricating microelectronic structures. The forming process can be easily regulated by adjusting parameters such as material flow rate, polymeric mixture composition, and nozzle width.

Water stability, an important property in biomedical applications, can be improved by incorporating an organic solvent and a cross-linking agent. The use of an organic solvent such as ethylene glycol or dimethyl sulfoxide (DMSO) in the synthesis of the polymeric material reduces the evaporation rate and enhances its stability. Furthermore, the addition of a suitable cross-linking agent promotes water stability in PEDOT:PSS chains through the hydrolysis and condensation of silane groups [14]. Another approach involves introducing DMSO and triton X into the polymer mixture, resulting in a biocompatible polymeric structure with high elasticity and electrical conductivity using the DIW technique [15]. In the design of a multilayer structure for a bionic photodetector that can be used as an eye implant, the PEDOT:PSS mixture is utilized as the transparent anode layer, while the poly (3-hexylthiophene) (P3HT):[6,6]-phenyl C61-butyric acid methyl ester (PCBM) serves as the photoactive layer. Additionally, silver nanoparticles are employed for metallic bonding, silicon acts as the insulating layer, and a eutectic gallium-indium mixture is used as the cathode [16]. Likewise, a flexible conductive polymeric material is produced through the co-processing of the PEDOT:PSS mixture with poly (N-acryloyl glycinamide-co-2-acrylamide-2-methylpropanesulfonic) PNAGA:PAMPS [17]. The combination of Nafion and PEDOT has proved a valuable material in the production of 3D organic electrochemical transistors [18].

The conductive composite structure of PEDOT-g-PLA was obtained using the DIW method, resulting in a visually appealing structure resembling body tissue. This composite exhibited high compatibility with cardiomyocytes and fibroblasts [19]. In a study focused on producing a biocomposite by combining tyramine-modified PVA with PEDOT, a scaffold was created with significantly shorter PEDOT chains within the PVA pores. The rapid bonding process, facilitated by the photochemical properties of ruthenium, led to the formation of a hydrogel structure with short chains. This unique structure exhibited electrical and physical properties that make it particularly suitable for pressure and temperature actuation. Like PEDOT:PSS, gelatin methacryloyl mixed with methyl cellulose/carrageenan or carbon methylcellulose has been used to produce high-conductivity biopolymer mixtures suitable for 3D printing. These materials possess high biocompatibility and allow for easy adjustment of properties such as conductivity and viscosity. They have been loaded with cells such as C2C12 or HEK-293, demonstrating their potential as functional alternatives in 3D tissue production [20]. Nanoparticles such as Sb_2Te_3 nanoflakes, multi-walled carbon nanotubes (MWCNTs), and Ag nanowires can be directly added during the polymerization stage to enhance the mechanical, electrical, and thermal properties of the PEDOT:PSS mixture. This enables the production of more optimal structures in a single-stage process. These structures are particularly effective in pressure and humidity detection [21]. PPy, another conductive polymer that cannot be used alone in 3D printing, can be combined with double-bond decorated chitosan. The dual structure is then subjected to acrylic acid polymerization, resulting in a material with strong electrical conductivity, suitable for 3D printing and medical device applications. Similarly, a mixture of PPy, nanocellulose, and poly(glycerol sebacate) can be used to create a 3D structure for controlled drug release [22]. It can be used in the production of 3D biomaterials with alginate alone. Alginate alone can also be utilized in the production of 3D biomaterials. Additionally, PPy, along with another conductive polymer, PSS, can be added to an alginate-gelatin mixture to construct a 3D scaffold for cartilage tissue implants [23]. It has been observed that the addition of PPy to PLA or PVA in nanostructure improves the electrical and mechanical properties in 3D printing [24]. MWCNTs offer a more advanced structure in the production of PPy-based biosensors [25]. The polymeric mixture obtained from PA and PCL results in a robust and highly conductive structure capable of loading adipose tissue-derived stem cells. This mixture can be produced through mechanical extrusion or chemical grafting methods [26]. Increasing the ratio of PANi added to the PCL/PANi nanofiber mixture leads to a decrease in average pore diameter as the fibers are separated into smaller pieces, resulting in significantly increased conductivity [27]. The PANi/reduced graphene oxide mixture exhibits desirable mechanical properties for 3D printing along with effective electrical conductivity [28].

19.3.3 Electrohydrodynamic Printing

In this method, an electric field is created between the tip of the printer and the surface on which the product is printed, and the material is transferred to the surface dissolved in a polarizable liquid. It offers advantages in the production of micro- and nano-sized products, as it provides higher resolution compared to techniques such as inkjet printing. In addition to the ink properties, parameters like voltage, pressure, and flow affect the feeding properties during production. The spray behavior depends on the applied voltage and the viscosity of the ink. Low voltage and low viscosity result in dripping behavior from the top of the Taylor cone, while high voltage and low viscosity create smaller droplets, leading to microdripping behavior. By increasing the flow rate when microdripping occurs, the ink flow becomes continuous. Increasing the voltage and viscosity generates a fine jet of liquid at the top of the Taylor cone. However, if the flow is increased to high values at this stage, an unstable output is observed. In the electrohydrodynamic technique for 3D production using a PEDOT:PSS mixture, poly(ethylene oxide) (PEO) was added to the mixture to adjust viscosity. This allowed the production of 3D structures with increasing thicknesses, reaching up to 100 layers. As the number of layers increased, a significant decrease in electrical resistance was observed. By incorporating PCL micro-sized parallel vertical walls between the nano-sized layers obtained from the same triple mixture, a multi-layered structure was formed. This structure showed potential as a substitute for the cardiac extracellular matrix. The impedance of this structure at physiological frequencies was found to be much lower than that of the pure PCL structure [29]. Furthermore, a promising organic field-effect transistor for electronic devices was developed by polymerizing PEDOT and PSS [30]. Another achievement involved the production of PPy-b-PCL, a biodegradable conductive polymer composite, using the electrohydrodynamic technique [31].

19.3.4 Light-Based Printing

In this technique, monomer or polymer precursors are contained and hardened by the effect of light, creating 3D shapes. In stereolithography (SLA), a laser beam is reflected on the liquid surface using a mirror, and the areas where the light is exposed are cured. Digital light processing (DLP) employs a device with millions of small mirrors to project light onto the liquid in the container. As a result, the SLA process progresses point by point, while DLP processes 2D images as slices. Selective laser sintering (SLS) is another light-based application. In this method, a polymer precursor that can be cross-linked with light is sintered by applying a high-energy laser beam to a powder mixture containing other materials to be formed. When it comes to using conductive polymers in light-based printing applications, it is necessary to disperse the polymer in a light-curable material to create an ink. For instance, the PEDOT:PSS mixture was prepared in a water-ethylene glycol mixture, and then mixed with poly(ethylene glycol) diacrylate containing a photoinitiator called bis(2,4,6-trimethylbenzoyl)phenylphosphineoxide to obtain a usable ink in SLA [32]. In the production of light-based 3D products, PEDOT:PSS has been incorporated into nanostructured electrically conductive hydrogels containing MWCNTs doped with a hydrophilic photoresist [33].

19.4 3D PRINTING OF CONDUCTIVE POLYMERS FOR MEDICAL PURPOSES

Various applications of structures produced using conductive polymers in 3D with cells have been documented in the literature. The PEDOT:PSS compound mixed with PEO was utilized to support the PCL matrix, resulting in the formation of a structure that facilitated the proliferation of H9C2 myoblasts and primary cardiomyocytes in tissues [29]. The composite of PPy with PLA extended the lifespan of fibroblasts by over 80%, while the composite of human embryonic stem cells-neural crest stem cells with PCL promoted the proliferation of human embryonic stem cells more effectively than the use of PCL alone [31]. The mixture of PCL and PANi exhibited a proliferation-enhancing effect on C2C12 mouse myoblasts and osteoblasts [27]. Several studies have addressed the improvement of communication between electroactive cells by leveraging the electrical conductivity of conductive polymers. In one study, PC12 cells were electrically stimulated by introducing the PPy:PVA structure into a collagen-mixed medium. This resulted in an increase in the formation of nerve cells that comprise neural networks, along with a 40% elongation in the length of the formed cells [34]. The 3D printing method, which involves incorporating cells into inks, is commonly used in constructing cell-laden scaffolds. For example, HEK-293 cells were used as bioinks in a mixture of methylcellulose/kappa-carrageenan/PEDOT:PSS combined with C2C12 cells, while immortalized dorsal root ganglion neuron cells were used with a mixture of GelMA/PEDOT:PSS. These applications aim to increase the lifespan and survival rate of cells or facilitate their transformation into nerve cells through electrical stimulation within the cell population [32]. Human neuronal stem cells were coated with a gel based on polysaccharides and printed on this gel using PEDOT:PSS. It was observed that the cells transformed into neurons and glial cells even without stimulation, but the stimulated cells exhibited strong connections to the underlying structure, with their structural extensions becoming more branched [13].

The PEDOT:PSS binary mixture has been utilized in sensors for measuring various metabolic parameters. Electrodes prepared with PEDOT:PSS have been extensively studied for electrocardiogram measurements, with a particular focus on presenting them as textile products. This polymeric structure offers high machinability, and its conductivity can be easily enhanced by using various dopants, making it a viable alternative to commercially used Ag/AgCl electrodes. Another application of this polymeric composite is in a respiratory rate measuring mask. This single-layer polymeric structure, which returns to its original state within a response time of 3 seconds, detects accelerated respiration. By creating a 3D structure that incorporates a poly(vinylidene fluoride-co-trifluoroethylene) layer between two PEDOT:PSS layers, sound waves can be converted into electric current. This enables differentiation between breathing and coughing, making the produced mask useful in diseases caused by viruses such as the coronavirus [35].

Artificial skin tissues produced using conductive polymers offer an advantage for electronic devices, such as robots, to mimic human skin. From a medical perspective, the use of electronic skin provides benefits not only in aesthetic applications but also in scenarios where sensors can be integrated internally within the body. These sensors can function as treatment devices that can be electrically manipulated externally. An artificial tendon with high conductivity up to 420 MJ/m^3, elasticity, and suitability for robotic applications was produced by reinforcing the PEDOT:PSS structure with single-walled carbon nanotubes (SWCNTs) and spider web. This tendon achieved force transfer while serving as a sensor [36]. For the same purpose, a highly flexible and strong supercapacitor was developed using a combination of the PEDOT:PSS mixture, polyurethane, and graphene. This material demonstrated only a 1% loss in capacitance ability when subjected to random bending, making it suitable for use in constantly moving organs like fingers [37]. In addition to PEDOT:PSS, PANi is a conductive and flexible polymer that can be utilized in artificial leather production. A mixture of PANi, poly(acrylic acid), and phytic acid was employed to create self-healing elastic leather pieces. The addition of phytic acid as a dopant increased the electrical conductivity by 600 times and quadrupled the toughness of the material [38]. An artificial skin with a temperature sensor function was developed by coating a polyurethane layer with PEDOT:PSS/graphene. The electrical resistance of the structure varied inversely with temperature, allowing for temperature measurement [39]. In another application, graphene oxide and silver were used as substitutes for graphene to produce an electrically more stable and measurable sensor, especially against mechanical effects [40].

19.5 PROSPECTS AND FUTURE DIRECTIONS

This chapter aims to provide a comprehensive understanding of 3D printing techniques used for conductive polymers. Also, the key ideas, benefits, and disadvantages of inkjet, extrusion, electrohydrodynamic and light-based printing are discussed, in addition to other printing processes used to fabricate conductive polymer structures. In-depth research has also been done on the use of 3D printing in the medical industry. Conductive polymers have been intensively investigated for their possible advantages and disadvantages in bioelectronics, tissue engineering, and the manufacture of medical devices. Conductive polymers are valuable in bioelectronics because of how adaptable they are. By using them to create wearable technology, biosensors, and neural interfaces, they can promote individualized healthcare and diagnostics. Conductive polymers are essential in tissue engineering because they help to build bioactive scaffolds that encourage tissue regeneration and cell proliferation. This opens new opportunities for regenerative medicine. Conductive polymers also help to create medical devices including implants, prosthetics, and medication delivery systems by improving their usefulness and compatibility. The future of conductive polymer 3D printing in the medical industry seems bright. Improvements in print resolution, material compatibility, and process optimization may result from continued research and development activities. This will make it possible to design more complex and precise conductive polymer structures that are specifically suited for medical purposes. The construction of complex hybrid structures with increased properties and functionalities is possible when conductive polymers are combined with other biomaterials. The ability to customize and produce medical devices on demand is another benefit of using 3D printing technologies. This enables the development of customized approaches that address the special requirements of various patients, improving treatment outcomes and increasing patient happiness. Finally, the application of conductive polymers and 3D printing in the medical sector offers enormous potential for improving medical technology. Conducive polymer 3D printing is poised to transform medical applications, advancing patient care, diagnostics, and regenerative medicine through ongoing research, collaboration, and technology improvements.

REFERENCES

[1] I. Michira, R. Akinyeye, P. Baker, E. Iwuoha, Synthesis and Characterization of Sulfonated Polyanilines and Application in Construction of a Diazinon Biosensor, Int. J. Polym. Mater. Polym. Biomater. 60 (2011) 469–489.
[2] M. J. Shin, J. G. Kim, J. S. Shin, Amperometric Cholesterol Biosensor Using Layer-by-Layer Adsorption Technique on Polyaniline-Coated Polyester Films, Int. J. Polym. Mater. Polym. Biomater. 62 (2013) 140–144.
[3] M. A. Booth, S. A. Harbison, J. Travas-Sejdic, Development of An Electrochemical Polypyrrole-Based DNA Sensor and Subsequent Studies on the Effects of Probe and Target Length on Performance, Biosens. Bioelectron. 28 (2011) 362–367.
[4] L. Zhang, H. Peng, P. A. Kilmartin, C. Soeller, J. Travas-Sejdic, Polymeric Acid Doped Polyaniline Nanotubes for Oligonucleotide Sensors, Electroanalysis. 19 (2007) 870–875.

[5] J. J. Langer, K. Langer, P. Barczyński, J. Warchoł, K. H. Bartkowiak, New "ON-OFF"-Type Nanobiodetector, Biosens. Bioelectron. 24 (2009) 2947–2949.
[6] F. Khan, Y. Dahman, A Novel Approach for the Utilization of Biocellulose Nanofibres in Polyurethane Nanocomposites for Potential Applications in Bone Tissue Implants, Des. Monomers Polym. 15 (2012) 1–29.
[7] M. R. Abidian, K. A. Ludwig, T. C. Marzullo, D. C. Martin, D. R. Kipke, Interfacing Conducting Polymer Nanotubes with the Central Nervous System: Chronic Neural Recording Using Poly(3,4-Ethylenedioxythiophene) Nanotubes, Advan. Mater. 21 (2009) 3764–3770.
[8] M. Y. Teo, N. Ravichandran, N. Kim, S. Kee, L. Stuart, K. C. Aw, J. Stringer, Direct Patterning of Highly Conductive PEDOT:PSS/Ionic Liquid Hydrogel via Microreactive Inkjet Printing, ACS Appl. Mater. Interfaces. 11 (2019) 37069–37076.
[9] A. S. Alshammari, M. Shkunov, S. R. P. Silva, Inkjet Printed PEDOT: PSS/MWCNT Nano-Composites with Aligned Carbon Nanotubes and Enhanced Conductivity, Phys. Status. Solidi. Rapid. Res. Lett. 8 (2014) 150–153.
[10] B. Weng, R. Shepherd, J. Chen, G. G. Wallace, Gemini Surfactant Doped Polypyrrole Nanodispersions: An Inkjet Printable Formulation, J. Mater. Chem. 21 (2011) 1918–1924.
[11] B. Weng, X. Liu, R. Shepherd, G. G. Wallace, Inkjet Printed Polypyrrole/Collagen Scaffold: A Combination of Spatial Control and Electrical Stimulation of PC12 Cells, Synth. Met. 162 (2012) 1375–1380.
[12] P. Patil, S. Patil, P. Kate, A. A. Kulkarni, Inkjet Printing of Silver Nanowires on Flexible Surfaces and Methodologies to Improve the Conductivity and Stability of the Printed Patterns, Nanoscale Adv. 3 (2021) 240–248.
[13] E. Tomaskovic-Crook, P. Zhang, A. Ahtiainen, H. Kaisvuo, C.-Y. Lee, S. Beirne, Z. Aqrawe, D. Svirskis, J. Hyttinen, G. G. Wallace, J. Travas-Sejdic, J. M. Crook, Human Neural Tissues from Neural Stem Cells Using Conductive Biogel and Printed Polymer Microelectrode Arrays for 3D Electrical Stimulation, Adv. Healthc. Mater. 8 (2019) 1–10.
[14] P. Zhang, N. Aydemir, M. Alkaisi, D. E. Williams, J. Travas-Sejdic, Direct Writing and Characterization of Three-Dimensional Conducting Polymer PEDOT Arrays, ACS Appl. Mater. Interfaces, 10 (2018) 11888–11895.
[15] S. Kee, M. A. Haque, D. Corzo, H. N. Alshareef, D. Baran, Self-Healing and Stretchable 3D-Printed Organic Thermoelectrics, Adv. Funct. Mater. 29 (2019) 1905426.
[16] S H. Park, R. Su, J. Jeong, S.-Z. Guo, K. Qiu, D. Joung, F. Meng, M. C. McAlpine, 3D Printed Polymer Photodetectors, Adv. Mater. 30 (2018) 1–8.
[17] Q. Wu, J. Wei, B. Xu, X. Liu, H. Wang, W. Wang, Q. Wang, W. Liu, A Robust, Highly Stretchable Supramolecular Polymer Conductive Hydrogel with Self-Healability and Thermo-Processability, Sci. Rep. 7 (2017) 1–11.
[18] A. I. Hofmann, I. Östergren, Y. Kim, S. Fauth, M. Craighero, M.-H. Yoon, A. Lund, C. Müller, All-Polymer Conducting Fibers and 3D Prints via Melt Processing and Templated Polymerization, ACS Appl. Mater. Interfaces. 12 (2020) 8713–8721.
[19] A. Dominguez-Alfaro, E. Gabirondo, N. Alegret, C. M. De León-Almazán, R. Hernandez, A. Vallejo-Illarramendi, M. Prato, D. Mecerreyes, 3D Printable Conducting and Biocompatible PEDOT-graft-PLA Copolymers by Direct Ink Writing, Macromol. Rapid Commun. 42 (2021) 1–8.
[20] P. Bao, Y. Lu, P. Tao, B. Liu, J. Li, X. Cui, 3D Printing PEDOT-CMC-Based High Areal Capacity Electrodes for Li-Ion Batteries, Ionics. 27 (2021) 2857–2865.
[21] X. He, R. He, Q. Lan, W. Wu, F. Duan, J. Xiao, M. Zhang, Q. Zeng, J. Wu, J. Liu, Screen-Printed Fabrication of PEDOT:PSS/Silver Nanowire Composite Films for Transparent Heaters, Materials. 10 (2017) 220.

[22] R. Ajdary, N. Z. Ezazi, A. M. Correia, M. Kemell, S. Huan, H. J. Ruskoaho, J. Hirvonen, H. A. Santos, O. J. Rojas, Multifunctional 3D-Printed Patches for Long-Term Drug Release Therapies after Myocardial Infarction, Adv. Funct. Mater. 30 (2020) 2003440.
[23] T. Distler, C. Polley, F. Shi, D. Schneidereit, M. D. Ashton, O. Friedrich, J. F. Kolb, J. G. Hardy, R. Detsch, H. Seitz, A. R. Boccaccini, Electrically Conductive and 3D-Printable Oxidized Alginate-Gelatin Polypyrrole:PSS Hydrogels for Tissue Engineering, Adv. Healthc. Mater. 10 (2021) 1–16.
[24] Y. Gu, Y. Zhang, Y. Shi, L. Zhang, X. Xu, 3D All Printing of Polypyrrole Nanotubes for High Mass Loading Flexible Supercapacitor, ChemistrySelect. 4 (2019) 10902–10906.
[25] J. H. Kim, S. Lee, M. Wajahat, H. Jeong, W. S. Chang, H. J. Jeong, J.-R. Yang, J. T. Kim, S. K. Seol, Three-Dimensional Printing of Highly Conductive Carbon Nanotube Microarchitectures with Fluid Ink, ACS Nano. 10 (2016) 8879–8887.
[26] A. Prasopthum, Z. Deng, I. M. Khan, Z. Yin, B. Guo, J. Yang, Three Dimensional Printed Degradable and Conductive Polymer Scaffolds Promote Chondrogenic Differentiation of Chondroprogenitor Cells, Biomater. Sci. 8 (2020) 4287–4298.
[27] J. Song, H. Gao, G. Zhu, X. Cao, X. Shi, Y. Wang, The Construction of Three-Dimensional Composite Fibrous Macrostructures with Nanotextures for Biomedical Applications, Biofabrication. 8 (2016) 1–14.
[28] Z. Wang, Q. Zhang, S. Long, Y. Luo, P. Yu, Z. Tan, J. Bai, B. Qu, Y. Yang, J. Shi, H. Zhou, Z.-Y. Xiao, W. Hong, H. Bai, Three-Dimensional Printing of Polyaniline/Reduced Graphene Oxide Composite for High-Performance Planar Supercapacitor, ACS Appl. Mater. Interfaces. 10 (2018) 10437–10444.
[29] Q. Lei, J. He, D. Li, Electrohydrodynamic 3D Printing of Layer-Specifically Oriented, Multiscale Conductive Scaffolds for Cardiac Tissue Engineering, Nanoscale. 11 (2019) 15195–15205.
[30] B. Lyu, S. Im, H. Jing, S. Lee, S. H. Kim, J. H. Kim, J. H. Cho, Work Function Engineering of Electrohydrodynamic-Jet-Printed PEDOT:PSS Electrodes for High-Performance Printed Electronics, ACS Appl. Mater. Interfaces. 12 (2020) 17799–17805.
[31] S. Vijayavenkataraman, S. Kannan, T. Cao, J. Y. H. Fuh, G. Sriram, W. F. Lu, 3D-Printed PCL/PPy Conductive Scaffolds as Three-Dimensional Porous Nerve Guide Conduits (NGCs) for Peripheral Nerve Injury Repair, Front. Bioeng. Biotechnol. 7 (2019) 1–14.
[32] D. N. Heo, S. J. Lee, R. Timsina, X. Qiu, N. J. Castro, L. G. Zhang, Development of 3D Printable Conductive Hydrogel with Crystallized PEDOT:PSS for Neural Tissue Engineering, Mater. Sci. Eng. C. 99 (2019) 582–590.
[33] Y. Tao, C. Wei, J. Liu, C. Deng, S. Cai, W. Xiong, Nanostructured Electrically Conductive Hydrogels Obtained: Via Ultrafast Laser Processing and Self-Assembly, Nanoscale. 11 (2019) 9176–9184.
[34] B. Weng, X. Liu, M. J. Higgins, R. Shepherd, G. Wallace, Fabrication and Characterization of Cytocompatible Polypyrrole Films Inkjet Printed from Nanoformulations Cytocompatible, Inkjet-Printed Polypyrrole Films, Small. 7 (2011) 3434–3438.
[35] W. Wang, K. Ouaras, A. L. Rutz, X. Li, M. Gerigk, T. E. Naegele, G. G. Malliaras, Y. Y. S. Huang, Inflight Fiber Printing Toward Array and 3D Optoelectronic and Sensing Architectures, Sci. Adv. 6 (2020) caba0931.
[36] L. Pan, F. Wang, Y. Cheng, W. R. Leow, Y.-W. Zhang, M. Wang, P. Cai, B. Ji, D. Li, X. Chen, A Supertough Electro-Tendon Based on Spider Silk Composites, Nat. Commun. 11 (2020) 1–9.
[37] F. Tehrani, M. Beltrán-Gastélum, K. Sheth, A. Karajic, L. Yin, R. Kumar, F. Soto, J. Kim, J. Wang, S. Barton, M. Mueller, J. Wang, Laser-Induced Graphene Composites for Printed, Stretchable, and Wearable Electronics, Adv. Mater. Technol. 4 (2019) 1–11.

[38] T. Wang, Y. Zhang, Q. Liu, W. Cheng, X. Wang, L. Pan, B. Xu, H. Xu, A Self-Healable, Highly Stretchable, and Solution Processable Conductive Polymer Composite for Ultrasensitive Strain and Pressure Sensing, Adv. Funct. Mater. 28 (2018) 1705551.
[39] T. Vuorinen, J. Niittynen, T. Kankkunen, T. M. Kraft, M. Mäntysalo, Inkjet-Printed Graphene/PEDOT:PSS Temperature Sensors on a Skin-Conformable Polyurethane Substrate, Sci. Rep. 6 (2016) 1–8.
[40] M. Soni, M. Bhattacharjee, M. Ntagios, R. Dahiya, Printed Temperature Sensor Based on PEDOT:PSS-Graphene Oxide Composite, IEEE Sens. J. 20 (2020) 7525–7531.

Index

Q

R

S

T

V

W

Z

For Product Safety Concerns and Information please contact our EU
representative GPSR@taylorandfrancis.com
Taylor & Francis Verlag GmbH, Kaufingerstraße 24, 80331 München, Germany

www.ingramcontent.com/pod-product-compliance
Lightning Source LLC
Chambersburg PA
CBHW060814220326
41598CB00022B/2615
* 9 7 8 1 0 3 2 5 4 2 6 2 1 *